Climate Governance across the Globe

This book takes an innovative approach to studying international climate governance by providing a critical analysis of climate leadership, pioneership and followership across the globe.

The volume assesses the interactions between climate leaders, pioneers and followers, across multilevel and/or polycentric climate governance contexts. Examining the state and sub-state levels in both the Global South and Global North, as well as regional, supranational EU and international climate governance levels, the authors explore 16 countries across Asia, Australasia, Europe, and Central and North America, plus the European Union. Each chapter employs a comprehensive and consistent framework for analyzing leadership and pioneership, as well as followership. The findings provide new insights into the strategies and actions of sub-state, state-level, and supranational leaders and pioneers.

This book will be of key interest to scholars, students and practitioners in environmental politics and climate change governance, as well as those interested in political elites, EU studies and, more broadly, comparative politics and international relations.

Rüdiger K.W. Wurzel is Professor of Comparative European Politics and Jean Monnet Chair in EU Studies at the University of Hull where he is Director of the Centre for European Union Studies (CEUS), UK.

Mikael Skou Andersen is Professor of Environmental Policy Analysis at Aarhus University, Denmark.

Paul Tobin is Senior Lecturer in Politics at the University of Manchester, UK.

Routledge Research in Comparative Politics

Climate Governance across the Globe

Pioneers, Leaders and Followers

**Edited by Rüdiger K.W. Wurzel,
Mikael Skou Andersen and Paul Tobin**

LONDON AND NEW YORK

First published 2021
by Routledge
2 Park Square, Milton Park, Abingdon, Oxon OX14 4RN

and by Routledge
52 Vanderbilt Avenue, New York, NY 10017

Routledge is an imprint of the Taylor & Francis Group, an Informa business

British Library Cataloguing-in-Publication Data
A catalogue record for this book is available from the British Library

Library of Congress Cataloging-in-Publication Data
A catalog record has been requested for this book

ISBN: 978-0-367-43436-6 (hbk)
ISBN: 978-0-367-65047-6 (pbk)
ISBN: 978-1-003-01424-9 (ebk)

Typeset in Times New Roman
by Deanta Global Publishing Services, Chennai, India

Contents

Figures

Tables

Contributors

Mikael Skou Andersen is Professor of Environmental Policy Analysis at Denmark's Aarhus University's Department of Environmental Science. He has undertaken research on carbon taxes for more than two decades, surveying their effectiveness in reducing emissions and the implications for economic performance as well as the political circumstances for their introduction. Mikael Skou Andersen is Editor of *Carbon-Energy Taxes: Lessons from Europe* (Oxford University Press, 2009) with Paul Ekins, University College London.

Pat Brereton is a Professor in the School of Communications at Dublin City University, Ireland and his latest book *Environmental Literacy and New Digital Audiences* has just been published by Routledge. He is co-PI on a project funded by the Irish EPA entitled Deepening Public Engagement on Climate Change: Lessons from the Citizens' Assembly.

Laura Devaney is a Postdoctoral Researcher in the School of Law and Government at Dublin City University, Ireland. She is a social scientist with international expertise across environmental governance arenas, including food, bioeconomy and transport governance. She positions her work at the research-policy interface.

Johann Dupuis is a Senior Researcher at the Swiss Graduate School of Public Administration (IDHEAP), University of Lausanne, Switzerland. He has a PhD in public administration and MA in political sciences. He is an expert in comparative politics, public policy and environmental governance. He has worked extensively on climate change policies.

Harald Fuhr is Professor of International Politics at the University of Potsdam, Germany. His current research activities focus on governance issues in developing countries, international development and climate change.

David Hall is Senior Researcher at The Policy Observatory, Auckland University of Technology, New Zealand. He has a DPhil in Politics from Oxford University with a focus on the politics of borders and climate change. He is Co-Chair of the Independent Advisory Group for Auckland Council's Climate

Action Plan and sits on the Technical Working Group for Aotearoa Circle's Sustainable Finance Forum. His policy-oriented work focuses on 'natural climate solutions' through the design of impact instruments. In 2019 he edited the collection, *A Careful Revolution: Towards a Low-Emissions Future*, on the transition to a low-emissions economy in New Zealand.

Julia Hertin is a Political Scientist who works at the German Advisory Council on the Environment (SRU), an interdisciplinary, independent advisory body funded by the German government.

Thomas Hickmann is a Postdoctoral Researcher and Lecturer at the Chair of International Politics of the University of Potsdam, Germany. In his research, he mainly deals with theories of international relations, global sustainability governance and climate policy-making.

Chris Höhne is a Research Associate and PhD candidate with a focus on international relations at the Technical University of Darmstadt, Germany. His research focuses on norm research and climate politics in the Global South with a special focus on India and Indonesia.

Karin Ingold is Professor at the University of Bern in Switzerland which is affiliated to the Institute of Political Science and Oeschger Centre for Climate Research. She is also group leader at the Department of Environmental Social Sciences at EAWAG. She is a scholar of the Advocacy Coalition Framework and other policy process theories.

Kirsten Jörgensen is Senior Lecturer at the Department of Political and Social Sciences at the Free University of Berlin, Germany. Her primary research interests include European and Indian environmental/climate policy. She has been a visiting lecturer at TERI University in New Delhi in India and Moscow State Institute of International Relations of the MFA of Russia, and is a Research Fellow at the Johns Hopkins University. She initiated the German-Indian Sustainability and Climate Change Dialogue student exchange and the Indian-European Multi-level Climate Governance Research Network. Since 2013 she has been a member of the Indo-German Expert Group on Green and Inclusive Economy.

Marlene Kammerer is a Postdoctoral Researcher at the Institute of Political Science and the Oeschger Centre for Climate Change research at the University of Bern, Switzerland. Her main research area is the intersection between international dynamics and national policy-making with a focus on climate change.

Markus Lederer is Professor of Political Science with a focus on international relations at the Technical University Darmstadt, Germany. His research focuses on global climate politics and green transformations in the Global South.

Xinlei Li is Associate Professor in the School of Political Science and Public Administration of Shandong University in China and Deputy Director of the Environmental Politics Research Institute. She specialises in studying energy and environment policy, with a particular focus on renewable energy. She has authored two books entitled *Renewable Energy Policy Change in China* and *Clean Energy Diplomacy: International Tendency and China's Path.*

Jeremy F.G. Moulton is an Associate Lecturer in the Department of Politics at the University of York, UK. He researches and publishes work on multilevel climate action in the European Union and on political myth.

Henrik Selin is Associate Professor of International Relations at the Pardee School of International Relations at Boston University, USA.

Giuseppina Siciliano is a Research Fellow at the School of Oriental and African Studies (SOAS), University of London, UK. She carries out research on the linkages between energy, the environment and development, with a particular focus on energy infrastructure development in Asia, Africa and Latin America, land use changes and the water-energy-food nexus. She is an Academic Tutor at SOAS and a Consultant for the Ministry of the Environment in Italy.

Fee Stehle is a Research Associate and PhD candidate at the Chair of International Politics of the University of Potsdam in Germany, specialising in international relations and public administration. Her research is concerned with climate governance and focused on polycentric climate governance in cities in Brazil and South Africa.

Sibyl Steuwer has carried out research on energy and climate policy at the Environmental Policy Research Center (FFU) at the Free University of Berlin, Germany for many years. She works at the Buildings Performance Institute Europe (BPIE).

Paul Tobin is Senior Lecturer in Politics at the University of Manchester, UK. He specialises in European and environmental public policy. He is leading a three-year project on polycentric climate governance. The support of the Economic and Social Research Council (ESRC) is gratefully acknowledged, having funded him via grant ES/S014500/1 during his involvement in this volume.

Diarmuid Torney is Assistant Professor in the School of Law and Government at Dublin City University, Ireland. He was a Member of the Expert Advisory Group to the Irish Citizens' Assembly on climate change. He is Co-Principle investigator on a project funded by the Irish EPA entitled Deepening Public Engagement on Climate Change: Lessons from the Citizens' Assembly.

Frauke Urban is Associate Professor in the Management of Sustainable Energy Systems at Royal Institute of Technology (KTH) in Stockholm, Sweden. She was Reader in Environment and Development at School of Oriental and African Studies (SOAS), University of London, UK; Research Fellow in Climate Change and Development at the University of Sussex, UK; and Research Fellow in Energy, Environment and Development at the University of Groningen, the Netherlands. Frauke works on low carbon energy transitions, renewable energy, technology transfer and cooperation, energy policy, climate policy, green economy and sustainable development.

Stacy D. VanDeveer is Professor of Global Governance and Human Security at the McCormack Graduate School of Policy and Global Studies at the University of Massachusetts Boston, USA. Among other publications, he co-authored *European Union and Environmental Governance* (Routledge 2015).

Alonso Villalobos is Associate Instructor in the Department of Political Science at the University of Costa Rica (UCR) and Social Researcher at OdD-UCR and INISA-UCR. His work is centered on Costa Rican climate change policy, avoided deforestation policies in Central America and development patterns in this region. He has been a consultant to organizations such as CATIE, FAO, IADB, IUCN, TNC and the Friedrich Ebert Foundation. He was a researcher for regional GEO reports from UNEP-LAC and human development reports from UNDP-Costa Rica. Villalobos received his PhD from the University of Potsdam and his master's degree from the University of Freiburg, both in Germany.

Rüdiger K.W. Wurzel is Professor of Comparative European Politics and Jean Monnet Chair in EU Studies at the University of Hull, UK, where he is Director of the Centre for European Union Studies (CEUS). He has published widely on environmental issues.

Foreword and acknowledgements

Climate change remains one of the most pressing, if not the most pressing, medium-to-long-term global challenges, despite the emergence of multiple additional global crises, including the COVID-19 crisis, refugee crisis and financial crisis. Environmental leaders and pioneers have long been identified as crucial agents of change who are of central importance to successful, innovative climate change measures at all levels of governance. Much of the existing literature on environmental and climate leaders and pioneers has focused primarily on countries in the Global North. We know relatively little about the climate leadership and pioneership in and by countries of the Global South. This volume tries to help close this gap in the literature, while also providing new empirical findings and novel analytical insights about climate leadership and pioneership in and by Global North countries.

The success of climate governance at multiple levels depends not only on leaders and pioneers but on followers which have remained largely under-researched. This volume also identifies followers and assesses occurrences of climate followership in various multilevel and polycentric climate governance structures. As explained in this volume, climate leaders and pioneers have arguably become even more important since the 2015 Paris Agreement, which is based on a bottom-up approach and demands that the acceding states put forward voluntary state-level pledges via Nationally Determined Contributions (NDCs). The Agreement refers explicitly to the need for and the exchange of information, experiences and best practices. Learning from innovative climate governance measures of climate leaders and pioneers and, if possible, emulating their best practices, have therefore become even more central to global climate efforts.

This edited volume grew out of an Innovations in Climate Governance (INOGOV) funded workshop on 'Pioneers and Leaders in Polycentric Climate Governance (PiLePoC)' at the University of Hull, UK which took place from 15 to 16 September 2016. We are grateful to INOGOV and its Chair Andrew Jordan. Rudi Wurzel would like to thank the British Academy (grant SG 131240), while Mikael Skou Andersen would like to thank Nordforsk (grant 82841) and Aarhus University's Interdisciplinary Climate Centre (iClimate) for financial support to language editing. Very early versions of some of the contributions to this volume were presented at the INOGOV workshop in Hull. We commissioned, however, also several new chapters especially on Global South countries after receiving extremely helpful comments from three anonymous reviewers for our book proposal. At Routledge we received excellent support from Andrew Taylor and Sophie Iddamalgoda throughout the process of completing this manuscript. We would like to thank all of our chapter authors for their insightful research, and their supportive and steadfast collaborative spirit, especially during the COVID-19 crisis.

Last but not least, Rudi would like to thank Ita, Alfred and Kelsey for their humour, patience and tolerance throughout the completion process of this book. Mikael would like to give thanks to his wife Rikke for her support and understanding, while from Paul, special thanks go to Ciara, who has supported him at every step of his academic career.

RKWW, MSA and PT

Abbreviations

ACCTS	Agreement on Climate Change, Trade and Sustainability
ANZUS	Australia, New Zealand, United States Security
APEC	Asian-Pacific Economic Cooperation
ASEAN	Association of Southeast Asian Nations
BASIC	Brazil, South Africa, India and China
BAU	Business as Usual
BREXIT	British exit (from the European Union)
C	Celsius
CAFE	Corporate Average Fuel Economy (United States)
CBDR	Common But Differentiated Responsibilities
CCAs	Climate Change Agreements (United Kingdom)
CCC	Climate Change Committee (United Kingdom)
CCL	Climate Change Levy (United Kingdom)
CCS	carbon capture and storage
CVER	Certified Voluntary Emission Reduction (China)
CDKN	Climate and Development Knowledge Network
CDM	Clean Development Mechanism
CDU	Christian Democratic Union (*Christlich Demokratische Union* – Germany)
CEO	Chief Executive Officer
CEES	Central and Eastern European States
CERs	Certified Emission Reductions
CMA	China Meteorological Administration
CO$_2$	carbon dioxide
CO$_2$eq	carbon dioxide equivalent
COP	Conference of the Parties
CPC	Congress of the Communist Party of China
CSU	Christian Social Union (*Christlich Soziale Union* – Germany)
DG	Directorate-General
DNPI	National Council on Climate Change (Indonesia)
EBRD	European Bank for Reconstruction and Development
EEA	European Economic Area

EEG	Renewable Energy Law (*Erneuerbare-Energien-Gesetz* – Germany)
EFG	Eco Forum Global
EIB	European Investment Bank
EPBD	Energy Performance of Buildings Directive (Germany)
EPA	Environmental Protection Agency (United States)
EPC	Energy Performance Certificates
ENCC	National Climate Change Strategy (Costa Rica)
ENGOs	Environmental Non-Governmental Organisations
EPDP	EU Energy Performance of Buildings Directive
ETS	Emissions Trading Scheme
EU	European Union
EV	electric vehicle
FDP	Free Democratic Party (*Freie Demokratische Partei* – Germany)
FC	Federal Council (Switzerland)
FoE	Friends of the Earth
FYPs	Five Year Plans (China)
G7	Group of Seven countries
G20	Group of 20 countries
G77	Group of 77 countries
GDP	gross domestic product
GEG	Energy in Buildings Law (*Gebäudeenergiegesetz* – Germany)
GHGE	greenhouse gas emissions
GNI	gross national income
GOI	Government of India
Gt	gigatonnes
GW	gigawatt
GWP	global warming potential
HFCs	hydrofluorocarbons
ICLEI	Local Governments for Sustainability
IEA	International Energy Agency
INDCs	Intended Nationally Determined Contributions
IPCC	Intergovernmental Panel on Climate Change
IRENA	International Renewable Energy Agency
JI	joint implementation
KBN	Kommunalbanken (Norway)
KfW	Development Bank (*Kreditanstalt für Wiederaufbau* – Germany)
LDCs	Least Developed Countries
LGFA	Local Government Financing Agency (New Zealand)
LULUCF	Land Use, Land-Use Changes and Forestry
MIDEPLAN	Ministry of National Planning and Economic Policy (Costa Rica)
MLG	multilevel governance
MP	Member of Parliament
MT	million tonne

MW	megawatt
NAP	National Allocation Plan
NAMAS	Nationally Appropriate Mitigation Actions
NAPCC	National Action Plan on Climate Change
NDCs	Nationally Determined Contributions
NGO	Non-governmental organisation
NIMBY	not in my backyard
NLGCC	National Leading Group on Climate Change (China)
NZ	New Zealand
NZD	New Zealand Dollar
nZEB	nearly zero energy building
NZU	New Zealand Unit
OECD	Organization for Economic Co-operation and Development
PAT	perform, achieve, trade
PES	payments for ecosystem services
PM	particulate matter
PPCA	Powering Past Coal Alliance
PPCDAm	Plan to Prevent and Control Deforestation in the Brazilian Amazon
PPP	public private partnership
PV	photovoltaics
RAN-GRK	National Action Plan for Greenhouse Gas Emissions Reductions
R&D	research and development
REDD+	Reducing Emissions from Deforestation and Forest Degradation
REM	REDD Early Movers Programme
REN21	Renewable Energy Policy Network for the 21st Century
RGGI	Regional Greenhouse Gas Initiative (United States of America)
RMA	Resource Management Act (New Zealand)
SAPCC	State Action Plans on Climate Change
SDGs	sustainable development goals
SDPC	State Development and Planning Commission (China)
SEM	Single European Market
SEPA	State Environmental Protection Administration
SHEA	Shanghai Carbon Emissions Allowance (China)
SPD	Social Democratic Party of Germany (*Sozialdemokratische Partei Deutschlands*)
SRI	socially responsible investment
SUV	sport utility vehicle
TCN	Transnational City Network
UN	United Nations
UNDP	United Nations Development Programme
UNEP	United Nations Environment Programme
UNESCO	United Nations Educational, Scientific and Cultural Organization
UNFCCC	United Nations Framework Convention on Climate Change

UPA	United Progressive Alliance
USA	United States of America
VAT	value-added tax
VGGS	Vietnam Green Growth Strategy
WTO	World Trade Organisation
WWF	World Wide Fund for Nature

Part 1
Introduction

1 Introduction

Climate governance across the globe: pioneers, leaders and followers

Rüdiger K.W. Wurzel, Mikael Skou Andersen and Paul Tobin

Introduction

In a global community of about 200 states and seven billion people, it could be easy to hide and not take action on climate change. Indeed, some have done exactly that. Yet, many choose to act, to achieve long-term benefits for themselves, for others and for those who are not yet born. Sometimes they do so in order to draw others to their cause; in other cases, they do so regardless of how their peers will behave. Urgent action is needed to mitigate climate change. Who leads? Who follows? How? When? Why?

Environmental leaders and pioneers have long been identified as pivotal actors for solving or at least mitigating environmental problems at both the domestic and international governance levels (e.g. Young, 1991; Underdal, 1994; Andersen and Liefferink, 1997; Jänicke, 2006; Liefferink and Wurzel, 2017). Leaders and pioneers can act as 'agents of change' (Liefferink and Wurzel, 2017) who are of central importance for successful climate change governance (e.g. Grubb and Gupta, 2000; Oberthür and Roche Kelly, 2008; Jordan *et al.*, 2012; Parker and Karlsson, 2010; Wurzel, Connelly and Liefferink, 2017; Wurzel, Liefferink and Torney, 2019). Much of the existing literature on leaders and pioneers in environmental governance in general and climate governance in particular has focused on countries in the Global North. We therefore know relatively little about the role environmental leaders and pioneers can play in relation to countries in the Global South. This volume tries to make a contribution towards closing this gap in the literature by offering a critical analysis of climate leadership and pioneership in countries of the Global South (see Chapters 2–6) and in jurisdictions in the Global North (see Chapters 7–13). The chapters in our volume cover a range of large and small countries from both the Global South and Global North. This breadth should allow us to undertake a more nuanced analysis and to draw more informed conclusions (see Chapter 14) on what factors help to explain how, when and why different types of countries offer what type of climate leadership and pioneership.

The 2015 Paris Agreement, which is based on a bottom-up approach and requests that states put forward voluntary national pledges in the form of Nationally Determined Contributions (NDCs), has arguably increased further the importance of climate leaders and pioneers. The Agreement refers explicitly to the need for leadership by stating that 'developed country Parties shall continue taking the lead' (article 4.4). In order to achieve 'rapid reductions' (article 4.1) in greenhouse gas emissions (GHGE), the Paris Agreement is complemented by a mechanism to facilitate 'the exchange of information, experiences and best practices' (adoption decision, items 33–34; see also Chapters 8 and 14). Learning from climate leaders' and pioneers' innovative climate governance measures and, if possible, emulating their best practices, has therefore become even more important.

The successful ratcheting up of ambitions and actions in international climate governance and domestic climate policy depends, however, not only on leaders and pioneers but also on followers. Therefore, the authors of the chapters in this volume were asked to identify, if possible, followers and assess occurrences of climate followership for their case countries. Up to now, the scholarly literature has paid relatively little attention to followers and their interactions with leaders and pioneers in multilevel and polycentric climate governance structures although there are important exceptions (e.g. Torney, 2014, 2015, 2019; Parker, Karlsson and Hjerpe, 2015; Wurzel, Liefferink and Torney, 2019).

Global climate change governance has always taken place within multilevel governance (MLG) structures that include at least the international, national and subnational levels as well as the supranational level for the currently 27 member states of the European Union (EU). While the 1997 Kyoto Protocol adopted a top-down 'targets-and-timetables' approach, the 2015 Paris Agreement rests on a bottom-up approach that relies on voluntary measures propped up by regular reviews and monitoring (e.g. Klein *et al.*, 2017). Many analysts have argued that the Paris Agreement has made global climate governance more *polycentric* (e.g. Ostrom, 2014; Oberthür, 2016a ; Jordan *et al.*, 2018). According to Elinor Ostrom (2010: 552), '[p]olycentric systems are characterized by multiple governing authorities at different scales rather than a monocentric unit ... Each unit within a polycentric system exercises considerable independence to make norms and rules within a specific domain'. However, as Liefferink and Wurzel (2018: 136) have argued, 'while polycentricity and monocentricity, which constitute opposite poles on the governance dimension, are useful heuristic analytical terms, they are rarely found (at least in their pure form)'. As we will explain in more detail below, although multilevel and polycentric governance approaches are not identical, they actually partially overlap, because both concepts emphasise the importance of multiple decision-making or governance centres.

As an analytical starting point, all chapters in this volume take Liefferink and Wurzel's (2017) differentiation that leaders normally actively seek to attract followers, while pioneers do not do so. Importantly, leaders have 'the explicit aim of leading others, and, if necessary, to push others in a follower position' (Liefferink and Wurzel, 2017: 953). Pioneers on the other hand focus primarily on their own actions. The French term *pionnier* refers to a foot soldier or soldier involved in digging trenches (Merriam-Webster online dictionary, 2020). Pioneers carry out activities which, depending on the specific circumstances and subsequent events 'in the field', may or may not help others to follow (Liefferink and Wurzel, 2017: 952–953).

Much of the existing literature tends to employ interchangeably a wide range of analytical terms to describe somewhat similar phenomena, such as leader, pioneer, entrepreneur, forerunner, front-runner, first mover, lead state, lead market, lead actor, pacesetter and trendsetter, to name only some of the most commonly used terms (Liefferink and Wurzel, 2013). The inflationary use of such a wide range of different analytical terms is 'making difficult the emergence of theory-guided cumulative empirical research on the actions and impact of leaders and pioneers, which are widely perceived as important agents of change' (Liefferink and Wurzel, 2017: 951). For this volume, we asked all chapter authors to make use only of the analytical terms *leader* and *pioneer*, and to follow Liefferink and Wurzel's (2017) differentiation, whereby leaders will actively seek to attract *followers*, while this is normally not the case for pioneers. Yet, leadership and pioneership can manifest in many different ways. To obtain clear insights into how countries play influential roles across the globe, we need to typologise their behaviours.

Different types of leadership and pioneership

There is broad agreement in the environmental governance literature that leaders usually exhibit not only one single type of leadership but different types of leadership, either simultaneously or sequentially (e.g. Burns, 1978; Young, 1991, 1999; Underdal, 1994, 1998; Grubb and Gupta, 2000; Parker, Karlsson and Hjerpe, 2015). However, different classifications exist regarding the exact *types* of leadership, which can be exhibited by different actors. For example, Young (1991) identifies structural, entrepreneurial and intellectual leadership, Underdal (1994) differentiates between coercive, instrumental and unilateral leadership, Grubb and Gupta (2000) distinguish between structural, instrumental and directional leadership, and Parker, Karlsson and Hjerpe (2015) demarcate structural, directional, idea-based and instrumental leadership. These concepts are distinct, but overlap with one another. Instead, the chapters in this volume all draw on the same four-fold leadership classification: (1) structural, (2) entrepreneurial, (3) cognitive and (4) exemplary leadership (see Liefferink and Wurzel, 2017; Wurzel, Connelly and Liefferink, 2017; Wurzel, Liefferink and Torney, 2019).

First, *structural* leadership is usually associated with military and/or economic power, especially by international relations (IR) scholars (e.g. Young, 1991, 1999; Underdal, 1994, 1998; Nye, 2008). For structural climate leadership, military power tends to be only of secondary importance – climate change cannot be solved by military means – while economic power is more central for being able to transform structural climate leadership capabilities into actual structural leadership. The size of the domestic market is an important source of structural power, which jurisdictions can try to transform into structural leadership. Young (1991: 288 and 289) has argued that structural leaders 'are experts in translating the possessions of material resources into bargaining leverage' while making use of 'the existence of asymmetries among participants or stakeholders in processes of institutional bargaining'. Although structural power and leadership are closely-related concepts, they are not identical (e.g. Young, 1991; Nye, 2008). Power is a necessary but not a sufficient condition for structural leadership. In his study on US presidents, Burns (1978: 19) has argued that '[a]ll leaders are actual or potential power holders, but not all power holders are leaders'. Burns' argument also applies for the analysis of the behaviour of states and their representatives in international negotiations. This conceptualisation helps to explain why powerful states (e.g. the USA and China) have sometimes failed to offer structural leadership in international climate governance, or have done so only intermittently (see Chapters 2 and 7). Oberthür (2016a: 83) has claimed that '[p]ower and power structures have become an increasingly prominent consideration in analyses of international climate policy in the 21st century' while pointing out 'the rise of climate change to high politics, great power politics and even geopolitics'.

An actor's relative contribution to a particular environmental problem may provide it with structural power. For example, China overtook the USA as the world's largest emitter of carbon dioxide (CO_2) emissions in the 2000s. By doing so, China increased its relevance (and thus arguably also its structural power) for any global climate regime. The bilateral agreement between Presidents Barack Obama and Xi Jinping in November 2014 constituted a milestone in the run-up to the 2015 Paris Agreement, not least because the USA and China constituted the two largest producers of GHGE (Bang and Schreurs, 2017; Li, 2017; Liefferink and Wurzel, 2018: 141). Conversely, one could argue that the EU's declining GHGE are reducing its *structural* or *systemic relevance* for international climate governance (see Chapter 8). However, systemic relevance can be achieved not only by contributing to a governance problem, but also by being able to offer solutions to collective action problems. Systemic relevance is important for a wide range of domestic politics and international governance issues because it provides the actor with considerable power resources that may be activated to try to influence positively (i.e. by offering leadership) or negatively (i.e. by vetoing or watering down) domestic policy-making and/or international governance which is aimed at solving collective problems. For example, Young (1991: 288) has argued that the USA was of systemic relevance for the Bretton Woods agreement because of the central importance of the US dollar, and Dyson (2014) has pointed out that the Economic and Monetary Union (EMU) within the EU would not have been viable without Germany, which constitutes the EU's largest economy.

Importantly, an actor using its structural power to influence domestic policy-making and/or international rule-making becomes a structural environmental leader only by mobilising its structural power in pursuit of strengthening collective goods (Young, 1991; Underdal 1998; Parker, Karlsson and Hjerpe, 2015). In other words, a state that vetoes and/or waters down climate protection measures does not act as a climate leader or pioneer, within this conceptualisation. We therefore follow Underdal's (1998: 101) argument that 'a leader is supposed to exercise what might be called "positive" influence, guiding rather than vetoing or obstructing collective action'. Similarly, Young (1991: 285) has defined leadership as 'the actions of individuals who endeavor to solve or circumvent the collective action problems that plague the efforts of parties seeking to reap joint gains in processes of institutional bargaining'. There is therefore 'a normative dimension which requires the leader/pioneer to facilitate rather than to veto ambitious environmental measures which help to solve collective action problems' (Liefferink and Wurzel, 2017: 957).

Secondly, *entrepreneurial* leadership requires negotiating and diplomatic skills for brokering compromise agreements (Young, 1991; Underdal, 1994, 1998). Young (1991: 295) has argued that 'entrepreneurs play key roles as facilitators of bargaining processes that can all too easily bog down or get diverted into blind alleys in the absence of skilful measures to keep them on track'. While entrepreneurial leaders will try to facilitate compromises acceptable to other actors, they themselves may have stakes and interests in the governance issues that are being negotiated. Large, powerful states usually have more diplomatic resources than smaller, less powerful countries. In the run-up to the Paris Climate Conference (Conference of the Parties – COP21) in late 2015, host nation France invested massive diplomatic resources to facilitate compromise solutions, which paved the way for the Paris Agreement (e.g. Bocquillon and Evrard, 2017; see also Chapter 8). However, smaller countries may sometimes find it easier to act as honest brokers for compromise agreements because their stakes and interests in particular solutions may be lower than those of large, powerful states. Their domestic constituencies tend to be more homogenous, and thus leave broader scope for negotiators in proposing possible and acceptable 'win-sets' in international negotiations, whereas larger countries can experience lock-ins due to heterogeneities in their base of interests at home (cf. Putnam, 1988; Andersen, 2019).

Thirdly, *cognitive* leadership involves putting forward innovative ideas and defining or redefining interests and problem perceptions. Cognitive leaders may propagate concepts to share with other states, such as 'sustainable development' and 'ecological modernisation'. While the concept of sustainable development postulates that equal attention should be paid to environmental, economic and social concerns, adherents to ecological modernisation argue that environmental measures are beneficial not only for the environment but also the economy, for instance, in form of the 'green' jobs created by a low carbon economy. Ecological modernisation in particular has been advocated predominantly by wealthier states, but another noteworthy concept that has been especially instrumental within the global climate governance community resulted from the cognitive leadership of Global South states. After much advocacy by countries that have historically produced fewer GHGE and have fewer resources with which to mitigate climate change (see Chapters 2 and 3), the principle of common but differentiated responsibilities (CBDR) underpinned the United Nations Framework Convention on Climate Change (UNFCCC) that was adopted at the 1992 Rio 'Earth Summit' (e.g. Rajamani, 2012). CBDR has since been enshrined also in the 1997 Kyoto Protocol and the 2015 Paris Agreement (e.g. Klein *et al.,* 2017).

Although powerful actors (e.g. large or resourceful states and transnational corporations) may find it easier to offer structural leadership, this is not necessarily the case for cognitive leadership. For example, smaller EU member states (such as Sweden, as well as, at times, Denmark and previously the Netherlands) are widely seen as having punched well above their structural leadership weight when it comes to being able to offer cognitive leadership (Andersen and Liefferink, 1997; Liefferink and Andersen, 1998; Andersen and Nielsen, 2017; see also Chapters 11, 12 and 13). Cognitive climate leadership usually relies strongly on scientific expertise and practical implementation knowledge (Liefferink and Wurzel, 2017: 959). The EU has been portrayed as a 'normative power' (Manners, 2002), which suggests that it relies more heavily on cognitive than on structural leadership (see Chapter 8) although Damro (2012) has pointed out that the EU frequently acts also as 'market power' when exerting structural climate leadership. Cognitive leadership may also include what Dyson (2014: 5) has called 'arguing power' which stems from the 'capacity to frame how policy issues ... are debated' and allows actors 'to set the normative standards of policy evaluation'. According to Dyson (2014), arguing power can more easily be established by actors who are of systemic relevance (i.e. have considerable structural power that they can also try to transform into structural leadership).

Cognitive leadership often requires considerable (financial, staff and time) resources especially if the generation of novel scientific findings and practical implementation knowledge is involved. However, cognitive leadership also includes the framing of principles such as the above-mentioned CBDR.

Cognitive leadership operates on a different timescale to structural and entrepreneurial leadership. While (military and) economic power can be used almost instantly or at least relatively quickly to transform structural leadership capacities into actual structural leadership, cognitive leadership often requires considerable time to achieve acceptance and become effective. Young (1991: 298) has argued that 'new ideas generally have to triumph over the entrenched mindsets or worldviews held by policymakers, so that the process of injecting new intellectual capital into policy streams is generally a time-consuming one'.

Fourthly, leadership by example, or *exemplary leadership*, is the intentional setting of examples for others, while unintentional example-setting is referred to as *exemplary pioneership* in this volume. *Intentional* exemplary leadership resembles what Grubb and Gupta (2000) have defined as directional leadership. Intentional exemplary leadership and directional leadership amount to what Liefferink and Wurzel (2017: 959) have called a *constructive pusher* position according to which a leader seeks to offer unconditional exemplary leadership. In other words, a constructive pusher will not make its climate leadership actions conditional on other actors taking the same or similar actions. *Constructive pushers* therefore unconditionally put forward domestic policies as models for other actors (cf. Liefferink and Andersen, 1998). Unlike *conditional pushers*, constructive pushers do not make conditional their own actions on the actions of other actors. As conditional and constructive pushers both seek to attract followers (although by different means) they can be subsumed under the umbrella term *leader*.

Pioneers may exhibit unintentional example-setting although they are not usually seeking to attract followers. As explained above, the actions of pioneers may be emulated by other actors despite this outcome not being actively pursued. The policy learning, diffusion and transfer literature offer an abundance of empirical examples of followers who emulated intentional or unintentional example setting by environmental leaders or pioneers (e.g. Tews, Busch and Jörgens, 2003; Tews, 2005; Jänicke, 2006).

Importantly, leaders usually combine different leadership types either simultaneously or sequentially. The specific mix of different types of environmental leadership offered by a particular actor can change over time and may vary across environmental issues and governance levels. This underscores the utility of employing a specific typology, as we do in this volume, in order to avoid seemingly enduring generic labels such as 'environmental leader' that fall to dust when scrutinised across time or policy areas.

Followers and followership

As pointed out above, environmental leaders and pioneers have received significantly more scholarly attention than followers, although there are notable exceptions (e.g. Torney, 2014, 2015, 2019; Parker, Karlsson and Hjerpe, 2015; for the general literature on leaders and followers see Rhodes and t'Hart, 2014). The dearth of studies on followers is at least partly due to the methodological and evidential challenges associated with empirically identifying followers and followership (Torney, 2019; Wurzel, Liefferink and Torney, 2019). Much of the policy transfer, diffusion and learning literature acknowledges that it is generally easier to identify actors who come up with policy innovations (i.e. act as their source). It is much harder to determine which actors have emulated the leaders/pioneers, or the mechanisms (e.g. transfer, diffusion and learning) through which emulation has taken place (e.g. Tews *et al.*, 2003; Tews, 2005).

There is a major challenge for studying followers and followership: states that adopt the same or similar climate governance measures may have gravitated towards such measures independently from each other, rather than by following the example of another state. The possibility of states adopting similar measures independently from each other has been captured well by Lowi's (1964) famous dictum that 'policy determines politics'. Decision-makers in states that possess similarly structured economic, social and political institutions and capacity tend to gravitate 'naturally' towards similar policy solutions. In line with this argument, Hoberg (1986: 358) has pointed out that '[m]any theorists have argued that there is only one best way to resolve a particular problem, and, since nations at similar levels of industrial development confront a common core of problems, responses will converge accordingly'.

However, although Lowi's dictum received considerable support especially among American political scientists, it gained much less traction among European political scientists, for whom Richardson's (1982) 'policy style' concept, which turned Lowi's dictum on its head by postulating in essence that 'politics determines policy', became a more influential concept. In other words, according to the policy style concept, states usually follow certain national path dependencies that may severely constrain the policy options available to domestic policy makers. Importantly, the 'policy determines politics' and the 'politics determines policy' schools of thought offer useful heuristic models that can rarely be found in pure form in real world domestic politics and international governance.

The contemporary debate about climate leaders/pioneers and followers and their preferred climate policy/governance solutions carries faint echoes of the 'policy determines politics' versus 'politics determines policy' scholarly debate. Jänicke (2006) has argued that there is frequently a high degree of 'conformism' according to which most countries converge towards the same preferred policy approaches, which echoes Lowi's dictum that 'politics determines policy' and implies that 'followers' will mostly 'copy' the climate mitigation measures taken and policy instruments applied by leaders or pioneers. In contrast, Rhodes and t'Hart (2014: 6) have pointed out '[t]here is now a growing body of thought and research that understands leadership as an interactive process between leaders and followers'. Seen from Rhodes and t'Hart's (2014) perspective, 'follower' countries can be expected to adapt and alter climate mitigation measures and policy instruments to their specific preconditions and their (national) institutions of governance. For example, energy efficient and low-carbon district heating is widely distributed in Europe's Nordic countries, that rely on it heavily not only because of their cold climate but also because of the presence of strong local authorities with planning powers and institutionalised sources of low-interest finance (see Chapter 11). Countries that seek to follow the Nordic countries' trail of collective heating systems may need to adapt such systems considerably, for example, to domestic traditions of private sector involvement and finance. Another example of the 'politics determines policy' patterns identified by Richardson and collaborators (1982) can be found in renewable energy policies. The 'feed-intariffs' approach pioneered in Germany was transformed by the state's followers, such as Sweden and the UK, into a more market-oriented system of renewable obligation certificates (Butler and Neuhoff, 2005; Söderholm, Ek and Pettersson, 2007).

In order to convincingly identify followers and followership, it is important not only to determine a purported leader/pioneer and an actor that has subsequently adopted similar policies or responses; one also has to establish that the actions of the purported follower were indeed caused by preceding actions carried out by leaders or pioneers (Torney, 2019; Wurzel, Liefferink and Torney, 2019: 12–13). Such requirements are inherently more complex, time-consuming and costly for researchers to fulfil, hence arguably the current lack of empirical examples within the current literature.

Followership and rapid policy learning from best practice are of central importance in multilevel and polycentric climate governance especially if, as postulated by the Paris Agreement, a bottom-up approach is adopted. The policy transfer, diffusion and learning literature covers effectively why and how pioneers may unintentionally, and leaders intentionally, attract followers (e.g. Tews *et al.*, 2003; see also Jänicke and Wurzel, 2019).

Why do some countries become climate leaders or pioneers?

Much of the state-centred comparative politics (CP) literature has focused on a wide variety of factors, which help to explain the actions of leaders and pioneers and their ambitions and motivations. Drivers for acting as environmental leaders and pioneers include a high level of problem pressure, high political salience of environmental issues and regulatory competition (e.g. Jänicke and Jacob, 2002; Liefferink *et al.*, 2009). The environmental capacity literature has identified, among others, institutional, politico-administrative, informational-cognitive and technological capacities as important factors for states being able to offer environmental leadership/pioneership (e.g. Jänicke, 2006).

High problem pressure, high political salience and regulatory competition are also important factors for explaining the actions of climate leaders and pioneers as well as followers (Wurzel, 2002). Creating and maintaining a 'green' public image appears to be particularly important for cities, regardless of whether they are affluent (e.g. Bulkeley and Betsill, 2005; Kern and Bulkeley, 2009) or structurally disadvantaged (Wurzel *et al.*, 2019).

Another way of looking at why actors strive to become leaders or pioneers is to investigate the way their 'green' ambitions are structured. While drawing on the distinction between leaders and pioneers, we assess four possible combinations of an actor's *internal* and *external ambitions* (von Prittwitz, 1984) on a scale ranging from 'low' to 'high'.

Liefferink and Wurzel (2017) distinguish in ideal-typical fashion between the following four *positions* that states may take up. First, low internal and low external ambitions lead actors to become *laggards* (or, at best, followers); second, the combination of high internal and low external ambitions turns actors into *pioneers* which try to 'go it alone'. Third, when the opposite is the case and there are low internal and high external ambitions, actors become *symbolic leaders* that fail to back up domestic climate action with their externally directed 'green' ambitions in foreign climate policy. Finally, as discussed above in the section on 'exemplary leadership', a state that marries high internal and high external ambitions is either a constructive or conditional pusher state. Constructive and conditional pusher states fall under the umbrella term *leaders*, which can then be manifested as one of the four types (structural, entrepreneurial, cognitive and exemplary).

Putting the distinction between the two concepts to one side for a moment, how do we define the 'ambitiousness' that is inherent to pioneership and leadership? In this volume, pioneership and leadership refer to actors who are either *first* to introduce and/or propagate a certain climate policy measure, or who exhibit the *highest* level of ambition. As Liefferink and Wurzel (2017: 956) have pointed out '[b]oth "the first in class" and "the best in class" can in principle be viewed as leaders or pioneers, although the motivations underlying their differing ambitions and the subsequent consequences may be different'. Importantly, Burns (2003: 26) has argued that '[f]ollowers might outstrip leaders. They might become leaders themselves'. Former followers who develop into leaders may become more innovative and ambitious than previous or existing leaders, who in turn may become followers. Sometimes, leaders become laggards, at least temporarily, or at best followers, because they downgrade their climate governance ambitions. For example, after an election, a new government may attribute significantly lower importance to climate protection; a scenario that happens in both Global South states (see especially Chapters 3, 5 and 6) and Global North countries (see especially Chapters 7, 10, 11 and 12). These examples show that it is also important to analyse the domestic politics of climate change. Yet IR-inspired studies of global climate governance sometimes ignore or at least downplay domestic politics, limiting their explanatory ability.

Moreover, all climate leaders have some blind spots (Wurzel, 2008). It is extremely difficult, if not impossible, for any single country to remain indefinitely a leader or pioneer across all climate governance issues. Having said that, some countries have shown considerable more staying power as climate leaders or pioneers than others. The chapters in this volume offer novel empirical findings and new analytical insights, which will help us to gain a better understanding of leaders, pioneers and followers from across the globe and how they act and interact with each over time. Both MLG and polycentric governance concepts are well geared towards capturing both policy and politics dimensions at different and/or multiple levels of climate governance.

Multilevel and polycentric governance

While focusing on different governance levels, MLG concepts usually emphasise the important role played by public institutions, such as the state, and supranational and subnational actors. The MLG concept was initially proposed to capture the complexities of federal states (e.g. USA) and quasi-federal jurisdictions, such as the EU (Marks, 1993). Following Hooghe and Marks (2003), the MLG approach has gradually been extended and applied to the dynamics of wider regional and international regimes. MLG concepts aim to grasp the subtle shift of the locus of decision-making to a wider spectrum of actors at different spatial and sectoral levels. Hooghe and Marks (2003) contrast the conventional state-centred approaches' assumption of a 'Russian doll' hierarchy between actors (one actor is smaller and embedded within the context of another), with the 'marble cake' conceptualisation of governance actors of MLG (whereby different actors interweave in myriad ways). For the latter concept, the demarcations of competences are partly overlapping and somewhat fluid, enabling various sub-state, sectoral, supranational and international actors to exploit MLG opportunity structures for initiative, interplay and influence. Importantly, this understanding applies to both (sub-state, sectoral, supranational and international) climate leaders/pioneers and to climate laggards at such levels. The MLG concept has frequently been criticised for being little more than a metaphor, lacking the analytical stringency required for disentangling causal factors, as part of a continuous and cumulative research programme (Jordan and Lieffeink, 2004; Blom-Hansen, 2005). Indeed, empirical applications of the MLG concept are wide-ranging and have sometimes relaxed the concept's implicit requirements. To narrow the conceptualisation of

MLG, we assume a specific definition in this volume of MLG. That is, MLG highlights 'above all the changing role and relevance of the traditional nation-state and the fading away of the Westphalian international system, which is gradually replaced by a more fluid politico-institutional order in which power and authority are redistributed from the state upwards but also downwards' (Tortola, 2017: 244), and possibly sideways, for example, towards business (e.g. Strange, 1988).

Seen through the MLG lens, the policy space provided for leadership (and followership) is hence considerably larger and wider than one would assume from the perspective of either traditional state-centred CP approaches, or a conventional IR perspective of a global climate regime with some 195 countries. The congregation of mayors from across the globe in the C40 network of 96 cities, in which 'the world's leading cities [are] taking bold climate action' (C40 Cities, 2020) reflects the significance of leadership exercised by subnational actors well beyond the framework of their respective regions and nation states. The structures and processes of MLG are of interest here, as we seek to understand how different types of actors employ various types of leadership, and for what purposes. Much of the existing climate governance literature has focused primarily on international climate negotiations, particularly within the UNFCCC. However, the Paris Agreement and simultaneous bottom-up developments are gradually widening the focus beyond state action to include also innovative subnational climate measures. Indeed, these initiatives may then be upscaled to the national level or transferred sideways to other subnational actors (e.g. through the C40 network) in an attempt to deliver the promised NDCs. It is especially during those delivery processes that the theoretical perspective offered by MLG provides useful analytical lenses for the assessment of leadership, pioneership and followership. The vacuum in global leadership that emerged after the announcement by President Trump that the USA will withdraw from the Paris Agreement at the end of 2020 has not yet been filled, with China being unable to do so (see Shen and Xie, 2019). Still, 'the mutual help among poor brothers' (ibid.: 717)[1(1)] that was provided by China's South–South climate fund serves as yet another illustration of the fluid processes that exist outside the core international climate governance structures.

When it comes to who are core actors, and what the relations are between leaders, pioneers and followers, MLG concepts are located somewhere in between polycentric and state-centric governance concepts. Broadly speaking, state-centred concepts argue that states are the most important actors. Polycentric governance concepts usually identify non-state actors within specific policy domains as core actors. Finally, MLG concepts tend to emphasise the importance of mutual dependencies between sub-state, state and supranational actors across different governance levels. However, as discussed above already, some variants of MLG concepts are quite close to polycentric governance concepts, regarding the importance they attach to non-state actors. In the climate governance literature, both MLG (e.g. Grubb and Gupta, 2000; Schreurs and Tiberghien, 2007; Jordan *et al.*, 2012) and polycentric governance (e.g. Ostrom, 2014; Homsy and Warner, 2014; Oberthür, 2016a; Morrison *et al.*, 2017; Jordan *et al.*, 2018) approaches have been widely used. Moreover, some studies have drawn on both concepts (e.g.

Wurzel, Connelly and Liefferink, 2017). The shift away from a top-down 'targets-and-timetables' climate governance approach, as embodied in the 1997 Kyoto Protocol, towards a bottom-up approach with voluntary pledges as enshrined in the 2015 Paris Agreement, has partly driven this development. Jordan and colleagues (2018: 4) have argued that developments in climate change governance 'appear to confirm the trend towards greater polycentricity'. Similarly, Oberthür (2016a: 81) has argued that the Paris 'Agreement recalibrates the role of the multilateral UN process as providing overall direction towards global decarbonisation, while leaving implementation to other international organisations, states and various non-state actors and initiatives'. Thus, these stances chime with the argument put forward by Selin and VanDeveer (Chapter 7) that the USA is experiencing a 'polycentric turn'.

Polycentric governance concepts share certain core presuppositions (such as multiple centres of decision-making) with MLG approaches, although conceptually they are *not* identical (e.g. Homsy and Warner, 2014; Wurzel *et al.*, 2017; Jordan *et al.*, 2018; Liefferink and Wurzel, 2018). By comparison with polycentric governance approaches, MLG concepts usually assume a stronger role for governmental (i.e. subnational, state and supranational) actors (Morrison *et al.*, 2017; Liefferink and Wurzel, 2018; Wurzel, Liefferink and Torney, 2019). As mentioned above, the MLG concept, which was initially developed for federal and quasi-federal political jurisdictions (e.g. Marks, 1993; Hooghe, 1996), emphasises the mutual dependency of subnational, national and supranational governmental actors. MLG concepts differ from state-centred approaches by rejecting the idea of traditional top-down *government* in favour of less hierarchical *governance*, which is an assumption that is shared by polycentric governance approaches. In contrast to MLG concepts, polycentric governance concepts attribute a higher degree of autonomy to non-state societal actors, such as business, NGOs and citizens, while putting greater emphasis on the role of functionally defined governance domains that are characterised by a high degree of self-organisation (e.g. Ostrom, 2010, 2014). According to Ostrom (2010: 552), who pioneered the concept of polycentric governance approaches alongside her co-authors, '[p]olycentric systems are characterized by multiple governing authorities at different scales rather than a monocentric unit … Each unit within a polycentric system exercises considerable independence to make norms and rules within a specific domain'. Her definition echoes a broad-based recognition that emerged in the wake of the failed climate negotiations at COP15 in Copenhagen in 2009: only by involving civil society in accelerating a low-carbon transition would it be possible to mitigate global warming and accomplish UNFCCC objectives.

According to Dorsch and Flachsland (2017) one of the advantages of polycentric governance is that it encourages experimentation and learning-by-doing at regional and local levels which, if successful, may lead to upscaling of innovative climate governance measures to higher climate governance levels (see also Ostrom, 2012, 2014; Kern, 2019).

In line with Wurzel, Liefferink and Torney (2019), we argue that polycentric governance approaches emphasise the importance of societal self-coordination within loosely coupled governance structures (e.g. Ostrom, 2012, 2014). Meanwhile, MLG advocates often posit that governmental actors (including supranational EU actors) must play an important, if not dominant, role for climate policy (e.g. Marks, 1993; Hooghe, 1996; Homsy and Warner, 2014). This distinction also has consequences for the conceptualisation of leadership and pioneership. Polycentricity can help us to understand why and how bottom-up self-governing initiatives emerge and flourish (Dorsch and Flachsland, 2017; Jordan *et al.*, 2018). Due to the relatively high autonomy of functionally specific (or domain specific) polycentric subsystems, it may, however, be difficult for leaders to attract followers from other functionally defined quasi-autonomous subsystems (see Liefferink and Wurzel, 2018).

Morrison *et al.* (2017: 2) have accused proponents of polycentricity of inadvertently ignoring 'not only different types of power at play but also how their distribution may affect both governance processes and environmental outcomes'. Similarly, Singleton (2017: 1000) has argued that '[p]ower is a concept that remains largely underdeveloped within Ostrom's work rendering her themes "curiously apolitical"' (Wall, 2014: 480). Interestingly, Lowi's (1964) argument that 'policy determines politics' also largely downplayed the importance of power and power asymmetries for decision-making.

In contrast, MLG concepts frequently adopt a 'top-down view of subnational actors' (Fairbrass and Jordan, 2004: 152), according to which national governmental actors and supranational EU actors have greater decision-making powers at their disposal. However, this status exists despite the mutual dependencies between national and supranational actors, as well as subnational governance actors.

Both MLG and polycentric governance approaches conceptualise 'the plurality of actors and levels and the complexity of their interactions not as obstacles but as an opportunity for innovation and interactive learning' (Jänicke, 2017: 118; see also Marks and Hooghe, 2004: 16; Ostrom, 2012, 2014). Ostrom (2014: 119) has advocated the adoption of 'a polycentric approach to the problem of climate change in order to gain the benefits at multiple scales as well as to encourage experimentation and learning from diverse policies adopted at multiple scales'. Similarly, Marks and Hooghe (2004: 16) argued that MLG structures 'facilitate innovation and experimentation'. Thus, these concepts proffer nuanced means for looking beyond traditional understandings of power and authority. As such, they are especially useful for analysing the complex webs of interactions that define climate governance, in which every actor is – to highly variegated degrees – both responsible for and vulnerable to the impacts of climate change.

The structure of this book

Following this introduction in Part I, our volume divides Part II into five chapters on Global South countries, namely China, India, Costa Rica and Vietnam, New Zealand, and Brazil and Indonesia. Part III shifts the focus onto Global North countries and jurisdictions. We examine the USA, EU, Germany, UK, four Nordic countries (Denmark,

Finland, Norway and Sweden), Ireland, and Switzerland. Finally, in Part IV, we use the conclusion to provide a comparative summary of the main novel empirical findings and new analytical insights put forward in the preceding chapters. There, we offer a critical comparative analysis of the core conceptual themes, namely types of climate leaders, pioneers and followers, and MLG and polycentric governance.

The chapters in this volume offer rich, detailed case studies of 16 countries and the EU, with the aim of gaining a better understanding of which climate change measures have been driven forward by leaders and pioneers, and how. The novel empirical findings and new analytical insights presented in the chapters should allow us to obtain a better comprehension of which climate innovations and best practices may have to be altered, if they are to be used by others. Indeed, in jurisdictions that have developed along certain path dependencies, even 'actors of change', such as leaders and pioneers, may find it challenging to alter their paths significantly. Yet, regardless of this difficulty, the threat of climate change demands transformational change and innovation, rather than merely transactional tweaks and adjustments. This volume provides an exploration of climate innovations across the globe. Time is running low for the kinds of systematic changes we need, but these chapters demonstrate that it is possible to build – but also to lose – momentum towards a more low-carbon future.

Acknowledgements

Rudi Wurzel would like to thank the British Academy for grant no. SG 131240. Mikael Skou Andersen gratefully acknowledges financial support from Nordforsk, Nordic Energy and Nordic Innovation for grant no. 82841. The support of the Economic and Social Research Council (ESRC), UK, is gratefully acknowledged by Paul Tobin, having funded him via grant no. ES/S014500/1. All three authors are grateful to the Innovations in Climate Governance (INOGOV) programme of COST which funded a workshop on 'Pioneers and Leaders in Polycentric Climate Governance' in Hull, UK.

Note

1 The expression cited by Shen and Xie (2019: 717) stems from Xie Zhenhua, who was China's lead negotiator at COP21 in Paris.

Bibliography

Andersen, M.S. (2019) 'The politics of carbon taxation: how varieties of policy style matter', *Environmental Politics*, 28(6): 1084–1104.

Andersen, M.S. and Liefferink, J.D. (eds.) (1997) *European environmental policy: the pioneers*, Manchester: Manchester University Press.

Andersen, M.S. and Nielsen, H.O. (2017) 'Denmark: small state with a big voice and bigger dilemas', In R. Wurzel, J. Connelly and D. Liefferink (eds.) *The European Union in international climate change politics*, London: Routledge, 83–97.

Bang, G. and Schreurs, M. (2017) 'The United States: the challenge of global climate leadership in a politically divided state'. In R. Wurzel, J. Connelly and D. Liefferink (eds.) *The European Union in International Climate Change Politics*, London: Routledge, 239–253.

Blom-Hansen, J. (2005) 'Principals, agents and the implementation of EU cohesion policy', *Journal of European Public Policy*, 12(4): 624–648.

Bocquillon, P. and Evrard, A. (2017) 'French climate policy: diplomacy in the service of symbolic leadership', In R. Wurzel, J. Connelly and D. Liefferink (eds.) *The European Union in international climate change politics*, London: Routledge, 98–113.

Bulkeley, H. and Betsill, M. (2005) 'Rethinking sustainable cities: multilevel governance and the "urban" politics of climate change', *Environmental Politics*, 14(1): 42–63.

Burns, J.M. (1978) *Leadership*, New York: Harper & Row.

Burns, J.M. (2003) *Transforming leadership*, New York: Grove Press.

Butler, L. and Neuhoff, K. (2005) *Comparison of feed in tariff, quota and auction mechanisms to support wind power development*, Working Papers in Economics 0503, Faculty of Economics, University of Cambridge.

C40 Cities (2020) https://twitter.com/c40cities (Accessed 14.7.2020).

Damro, C. (2012) 'Market power Europe', *Journal of European Public Policy*, 19(5): 682–699.

Dorsch, M.J. and Flachsland, C. (2017) 'A polycentric approach to global climate governance', *Global Environmental Politics*, 17(2): 45–64.

Dyson, K. (2014) *States, debt and power. 'Saints' and 'sinners' in European history and integration*, Oxford: Oxford University Press.

Fairbrass, J. and Jordan, A. (2004) 'Multi-level governance and environmental policy', In I. Bache and M. Flinders (eds.) *Multi-level governance*, Oxford: Oxford University Press, 147–164.

Grubb, M. and Gupta, J. (2000) 'Climate change, leadership and the EU', In J. Gupta and M. Grubb (eds.) *Climate change and European leadership*, Dordrecht: Kluwer, 3–14.

Hoberg, G. (1986) 'Technology, political structure and social regulation', *Comparative Politics*, 18(3): 357–376.

Homsy, G.C. and Warner, M.E. (2014) 'Cities and sustainability: polycentric action and multilevel governance', *Urban Affairs Review*, 49(1): 1–28.

Hooghe, L. (ed.) (1996) *Multi-level governance and European integration*, Oxford: Clarendon Press.

Hooghe, L. and Marks, G. (2003) 'Unraveling the central state, but how? Types of multi-level governance', *American Political Science Review*, 97(2): 233–243.

Jänicke, M. (2006) 'Trend setters in environmental policy: the character and role of pioneer countries', In M. Jänicke and K. Jacob (eds.) *Environmental governance in global perspective*, Berlin: Environmental Policy Research Centre, 51–66.

Jänicke, M. (2017) 'The multi-level system of global climate governance – the model and its current state', *Environmental Policy and Governance*, 27: 108–121.

Jänicke, M. and Jacob, K. (2002) *Ecological modernisation and the creation of lead markets*, FFU Report 02-03, Berlin: Environmental Policy Research Centre.

Jänicke, M. and Wurzel, R. (2019) 'Leadership and lesson-drawing in the European Union's multilevel climate governance system', *Environmental Politics*, 28(1): 22–42.

Jordan, A. and Liefferink, D. (eds) (2004) *Environmental policy in Europe: The Europeanization of national environmental policy*, London: Routledge.

Jordan, A. *et al.* (2012) 'Understanding the paradoxes of multilevel governing: climate change policy in the European Union', *Global Environmental Change*, 12(2): 43–66.

Jordan, A., Huitema, D., van Hasselt, H. and Forster, J. (eds.) (2018) *Governing climate change: polycentricity in action*, Cambridge: Cambridge University Press.

Kern, K. (2019) 'Cities a leaders in multilevel climate governance: embedded upscaling of local experiments in Europe', *Environmental Politics*, 28(1): 125–145.

Kern, K. and Bulkeley, H. (2009) 'Cities, Europeanization and multi-level governance: governing climate change through transnational municipal networks', *Journal of Common Market Studies*, 47(2): 309–332.

Klein, D., Carazo, M.P., Doelle, M., Bulmer, J. and Higham, A. (eds.) (2017) *The Paris agreement on climate change: analysis and commentary*, Oxford: Oxford University Press.

Li, X. (2017) 'China: from a marginalized follower to an emerging leader in climate politics', In R. Wurzel, J. Connelly and D. Liefferink (eds.) *The European Union in international climate change politics*, London: Routledge, 239–253.

Liefferink, D. and Andersen, M.S. (1998) 'Strategies of the "green" member states in EU environmental policy-making', *Journal of European Public Policy*, 5(2): 254–270.

Liefferink, D., Arts, B., Kamstra, J. and Ooijevaar, J. (2009) 'Leaders and laggards in environmental policy', *Journal of European Public Policy*, 16(5): 677–700.

Liefferink, D. and Wurzel, R.K.W. (2013) Environmental leaders and pioneers: towards a better conceptualisation, Paper presented at the UACES 43rd Annual Conference, Leeds, 2–4 September.

Liefferink, D. and Wurzel, R.K.W. (2017) 'Environmental leaders and pioneers: agents of change?', *Journal of European Public Policy*, 24(7): 651–668.

Liefferink, D. and Wurzel, R.K.W. (2018) 'Leaders and pioneers in polycentric governance', In A. Jordan *et al.* (eds.) *Governing climate change: polycentricity in action*, Cambridge: Cambridge University Press, 135–151.

Liefferink, D., Boezeman, D. and de Coninck, H. (2017) 'The Netherlands: a case of fading leadership', In R. Wurzel, J. Connelly and D. Liefferink (eds.) *The European Union in international climate change politics*, London: Routledge, 131–144.

Lowi, T. (1964) 'American business, public policy, case studies and political theory', *World Politics*, 16, 677–715.

Manners, I. (2002) 'Normative power Europe: a contradiction in terms', *Journal of European Public Policy*, 40(2): 235–258.

Merriam-Webster Online Dictionary (2020) 'Pioneer', *Merriam-Webster.com*, https://www.merriam-webster.com/dictionary/pioneer (Accessed 14.7.2020).

Marks, G. (1993) 'Structural policy and multi-level governance in the EC', In A. Cafruny and G. Rosenthal (eds.) *The state of the European community*, Boulder: Lynne Rienner, 391–411.

Marks, G. and Hooghe, L. (2004) 'Contrasting visions of multi-level governance', In I. Bache and M. Flinders (eds.) *Multi-level governance*, Oxford: Oxford University Press, 15–30.

Morrison, T.H., Adger, W.N., Brown, K., Lemos, M.C., Huitema, D. and Hughes, T.P. (2017) 'Mitigation and adaptation in polycentric systems: sources of power in the pursuit of collective goals', *WIREs Climate Change*, 7: 1–16, doi: 10.1002/wcc.479.

Nye, J. (2008) *The powers to lead*, Oxford: Oxford University Press.

Oberthür, S. (2016a) 'Reflections on global climate politics post Paris: power, interests and polycentricity', *The International Spectator*, 51(4): 80–94.

Oberthür, S. (2016b) 'Where to go from Paris? The European Union in climate geopolitics', *Global Affairs*, 2(2): 119–130.

Oberthür, S. and Roche Kelly, C. (2008) 'EU leadership in international climate policy: achievements and challenges', *The International Spectator*, 43(3): 35–50.

Ostrom, E. (2010) 'Polycentric systems for coping with collective action and global environmental change', *Global Environmental Change*, 20(4): 550–557.

Ostrom, E. (2012) 'The future of the commons: beyond market failure and government regulation', In E. Ostrom *et al.* (eds.) *The future of the commons*, London: Institute of Economic Affairs, 68–83.

Ostrom, E. (2014) 'A polycentric approach for coping with climate change', *Annals of Economics and Finance*, 15(1): 97–134.

Parker, C.F. and Karlsson, C. (2010) 'Climate change and the European Union's leadership moment: an inconvenient truth?', *Journal of Common Market Studies*, 48(4): 923–943.

Parker, C.F., Karlsson, C. and Hjerpe, M. (2015) 'Climate change leaders and followers. Leadership recognition and selection in UNFCCC negotiations', *International Relations*, 29(4): 434–454.

von Prittwitz, V. (1984) *Umweltaußenpolitik*, Frankfurt: Campus Verlag.

Putnam, R. (1988) 'Diplomacy and domestic politics: the logic of two-level games', *International Organization*, 42(3): 427–460.

Rajamani, L. (2012) 'The reach and limits of the principle of common but differentiated responsibilities and respective capabilities in the climate change regime', In N.K. Dabash (ed.) *Handbook of climate change and India: development, politics and governance*, London: Earthscan, 118.

Rhodes, R.A.W. and 't Hart, P. (2014) 'Puzzles of political leadership', In R.A.W. Rhodes and P. 't Hart (eds.) *The handbook of political leadership*, Oxford: Oxford University Press, 1–21.

Richardson, J. (ed.) (1982) *Policy styles in Western Europe*, London: Allen and Unwin.

Schreurs, M.A. and Tiberghien, Y. (2007) 'Multi-level reinforcement: explaining European Union leadership in climate change mitigation', *Global Environmental Politics*, 7(4): 19–46.

Shen, W. and Xie, L. (2019) 'Can China lead in multilateral environmental negotiations? Internal politics, self-depiction, and China's contribution in climate change regime and Mekong governance', *Eurasian Geography and Economics*, 59(5–6): 708–732.

Singleton, B.E. (2017) 'What is missing from Ostrom? Combining design principles with the theory of sociocultural viability', *Environmental Politics*, 26(6): 994–1014.

Söderholm, P., Ek, C. and Pettersson, M. (2007) Wind power development in Sweden: global policies and local obstacles, *Renewable and Sustainable Energy Reviews*, 11(3): 365–400.

Strange, S. (1988) *States and markets: an introduction to international political economy*, London: Pinter.

Tews, K. (2005) 'The diffusion of environmental policy innovations: cornerstones of an analytical framework', *Environmental Policy and Governance*, 15(2): 63–79.

Tews, K., Busch, P. and Jörgens, H. (2003) 'The diffusion of new environmental policy instruments', *European Journal of Political Research*, 42(4): 569–600.

Torney, D. (2014) 'External perceptions of and EU foreign policy effectiveness: the case of climate change', *Journal of Common Market Studies*, 52(6): 1358–1373.

Torney, D. (2015) *European Climate Leadership in Question*, Cambridge: MIT Press.

Torney, D. (2019) 'Follow the leader? Conceptualising the relationship between leaders and followers in polycentric climate governance', *Environmental Politics*, 28(1): 167–182.

Tortola, P.D. (2017) 'Clarifying multilevel governance', *European Journal of Political Research*, 56(2): 234–250.

Underdal, A. (1994) 'Leadership theory: rediscovering the arts of management', In W.I. Zartman (ed.) *International multilateral negotiation*, San Francisco: Jossey-Bass, 178–197.

Underdal, A. (1998) 'Leadership in international environmental negotiations: designing feasible solutions', In A. Underdal (eds.) *The politics of international environmental management*, Dordrecht: Kluwer, 101–127.

Wall, D. (2014) *The sustainable economics of Elinor Ostrom*, London: Routledge.

Weidner, H., Jänicke, M. and Jörgens, H. (eds.) (2002) *Capacity building in national environmental policy*, Berlin: Springer.

Wurzel, R.K.W. (2002) *Environmental policy-making in Britain, Germany and the European Union*, Manchester: Manchester University Press.

Wurzel, R.K.W. (2008) 'Environmental policy: EU actors, leader and Laggard States', In J. Hayward (ed.) *Leaderless Europe*, Oxford: Oxford University Press, 66–88.

Wurzel, R.K.W., Connelly, J. and Liefferink, D. (eds.) (2017) *The European Union in international climate change politics*, London: Routledge.

Wurzel, R.K.W., Liefferink, D. and Torney, D. (2019) 'Climate pioneership and leadership in structurally disadvantaged maritime port cities', *Environmental Politics*, 28(1): 1–21.

Wurzel, R.K.W., Moulton, J., Osthorst, W., Mederake, L., Deutz, P. and Jonas, A. (2019) 'Climate pioneership and leadership in structurally disadvantaged Maritime Port cities', *Environmental Politics*, 28(1): 146–166.

Young, O.R. (1991) 'Political leadership and regime formation: on the development of institutions in international society', *International Organization*, 45(3): 281–309.

Young, O.R. (1999) 'Regime effectiveness: taking stock', In O.R. Young (ed.) *The effectiveness of international environmental regimes*, Cambridge: MIT Press, 249–279.

Part 2
Global South

2 China

Emerging low-carbon pioneers at city level

Xinlei Li

Introduction

China has complex climatic conditions and a fragile ecological environment which has frequently suffered natural disasters. In 2018, China's total carbon dioxide (CO_2) emissions reached 10 billion tonnes, which accounts for about 30% of global CO_2 emissions. Climate change has become a politically more salient issue together with related domestic environmental issues. In recent years, the increased frequency of extreme environmental events (such as large-scale particulate matter [PM] 2.5 smog pollution in the northern parts of China) has brought domestic environmental problems to the fore along with major international climate governance issues. The government has recognised that incorporating climate change into its national sustainable development strategy and overall socio-economic development planning is conducive to establishing a resource-saving and environment-friendly society.

China's climate change policy is not only a domestic issue, but has significant repercussions for the rest of the world. China's position and interest in climate politics has achieved increased global attention since China has become the world's largest emitter of greenhouse gas emissions (GHGE) in 2007. Initially China took part in the international climate negotiations mainly as a foreign policy requirement, with the aim of breaking its diplomatic isolation after the 1989 Tiananmen incident. It allied itself with developing countries in dealing with international issues. China has evolved from an observer in the 1987 Montreal Protocol, a marginalised participant in the 1992 United National Framework Convention on Climate Change (UNFCCC), an active G77 leader in the 1997 Kyoto Protocol (Heggelund, 2007), a major player at the 2009 Copenhagen climate conference to an emerging leader at the 2015 Paris Climate Change Conference.

Under the Paris Agreement, China pledged that its emissions will peak by 2030 and that it will reduce its CO_2 emissions per unit of gross domestic product (GDP) by 60–65% by 2030 (compared to 2005). To fulfil these ambitious commitments in the most cost-effective way, policy makers seek to gather increasingly specific, subnational information about sources of CO_2, their reduction potential and the economic implications for possible policies. More and more pioneering cities are emerging in China which has embarked on a course of low-carbon transition to enhance sustainability. Since China has great weight in global climate governance, its low-carbon transition solution model, if successful, will have a significant influence on future global climate governance approaches.

For the Chinese government, cities have become an important priority for low carbon development and climate actions (Lester, 2006). By late 2019, China's urbanisation rate reached a record 60%. It is expected to rise to 75% by 2050. In the late 2010s, cities produced almost 85% of China's CO_2 emissions. Urban direct energy use per capita is estimated to be three times higher than that of rural areas and indirect energy use (e.g. through infrastructure and consumption of goods) is even higher.

The Chinese government has realised that although cities are part of the climate change problem they will also need to be part of the solution. Any international and national climate policy will ultimately be implemented at the local level, the most important of which is the city. Cities have sufficient funds and human and technology resources to become low-carbon pioneers (Bulkeley and Betsill, 2005). Furthermore, many cities are important economic and financial centres which are at the forefront of high-tech, high value-added and independent industrial innovation capabilities. Global cities attract talent, technology, capital, and information. China is a large and economically diverse country that has rapidly industrialised and urbanised although through unbalanced development. The cities in the southeast coastal areas, which exhibit a high degree of development, have the capacity and willingness to improve their green industry and low-carbon transition. They are the main drivers for multilayered climate governance, and have gradually emerged as significant pioneering low-carbon actors. Sub-national actors like cities have become the focus of global climate governance because they can provide creative climate policies which could be upscaled to the national and/or global climate governance level. Cities have also created global governance networks through which innovative ideas and practices have been diffused and transferred (e.g. Homsy and Warner, 2013).

This chapter examines climate leadership and pioneership, primarily at the level of cities. The next section introduces China's changing climate policy and assesses the differing roles which the country has played in international climate governance. The section which follows focuses on the role of city governments in multilevel climate governance and on the mechanisms which support cities' low-carbon efforts. The third section examines Shanghai. The penultimate section explores the potential of China's pioneering cities in international climate co-operation. Finally, analysis of the empirical findings will be based on the theoretical concepts used in this chapter.

China's changing international climate policy

China's participation in global climate change governance served as an ideal opportunity for trying to heal the diplomatic rift with Western countries and to regain its international position in the early 1990s (Chen, 2009: 106–107). China initially lacked a comprehensive understanding of the global climate change regime. Its limited recognition of the climate change issue was reflected in the institutional composition of the National Group of Co-ordination on Climate Change established in September 1990. China aimed to co-ordinate diplomatic negotiations but failed to include officials in charge of economic development (Yan and Xiao, 2010: 83). The main leading institutions of this delegation were from the China Meteorological Administration (CMA) and the State Environmental Protection Administration (SEPA). Although China did not initially recognise the impact of climate change on its domestic energy regulation and economic development, the country perceived joining the UNFCCC as an opportunity to reinforce its diplomatic relations with other developing countries. These efforts promoted the formation of the bloc of the G77 plus China. Since 1991, China has aligned itself closely with India. Both countries have emphasised the historical responsibility of Annex-1 countries (i.e. industrialised countries) and strongly opposed legally binding emissions reduction targets for developing countries while insisting that industrialised countries should provide technology and financial aid to developing countries (Chen, 2009: 104; see also Chapter 4 in this volume).

The 1997 Kyoto Protocol became the first international treaty to set legally binding GHGE reduction targets for industrialised countries. Along with its increased involvement in the international climate change negotiations, China has gradually realised that the Kyoto Protocol not only relates to environmental concerns, but also has great influence on national economic growth. Consequently, climate change gained more attention domestically which, in turn, led to institutional changes in the bureaucratic system. In 1998, the main responsibility for climate change negotiations shifted to the most powerful agency, the State Development and Planning Commission (SDPC). It symbolised that climate change had entered the political mainstream in China. In the same year, SDPC representatives signed the Kyoto Protocol and promoted an attitudinal change in relation to the three flexible mechanisms in the Kyoto Protocol – Clean Development Mechanism (CDM), Joint Implementation (JI) and emissions trading. In around 2001, China gradually changed its sceptical attitude towards the CDM which has stimulated the initial development of large-scale renewable energy (e.g. Heggelund, 2007).

China has been subject to increasing international pressure after it surpassed the US to become the largest GHGE emitter in 2007. The 2007 Bali Climate Change Conference, which was the 13th conference of the parties (COP13) to the UNFCCC, produced the so-called Bali Roadmap that outlined the core design principles of a post-Kyoto climate regime. The Bali conference constituted a watershed because developing countries showed a willingness to discuss voluntary reduction commitments for the first time (Liang, 2010: 68). In June 2007, China announced its first global warming policy initiative (*China's National Climate Change Programme*) while the Climate Change Coordination Group under the National Development and Reform Commission (NDRC), which was the successor of SDPC, was elevated to the National Leading Group on Climate Change (NLGCC) directly under the State Council and was led by former Premier Wen Jiabao (Hallding *et al.*, 2009: 125–126).

In the run up to the 2009 Copenhagen Climate Change Conference, pressure on large developing countries mounted. China first announced its voluntary national climate mitigation action targets – in tandem with pledges made by US – in a declaration to a United Nations (UN) climate summit of heads of states in New York in September 2009. China pledged to cut carbon intensity by 40%–45% per unit of GDP by 2020 (compared to 2005), and to aim for 15% energy consumption from non-fossil fuels by 2020. For the first time it was China instead of the US that was blamed for dragging its feet in the international climate negotiations and for contributing to the conference's 'disappointing outcome' (Zhang, 2010: 239–240). Following the disappointing Copenhagen conference and the struggles to save the multilateral climate regime at the 2010 Cancún Climate Change Conference (COP16), the 2011 Durban Climate Change Conference (COP17) was an important turning point which paved the way for the post-Kyoto framework (IISD, 2011). The establishment of the Ad Hoc Working Group on the Durban Platform for Enhanced Action demonstrated the strong pressure for designing a legally binding framework to cover all COP member countries by 2015. China began to pay more attention to entrepreneurial leadership and cognitive leadership especially since the Durban conference. Xie Zhenhua, vice chairman of the NDRC and the leader of the Chinese delegation, declared that China should negotiate a legally binding document after 2020, although with conditions. This declaration was interpreted as China taking an active stance in the global negotiations by promising to put in place legally binding carbon reduction plans (Chinafaqs, 2011). In order to boost its cognitive leadership, China hosted its own pavilion at the UN climate change conferences and acted in a more transparent manner during the talks by, for example, holding more regular meetings with subnational actors, NGOs and journalists (Geall and Hui, 2016).

Since President Xi Jinping came to power in 2013, there has been a significant change in Chinese climate diplomacy: the doctrine of 'hide one's capacities and bide one's time' was replaced with the doctrine of 'diligent and ambitious diplomacy' (Ren, 2014). The new diplomatic style symbolised that China was beginning to perceive itself as a rising big power which could play a more central role in the international arena while taking on its international responsibilities. In 2013, China introduced its first initiatives to cap the use of coal, aiming to restrict its share in the national energy mix to 65% by 2017. According to *China Energy Outlook 2030,* China's GHGE may peak by 2025, five years ahead of the date it has pledged to the UN (Chinapower, 2016). Especially after the 2014 Lima Climate Change Conference (COP20), China became 'less defensive and more inviting, trying to take on a leadership role' (Soutar, 2015; see also Dong, 2017). China submitted its Intended Nationally Determined Contribution (INDC) according to which it planned to reduce the carbon intensity of its economy by 60–65% per unit of GDP by 2030 (compared to 2005), and repeated a previously announced aim that non-fossil fuel should make up 20% of its primary energy supply.

Compared with the Copenhagen conference negotiations, China played a positive, leading role in the Paris Agreement negotiations. For example, it offered entrepreneurial and structural leadership to enhance bilateral co-operation with the US and EU. China showed emerging cognitive leadership while trying to link the important Chinese concept of 'ecological civilisation' with sustainable development and low-carbon ideas. Importantly, the Paris experience has become a breakthrough in China's climate diplomatic transformation (Dong, 2017). The 2015 Paris Agreement adopted a new INDC-led mechanism and, for the first time, agreed on cuts in GHGE also for developing countries. It also stipulated the principle of common but differentiated responsibilities (CBDR), and established the legal framework for INDCs and international institutions to combat climate change post-2020. However, major disagreements between developed and developing countries have remained unresolved regarding the principle of CBDR, funding and adaptation measures. The developing countries – especially major emerging developing powers such as the BASICs (Brazil, South Africa, India and China) – will increasingly take on greater responsibilities in global climate governance. Against this background, China has become more demonstrative and has increasingly taken on a leading role amongst followers while still being cautious about acting as a possible climate leader. China has tried to link the 2030 Sustainable Development Goals with the climate change negotiations in international settings and to use this issue-linkage to promote domestic economic reforms (Dong, 2017). As a leading country in renewable energy, China began to promote 'clean energy' South–South co-operation. Through exemplary leadership (Liefferink and Wurzel, 2017), China shared its renewable energy best practices and project experience with other developing countries while offering financial aid, technology transfer and personnel training.

In the 2017 report of the 19th National Congress of the Communist Party of China (CPC), President Xi announced that China will be 'speeding up reform of the system for developing an ecological civilisation, and building a beautiful China … and become an important participant, contributor and leader of global ecological civilisation construction'(Xinhuanet, 2017). In March 2018, the National People's Congress amended the constitution to incorporate the construction of 'beautiful China' and 'ecological civilisation'. The construction of 'ecological civilisation' was promoted to a national core strategy. These political low-carbon transition efforts illustrate China's ambitious CO_2 emissions targets. However, the negative economic impact of Sino–US trade disputes has somewhat weakened China's attention to climate issues. In addition, the severe economic recession caused by the COVID-19 pandemic has also reduced the focus on climate change as an issue on the political agenda. At the same time, it cannot be ignored that the impact of extreme weather events triggered by climate change is ushering in an era of compound crises. Southern China has experienced the most intense, widest ranging and longest lasting flood since May 2020. More than eight million people over 11 provinces have been affected.

Multilevel climate governance in China and pioneering low carbon cities

Acknowledging the transnational nature of climate change, the 2015 Paris Agreement has encouraged cities, private actors and other stakeholders to participate in global climate governance while showcasing their climate actions. In contrast to the Kyoto Protocol's top-down 'targets and timetables' approach, the Paris Agreement strongly relies on a bottom-up approach, which encourages the involvement of local stakeholders (including cities) and voluntary measures such as INDCs. As the limitations of global climate governance based on international co-operation between states have become more apparent (e.g. at the 2009 Copenhagen Climate Change Conference), the emphasis has shifted towards transnational climate co-operation at the sub-state level and particularly in cities. Since the 2000s, cities have made increasingly significant contributions to global climate governance. For example, they attended some COPs, established transnational local government climate networks and influenced their countries' climate policies. They have therefore acted as catalysts for the global climate governance development. For instance, host cities of COPs have played an iconic role in climate governance (e.g. Kyoto for the Kyoto Protocol and Paris for the Paris Agreement).

China has one of world's fastest growing urbanisation rates. Cities contribute an estimated 70% of the energy-related GHGE and are therefore crucial for China's carbon reduction targets (Li, 2015). Cities have gradually received more attention also in Chinese climate change policy. The role of cities in combating climate change has changed because of the transformation of the global climate governance regime. Cities can act as climate leaders in multiple ways. Cities can offer cognitive leadership in the form of new, innovative ideas which they can spread across city climate networks, thus offering also entrepreneurial leadership. They can also establish carbon trading markets in cities. Examples include carbon emissions trading schemes (ETS) in Shenzhen since 2011 and pilot schemes in Beijing, Shanghai, Tianjin and Chongqing since 2013. These sub-national ETS pilots explore replicable and generalisable emission accounting standards and quota allocation to guide the long-term development of the national carbon market. Moreover, global city climate networks can influence the international climate negotiations by promoting capacity building at the urban level and by putting into practice new ideas to combat climate change.

In China, the signalling mechanism and 'learning by doing' are the two main dynamic mechanisms for cities to participate in climate governance (Li, 2016). The top–down signalling mechanism is based on the idea that climate leaders making (voluntary) commitments to reduce GHGE in the international climate negotiations will raise their national reputation and reduce the international pressure on China. It will send positive signals to the domestic audience and make easier local policy implementation efforts. China's state-over-society structure enhanced the effect of the signalling mechanisms, especially after national elites had reached a general consensus on a climate mitigation commitment. Historically speaking, China has a top–down approach to designing and implementing climate policies which set long-term, consistent climate targets in the Five Year Plans (FYPs). It has been spurred on by the political and economic resource reallocation at the domestic level (Schuman and Lin, 2012: 102), although increasingly competitive low-carbon development is emerging in a bottom-up way from cities. In order to be able to implement the voluntary national carbon-intensity reduction targets and the ambitious renewable energy targets, the central government devolved the national targets to the provincial and city state levels. It did so through mandatory indicator/index allocation of the national total energy conservation and emission reduction targets which were then broken down to subnational level targets. Regional (i.e. provincial and city state) competition in China has stimulated the emergence of low-carbon pioneering provinces and cities (see below). Furthermore, international commitments have promoted the greening of standards in the local official performance assessment system. Emissions reduction and renewable energy development requirements have become key evaluation standards to measure local officials' political performance (Li, 2016). The State Council will carry out annually a complete assessment and evaluation of energy production, distribution and consumption as well as renewable energy development at the local level. The performance of local officials is assessed through the combination of the enforcement of five-year targets and annual targets (Li, 2016).

China has a 'learning by doing' tradition which can be traced to the Deng Xiaoping era that began in the late 1970s. Deng encouraged incremental policy reform and innovation while relying on trial and error experimentation. In 1979, Deng proposed the creation of special economic zones with favourable economic policies in order to speed up the market economy reform trials with the aim of using the experience gained in other areas in China. The learning-by-doing pattern has been adopted to avoid unnecessary mistakes and setbacks. Small-scale pilot programmes and demonstration projects have been seen as better approaches to carrying out innovative low-carbon policies because they allow for the correction of errors in a timely manner and facilitate the accumulation of experience through learning by doing. The successful performance achieved in the pilot programmes boosted national confidence in nationwide expansion. Low-carbon pilot cities and carbon-trading pilot cities are two important examples (Li, 2016). With preferential policies and resources allocation, these pilot cities or provinces have been allocated structural leadership capacities to implement innovative low-carbon practices. This is a typically Chinese way to promote subnational cognitive and exemplary climate leadership.

The NDRC, which is China's top planning agency responsible for formulating and implementing national economic and social development strategies, launched the Low-Carbon Pilot Cities and Provinces Project in August 2010. Five provinces (Guangdong, Liaoning, Hubei, Shaanxi and Yunnan) and eight cities (Tianjin, Chongqing, Shenzhen, Xiamen, Hangzhou, Nanchang, Guiyang and Baoding) have acted as low-carbon pilots to demonstrate the feasibility of low-carbon development plans and renewable energy application at the urban governance level. These pilot cities and provinces were selected based on geographic, social and economic diversity and representativeness; existing and/or preparation work in low-carbon development; and a demonstrated interest by the local regions to become pilot locations. A further 16 cities were named in the second batch of pilot cities in February 2012; they include Beijing, Kunming, Xi'an, Ningbo, Guangzhou, Shenyang, Harbin, Huai'an, Yantai, Haikou, Chengdu, Qingdao,

Zhuzhou, Bengbu, Shiyan and Jiyuan. Lessons learned from successful trials will subsequently be used for large-scale, low-carbon reforms and the integration of innovative policy instruments into the national climate policy. Each pilot city was asked by the NDRC to develop and propose a low carbon development plan, formulate supporting policies, develop low carbon industry, establish CO_2 emission statistics and data management systems, and encourage low carbon lifestyles and consumption (Khanna, Fridley and Hong, 2014).

In 2017, the NDRC launched the third batch of 45 pilot cities from 22 provinces. These pilot areas account for about 40% and 60% of China's population and GDP, respectively. Since 2018, China has further expanded the low-carbon pilot schemes to more than 100 cities (NDRC, 2017). In contrast to the first and second batch, the NDRC has stipulated for the third batch that the overall targets of the low-carbon city plans should include both carbon emission and energy consumption reduction targets which correspond with 13th FYP and 2020 targets. Additionally, roadmaps of the low-carbon city plans need to be formulated and sector-based targets should be adopted on the basis of the overall targets with the aim of enabling better implementation, performance evaluation and policy adjustment. The NDRC therefore requires different low-carbon cities to work out specific low-carbon development plans according to their carbon emission peak target and pilot construction target. The performance of these pilots depends largely on whether their low-carbon development in the annual plan for regional economic and social development can be maintained in the long run. This will require the adjustment of industrial structure and the optimisation of energy structure, conservation and consumption. New pilot cites are expected to become leading actors which establish assessment systems for GHGE objectives and allocate emission reduction tasks to their lower administrative regions and key enterprises. In order to support these pilots, the local environmental protection system has undergone vertical reforms launched in 2018. This reform means that local environmental protection bureaus will acquire more real power to promote environmental supervision on low-carbon development. The leaders of county and district environmental protection bureaus will be appointed and removed by municipal or provincial governments. The aim is to make the local environmental protection department an independent system which can remove the economic development orientation ('GDPism') of local government. In the long term, the leading low-carbon pilots and especially their environmental protection and sustainable development bureaus, will need more powers to formulate the composition and assessment methods of carbon emission indicators in the region so that they can track the completion of emission reduction tasks set for each responsible body.

The establishment of market mechanisms in support of low carbon development is still at the trial stage. In October 2005, carbon trading under the CDM began in China. However, local stakeholder consultation and participation of CDM is still quite weak and lacks good practice guidance (Dong and Olsen, 2015). It was administered at the national level by the NDRC in a top-down approach. In 2008, several environmental and carbon trading schemes were established including the Tianjin Climate Exchange, China Beijing Environment Exchange and Shanghai Environment and Energy Exchange. From 2009 to 2010, additional environmental and carbon trade exchanges were created, covering Wuhan, Hangzhou, Kunming and Guiyang. In August 2010, the NDRC encouraged low-carbon pilot provinces and cities to include carbon trading as part of the overall development strategy. Having drawn lessons from the European Union (EU) ETS, in October 2011, China issued the Notice on Carbon Trading Trials Scheme, which initiated city-state and provincial-level ETS trials (Wang, 2012). Five cities (Beijing, Tianjin, Shanghai, Chongqing and Shenzhen) and two provinces (Hubei and Guangdong) were chosen for these pilots. Based on the experience of these ETS pilots in city-states and provinces, the Chinese government launched a national-level ETS in 2017. The new ETS will monitor and control national CO_2 emissions and energy consumption at the local government and firm level so that an emission peak can be reached by 2030. The first stage of the national carbon emission trading market will cover key industries, such as petrochemicals, chemicals, building materials, steel, non-ferrous metals, paper, electricity and aviation. Due to resistance from conventional energy actors, the coverage of the ETS was eventually reduced to only one sector, namely the power generation industry. The threshold for the power generation industry to be included in the national carbon market is set at 26,000 tonnes of CO_2-equivalent per year. More than 1,700 enterprises are included and the scale of emissions trading is more than 3 billion renminbi (RMB). The city pilots have also allowed local governments to experiment with and develop tailored local pathways of low carbon urban development instead of all cities having to follow a generic top-down mandated low carbon action plan. This has resulted in a rising tendency towards combining top–down and bottom–up approaches which provides more room for city actors to compete with each other for a leading role in domestic low-carbon development and local climate governance (see Table 2.1).

With increasing public air pollution awareness and the salience of climate change issues at the local level, the move towards a bottom–up approach for low-carbon initiatives has been promoted by multiple actors including local authorities, think tanks, environmental NGOs and renewable energy industry associations (Li, 2016). Even though they were not included in the national-level pilots, small cities like Rizhao, Dezhou and Weihai have put forward their own low-carbon and ETS development blueprint with support from civil society. By the end of 2010, more than 30 autonomous regions and cities in China have started to prepare for provincial-level climate change action plans. Some of them have been included in the third batch of the national low-carbon city list. Regardless of whether cities have been included in the national low-carbon cities list, they can choose their own approaches for climate mitigation and adaption, to improve their competitive position and to diffuse best practices with the support of the national government.

Shanghai's leading role

Shanghai, which is China's leader in economic and social development, is playing an increasingly important role in low-carbon development. Shanghai has

Table 2.1 Third batch of low-carbon cities pilot list.

Province	No.	City	Peak Year	Differentiated Innovation Focus
Inner Mongolia	1	Wuhai	2025	1. Establish carbon management system 2. Explore direct reporting system for GHGE for key stakeholders 3. Establish low-carbon technology innovation mechanism 4. Promote modern low-carbon agriculture 5. Establish a low-carbon and ecological civilisation evaluation mechanism
Liaoning	2	Shenyang	2027	1. Establish online monitoring system for carbon emissions in key energy consuming enterprises 2. Improve the central carbon emissions management platform
	3	Dalian	2025	1. Develop and promote technical standards for the evaluation of low-carbon product certification 2. Establish a 'carbon marking' system 3. Establish a green, low-carbon supply chain system
	4	Chaoyang	2025	1. Establish total carbon emission control system 2. Establish a low-carbon transport operation system
Heilongjiang	5	Xunke	2024	1. Explore the development model and support low-carbon agriculture
Jiangsu	6	Nanjing	2022	1. Establish 'double control' system for total carbon emissions and intensity 2. Establish a system for paid use of carbon emission rights 3. Establish a low-carbon integrated management system
	7	Changzhou	2023	1. Establish a total carbon emission control system 2. Establish a low-carbon model enterprise creation system 3. Establish mechanisms to promote green building and technology optimisation

Table 2.1 (Continued)

Province	No.	City	Peak Year	Differentiated Innovation Focus
Zhejiang	8	Jiaxing	2023	Explore the innovation of synergy system for low-carbon development
	9	Jinhua	2020	Explore the evaluation system of target responsibility for reducing emissions in key energy-using enterprises
	10	Quzhou	2022	1. Establish carbon productivity assessment mechanism 2. Explore regional carbon assessment and project carbon emissions assessment 3. Establish innovative photo-voltaic poverty alleviation mechanism
Anhui	11	Hefei	2024	1. Establish carbon data management system 2. Explore low-carbon product and technology extension system
	12	Huaibei	2025	1. Establish carbon approval access mechanisms for new projects 2. Establish assessment (objective) mechanism 3. Establish energy-saving and carbon-reducing monitoring mechanism 4. Explore innovation of carbon financial system 5. Promote low-carbon key technological innovation
	13	Huangshan	2020	1. Implement total quantity control and decomposition mechanism 2. Develop low-carbon and smart tourism
	14	Liuan	2030	1. Develop low-carbon development performance evaluation and assessment 2. Improve green, low-carbon and eco-protection market system
	15	Xuancheng	2025	Explore low carbon technology and product promotion system innovation
Fujian	16	Sanming	2027	1. Establish carbon data management mechanism 2. Explore mechanism of forest carbon sink compensation

(Continued)

Table 2.1 (Continued)

Province	No.	City	Peak Year	Differentiated Innovation Focus
Jiangxi	17	Gongqin-gcheng	2027	Establish low-carbon city planning system
	18	Jian	2023	Explore low-carbon community and carbon neutralisation demonstration projects in rural areas
	19	Fuzhou	2026	Establish carbon neutralisation demonstration zone in Zixi County
Shandong	20	Jinan	2025	1. Explore carbon emission data management system 2. Explore the total carbon emission control system 3. Explore the carbon assessment system for major projects
	21	Yantai	2017	1. Explore the total carbon emission control system 2. Explore carbon emission evaluation system for fixed assets investment projects 3. Develop low-carbon technology promotion catalogue
	22	Weifang	2025	1. Establish a 'four-carbon-in-one' system 2. Establish a carbon data information platform
Hubei	23	Changyang Tujia	2023	Create carbon neutralisation demonstration project in Qingjiang Gallery tourism area, Changyang Innovation Industrial Park
Hunan	24	Changsha	2025	1. Pilot 'Three Synergy' development mechanism 2. Establish carbon integration system
	25	Zhuzhou	2025	1. Promote low carbon transformation of urban old industrial base 2. Create low-carbon smart transport system
	26	Xiangtan	2028	1. Explore the low-carbon transformation model of the old industrial base
	27	Chenzhou	2027	1. Establish a green financial system

Table 2.1 (Continued)

Province	No.	City	Peak Year	Differentiated Innovation Focus
Guangdong	28	Zhongshan	2023–2025	Deepen the carbon inclusion system
Guangxi	29	Liuzhou	2026	1. Establish cross-sectoral collaborative carbon data management system 2. Establish total carbon emission control system 3. Establish regular working mechanism for GHGE inventory preparation
Hainan	30	Sanya	2025	Select independent island area to create a carbon neutralisation demonstration project
Li and Miao Autonomous County	31	Qiongzhong	2025	1. Establish low-carbon rural tourism development model 2. Explore low-carbon poverty alleviation models and systems
Sichuan	32	Chengdu	Before 2025	1. Implement the 'Low-Carbon Benefit Tianfu' Project 2. Explore carbon emission peak tracking system
Yunnan	33	Yuxi	2028	1. Establish an early warning mechanism for monitoring and analysis of emission data reporting in key enterprises 2. Develop specifications for statistical analysis of community emission data
	34	Pu'er	2025	Establish statistical management system for GHGE
Xizang	35	Lasa	2024	Create carbon neutralisation demonstration projects
Shanxi	36	Ankang	2028	1. Implement 'multi-planning in one' in pilot 2. Establish ecological compensation mechanism for carbon sinks 3. Establish a poverty alleviation mechanism for low carbon industries
Gansu	37	Lanzhou	2025	Explore multi-domain co-construction of low-carbon cities
	38	Dunhuang	2019	Establish cross-sectoral development and work management platform
Qinghai	39	Xining	2025	Establish resident living carbon integration system

(*Continued*)

Table 2.1 (Continued)

Province	No.	City	Peak Year	Differentiated Innovation Focus
Ningxia	40	Yinchuan	2025	1. Improve preferential policies and incentives for low-carbon technology and product promotion 2. Promote low-carbon technology and product platform 3. Establish mechanisms to explore, evaluate and promote low-carbon products and technologies
	41	Wuzhong	2020	Establish carbon neutralisation demonstration project in Jinji Industrial Park
Xinjiang	42	Changji	2025	1. Create linkage mechanism for total carbon emission control 2. Establish a carbon emission data management platform and database 3. Establish fixed assets investment carbon emission evaluation system
	43	Yining	2021	1. Develop low-carbon, green demonstration in government 2. Explore the creation of a low-carbon technology extension service platform 3. Establish carbon sink compensation mechanism
	44	Hetian	2025	1. Establish total carbon emission control system 2. Establish total carbon emission assessment and management system 3. Establish carbon assessment system for major construction projects 4. Create integrated carbon emission management service platform
Xinjiang	45	Alar City	2025	1. Explore total control and carbon data management systems 2. Promote low-carbon products and technologies 3. Explore the carbon assessment system for new projects

Source: Environmental Supervision Network (2019).

developed its 'Shanghai 2035 outstanding global city' target which emphasises innovation, humanity and ecology (SEEEX, 2019). In December 2013, the UN General Assembly decided to make 31 October the World City Day. It was initiated by the Shanghai declaration at the 2010 Shanghai World Expo Summit Forum. This is the first time that China has successfully promoted the establishment of an international day at the UN level. As part of the legacy of the 2010 Shanghai World Expo, the Annual World City Day Conference plays an active role in promoting international communication and co-operation on the living environment and green development by cities which are part of rapid global urbanisation. Annual conference reports (or Shanghai manual) on best city-level low-carbon practices were issued in 2011, 2016, 2017 and 2018. These initiatives illustrate the cognitive leadership ambitions of Shanghai which aim to establish a sustainable green future.

The Shanghai government recognises that low-carbon development is not only a crucial requirement for national sustainable development, but also an effective way of undertaking economic transformation and improving competitiveness. Shanghai is a leading low-carbon city which is one of a few cities that has adopted both fiscal/tax incentives and a local ETS to promote the low-carbon economy. The city has also adopted financial and technological support for industrial energy efficiency, transportation, energy saving, new energy and recycling. Shanghai is a thriving shipping, financial and economic centre showcasing the outward looking, modern China to the rest of the world. The city has been deeply involved in the low-carbon revolution as a way of life that promotes national interests and values. In 2011, Shanghai carried out a pilot project in accordance with the requirements of the NDRC. As one of the earliest carbon trading pilot areas in China, Shanghai officially launched its ETS on 26 November 2013. In December 2017, the NDRC issued the national ETS market construction plan (only for the power generation industry), launching the market construction of the national ETS. The plan made clear that Shanghai will take the lead in the construction, operation and maintenance of the unified national ETS. Following the learning-by-doing pattern, China's ETS has been gradually promoted from local pilots to a unified national market which has been helped a lot by Shanghai's leading role. By late 2019, the Shanghai carbon trading market covered more than 20 industries and more than 300 entities.

In July 2012, Shanghai issued the Opinions of the Shanghai People's Government on the Implementation of the Pilot Work on Carbon Emissions Trading in the City (SEEEX, 2019). In the same year, the Shanghai Development and Reform Commission also issued Guidelines for Accounting and Reporting of Greenhouse Gas Emissions in Shanghai (Trial) and sub-sector guidelines. In preparation, Shanghai issued the Shanghai for Carbon Emission Management Trial Measures in the form of government orders, which clearly defined the various elements and legal responsibilities of the carbon ETS that introduced quota allocation management plans and interim quota registration management. At the end of this process, a relatively complete policy system of 'government regulations + normative documents + supporting documents' had been formed. It shows that

Shanghai not only offered structural and entrepreneurial leadership by making full use of its local government powers to mobilise various resources to support the ETS, but also displayed cognitive leadership by experimenting with an innovative policy instrument and drafting normative documents.

The Shanghai Carbon ETS was officially launched in 2013. According to the city government's document Measures for the Implementation of Carbon Emission Management in Shanghai (SEEEX, 2019), the Shanghai Development and Reform Commission will regulate the details (e.g. formulate allocation plans, determine emission allowances, establish monitoring, etc.). It has established the Shanghai Environment and Energy Exchange as the trading platform (SEEEX, 2019). At the initial design stage of the ETS in Shanghai, the role of the market mechanism to reduce GHGE was the main focal point. Accordingly, the relevant policies and management models were formulated to form a relatively complete market management system (CNEEEX, 2019). Shanghai has become a pioneering actor for leading the system design. The trading varieties of the Shanghai carbon market mainly include the Shanghai Carbon Emissions Allowance (SHEA) and Certified Voluntary Emission Reduction (CVER). Eligible enterprises and institutions can be listed on the trading platform after becoming a member of the Shanghai Environment and Energy Exchange. By late 2019, the total turnover of the nine carbon markets in China amounted to 495 million tons. The total quota was 309 million tons and the total CVER amounted to 185 million tonnes. Shanghai's CVER turnover was the highest in the country (CNEEEX, 2019), which illustrates Shanghai's ambitious leading action in low-carbon development (See Figure 2.1).

China's ETS, which differs from the EU ETS, takes into account the country's developing status. Shanghai has offered structural leadership for innovative carbon financial products. Shanghai took the lead in launching CVER trading for which it ranks first in China according to trading volume. In June 2015, the Shanghai Environment and Energy Exchange issued the Shanghai Environment and Energy Exchange Carbon Trading Business Rules (Trial). In August 2015 it completed the first single carbon trading business. Shanghai has innovated with a carbon quota repurchase business. In March 2016, the Industrial Bank's Shanghai branch, Spring Airlines Co. Ltd. and Shanghai Confidant Carbon Asset Management Company signed a carbon quota asset sale repurchase contract at the Shanghai Environment and Energy Exchange to complete a 500,000 tonnes carbon credit. This marked the arrival of Shanghai's first carbon allowance asset sales and repurchase business. Shanghai's cognitive leadership helped to diffuse best practices and to promote the capacity training by establishing, in May 2016, the national carbon market capacity building (Shanghai) centre which holds professional conferences related to ETS. More than 20,000 people from nearly 30 cities in 11 provinces and autonomous regions have been trained in Shanghai.

Shanghai aims to develop into an outstanding global city in 2040 (Shanghai 2040 target), which functions as a key international economic, financial, trade, shipping and technological innovation centre. However, the low-carbon economy in Shanghai is still relatively weak. The city still has excessive growth, high energy consumption, a heavy industrial structure, scarcity of green capital, and institutional barriers which make it difficult to deal with climate change (SEEEX, 2019).

China's pioneer cities in international co-operation

Initially, China's development of low-carbon cities was largely driven by external actors. In 2008, the World Wide Fund for Nature(WWF) launched its pilot low-carbon city development programme. It selected Baoding and Shanghai as the first two pilot cities in recognition for their leadership on local low-carbon development. In October 2008, the UN Development Programme (UNDP), Norway and the EU jointly launched a project to support Chinese provincial climate change programmes and projects (see also Chapters 8 and 11 in this volume). The UK's Strategic Programme Fund (SPF) has provided support to Jilin City, Nanchang, Chongqing and Guangdong province in its low-carbon city development research and planning (see also Chapter 10 in this volume). In June 2010, the Switzerland–China Low Carbon Cities Project was launched, for which Yinchuan, Beijing Dongcheng District, Dezhou and Meishan were selected as cities for a pilot on city management, a low-carbon economy, transportation and green buildings (see also Chapter 13 in this volume).

China's subnational pilot actors have become more active and open to the international co-operation with foreign cities. China's transnational city co-operation in sustainable development is gradually evolving from externally-driven to internally-driven co-operation. China's subnational actors have launched bilateral urban co-operation with Germany, France, the UK, Switzerland, Finland and other European countries. They have launched co-operation projects such as the China–EU Mayors' Forum and the China–EU Low Carbon Eco City Platform. In 2012, China promoted a declaration on China–EU urbanisation partnership including an annual China–EU urbanisation forum which guides the partnership. The forum is held alternately in China and the EU and its results are presented to the leaders of China and the EU (see also Chapter 8 in this volume).

In June 2016, the first Asian-Pacific Economic Cooperation (APEC) High Level Forum on urbanisation (and inclusive growth) was held in Ningbo. It was hosted by the China urban and small town reform and development centre of the NDRC. The forum is the first city co-operation activity under the APEC framework. Staging it showed structural and entrepreneurial leadership by China at the subnational level. At the 18th China–ASEAN leaders meeting, China put forward a co-operation initiative to establish the China–ASEAN eco-friendly city development partnership. This partnership will become an important starting point for developing Belt and Road Initiative environmental co-operation and for participating in the regional urban sustainable development co-operation dialogue (ORAPE, 2016). The government's Green Belt and Road Initiative needs more consistent and comprehensive public disclosure and dissemination of information to allow the public to participate and oversee the implementation of the low-carbon city plans.

Due to the increased role of cities in multilevel global climate governance, China's local actors have paid more attention to urban climate diplomacy and international climate co-operation among cities. Without necessarily having been authorised by national governments, the transnational climate cities networks are treated as relatively independent subnational climate governance networks. This has increased the structural leadership potential of subnational actors. In recent years, Chinese cities have more actively participated in global climate networks, such as Local Governments for Sustainability (ICLEI), C40 Cities Climate Leadership Group and the World Alliance for Low Carbon Cities (WALCC). Before 2013, Chinese cities were cautious and conservative about transnational networks. Mainly due to the promotion of the Belt and Road Initiative and the need for continued openness, more cities have joined transnational local government networks. ICLEI is a city network of more than 1,750 regions and local governments around the world, committed to improving the environment for sustainable urban development. Since 2016, nearly 20 cities (e.g. Shenyang, Jilin, Tianjin, Baoding and Yangzhou) have joined ICLEI. In March 2018, the Beijing Representative Office of the East Asia Secretariat of ICLEI was officially established to support local governments in building low-carbon, resilient and ecologically-friendly cities, and to develop a green and circular economy. Beijing, Shanghai, Hong Kong, Shenzhen, Wuhan, Guangzhou, Nanjing, Qingdao, Dalian, Chengdu, Fuzhou and Hangzhou have become members of the C40 World Metropolis Pioneer Group (C40, 2019).

Some Chinese cities have begun bilateral co-operation with foreign cities with the aim of initiating larger transnational local government networks. For instance, Shenzhen became a founding member of the WALCC which was established in October 2011. Professor Kang Yufei, who is the Dean of Shenzhen Graduate School of Tsinghua University, became the president of the Council in which local cities from China, Sweden, Finland and other countries participate. The WALCC annually holds a world Low Carbon Cities Alliance conference in a Chinese or Nordic city. The main areas of co-operation include urban planning, green building, low-carbon transportation, and renewable resources and energy. The alliance has carried out some demonstration projects including electric vehicle technology projects (ORAPE, 2016).

Cities in China have played a leading role in establishing high-level eco-forums for subnational intergovernmental and cross-industry green development co-operation. Founded in a bottom–up or polycentric fashion in 2009, the ecological civilisation Guiyang conference was officially upgraded to the Eco Forum Global (EFG) in July 2013. It has provided cognitive leadership to promote local ecological development. The biggest event for EFG is the Eco Forum Global Annual Conference Guiyang, which is held annually in Guiyang in July and provides a high-level platform for the exchange of cross-border, cross-industry and cross-discipline ideas, knowledge and experience. It facilitates domestic and international co-operation between governmental and societal actors. The forum has promoted the continuous embedding of a green agenda at the international, regional and industrial levels (EFG, 2019). The EFG has also focused on building a global partnership network through co-operation with other international actors, such as the UNDP, the United Nations Educational, Scientific and Cultural Organization (UNESCO) and other UN institutions (EFG, 2019). In 2016, Guizhou, Fujian and Jiangxi provinces were approved by the central government to build an Ecological Civilisation Pioneer Zone. Consequently, Guizhou has become a comprehensive testing platform for carrying out the ecological civilisation reform which resembles a bottom–up approach combined with a top–down multilevel government approach. It was first proposed at the local level and then supported by the central government to promote the diffusion of a successful model. In 2018, UN Secretary-General António Guterres sent a video message to China's 10th Eco Forum Global Annual Conference.

Conclusion

This chapter has assessed China's climate leadership and pioneership while focusing on the national and subnational governance levels. It examined the role which China's cities have played in international climate governance while focusing on emerging low-carbon cities and the part which they have played in China's domestic climate politics.

China has become a global climate power. As the world's largest emitter, it is a potential veto player in the international climate negotiations. China has considerable structural power which it has used for offering structural climate leadership only since the 21st century. China's role in international climate governance has evolved over time. It was an observer in the Montreal Protocol negotiations, a marginalised participant in the UNFCCC, an active leader of the G77 in the Kyoto Protocol negotiations, a major player in the Copenhagen Climate Change Conference (COP15) and an emerging leader in the 2015 Paris Climate Change Conference. Over time, China's climate diplomacy has become more demonstrative and confident while exhibiting bouts of structural and entrepreneurial climate leadership. Before the 2015 Paris Climate conference, China promoted several bilateral climate agreements with big emitters like the US, EU, India and Brazil (Xinhuanet, 2015). China has aligned itself with developing country coalitions while emphasising the principle of respective capabilities based on a weakened CBDR principle. However, China has admitted that major developing countries will increasingly have to accept wider responsibilities in global climate governance, and that ambitious international climate commitments can act as drivers for domestic low-carbon transformation.

In China, cities are treated as the main battleground for global climate change. Urban climate innovations can play a unique role in global climate governance and cities are at the forefront of climate change. Due to the signalling mechanism and learning-by-doing pattern, the role of China's subnational actors at the provincial and city levels is strongly shaped and directed by the national government in domestic climate policy. By 2019, at least 69 cities were listed as national pioneering low-carbon cities. China has combined a top–down approach, which is led by the central government, with a bottom–up approach that is stimulated by

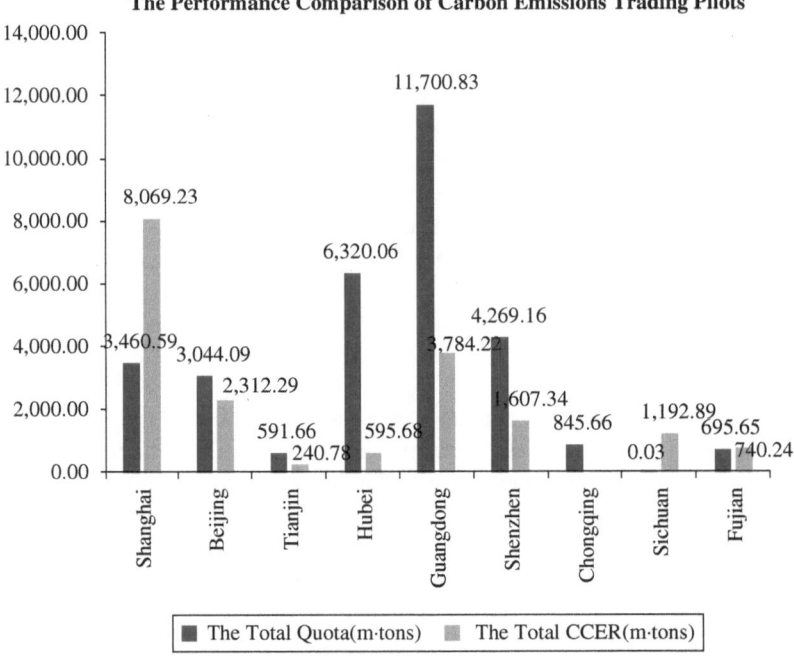

The Performance Comparison of Carbon Emissions Trading Pilots

Figure 2.1 Performance comparison of carbon emissions trading pilots. Note: Shenzhen is a special economic zone. The rest are municipalities. The columns in blue refer to quota turnover while the columns in red show the total turnover. *Source:* Shanghai Carbon Market Report (2018).

regional green development competition. More cities are eager to join the competition for low-carbon pilots. The central government has not set strict and detailed regulations for low-carbon pilots but encourages local innovations. However, the multitude of parallel local climate governance programmes has created complexity, confusion and overlaps in the development of low-carbon cities. Some cities belong to several pilot programmes. While it might be beneficial to receive technical and financial support through different programmes, the overlap of programmes has resulted in an unclear focus, repetitive planning processes and ineffective implementation of low carbon development planning. The emerging local low-carbon actors are not only involved in a green competition at the domestic level, but also try to play a more active role at the international level. As part of the large-scale development of low-carbon cities in China, Shanghai has increasingly shown cognitive and/or entrepreneurial leadership. The top–down approach of developing low-carbon action plans for local governments is still dominant. It limits the input from stakeholders and the public and thus makes it difficult to develop a comprehensive plan that adequately represents the various aspects of urban development. Third-party monitoring and evaluation are needed at the different levels of government (Khanna, Fridley and Hong, 2014).

Bibliography

Bulkeley, H. and Betsill, M. (2005) 'Rethinking sustainable cities: multilevel governance and the "urban" politics of climate change', *Environmental Politics*, 14(1): 42–63.

C40 (2019) *Member cities of C40*, https://www.c40.org/citiesn (Accessed 17 October 2019).

Chen, G. (2009) *Politics of China's environmental protection: problems and progress*, Singapore: World Scientific Publishing.

Chinafaqs (2011) *Propelling the Durban climate talks*, http://www.chinafaqs.org/blog-posts/propelling-durban-climate-talks-china-announces-willingness-consider-legally -binding-comm (Accessed 17 November 2018).

Chinapower (2016) *The issuing of China energy outlook 2030*, http://www.chinapower .com.cn/finance/20160322/21414.html (Accessed 27 May 2019).

CNEEEX (2019) *Shanghai environment and energy exchange*, http://www.cneeex.com/ (Accessed 17 July 2019).

Dong, L. (2017) 'Bound to lead? Rethinking China's role after Paris in UNFCCC negotiations', *Chinese Journal of Population Resources and Environment*, 15(1): 32–38.

Dong, Y. and Olsen, K.H. (2015) 'Stakeholder participation in CDM and new climate mitigation mechanisms: China CDM case study', *Climate Policy*, 17(2): 1–18.

EFG (2019) *Who we are*, http://www.efglobal.org/whoweare (Accessed 17 July 2019).

Environmental Supervision Network (2019) http://www.envsv.com/huanbaowangzhi/hjzl/ 5016.html (Accessed 27 February 2019).

Geall, S. and Hui, L.N. (2016) 'Chin's low carbon future offers global opportunities', *Chinadialogue Brief*, https://steps-centre.org/wp-content/uploads/china_low:carbon _opportunities_v5.pdf

Ha, S., Hale, T. and Ogden, P.(2016) *Ahead of the Paris treaty signing ceremony, it's "All hands on Deck" for climate finance*, https://www.chinadialogue.net/article/show /single/en/8792-Ahead-of-the-Paris-Treaty-signing-ceremony-it-s-all-hands-on-deck-f or-climate-finance> (Accessed 24 May 2019).

Hallding, K., Han, G., and Olsson, M. (2009) 'China's Climate- and Energy-Security Dilemma: Shaping a New Path of Economic Growth', *Journal of Current Chinese Affairs*, 38(3): 119–134.

Heggelund, G. (2007) 'China's climate change policy: domestic and international developments', *Asian Perspectives*, 31(2): 155–191.

Homsy, G.C. and Warner, M.E. (2013) 'Cities and sustainability: polycentric action and multilevel governance', *Urban Affairs Review*, 49(1): 1–28.

IISD (2011) *Summary of the Durban climate change conference*, http://www.iisd.ca/vol12 /enb12534e.html (Accessed 23 March 2019).

Khanna, N., Fridley, D. and Hong, L. (2014) 'China's pilot low-carbon city initiative: a comparative assessment of national goals and local plans', *Sustainable Cities and Society*, 1: 110–121.

Lester, R.B. (2006) *Plan B 2.0: rescuing a planet under stress and a civilization in trouble*, Beijing: Oriental Publishing House.

Liang, W. (2010) 'Changing Climate? China's New Interest in Global Climate Change Negotiations', in J. J. Kassiola and S. Guo, (eds.), *China's Environmental Crisis*, Basingstoke and New York: Palgrave Macmillan, 61–84.

Li, X. (2015) 'The action logic of transnational municipal networks (TMNs) in the global climate governance', *International Review*, 5: 104–108.

Li, X. (2016) *Renewable energy policy change in China: interlinking the climate change regime and domestic pro-renewable energy coalition*, China: Shandong University Press.

Liefferink, D. and Wurzel, R.K.W. (2017) 'Environmental leaders and pioneers: agents of change?', *Journal of European Public Policy*, 24(7): 651–668.

Lu 上海碳排放交易试点实践经验及启示文章来源:电力决策与舆情参考陆冰清 2019-08-11 12:47 http://www.tanpaifang.com/tanjiaoyi/2019/0811/65120.html

NDRC (2017) *The pilot work of the third batch of national low carbon cities*, http://www.gov.cn/xinwen/2017-01/24/content_5162933.htm (Accessed 18 May 2019).

ORAPE (2016) *Experience and enlightenment of international -eration in urban sustainable development*, Report of Observation and Research on Asia-Pacific Environment, http://mil.news.sina.com.cn/2014-01-15/1430760216.html (Accessed 17 April 2017).

Ren, W. (2014) 'The 3.0 Version of China's Diplomacy'. 15.01.2014, available at http://mil.news.sina.com.cn/2014-01-15/1430760216.html (accessed 27.12.2019).

Schuman, S. and Lin, A. (2012) 'China's renewable energy law and its impact on renewable power in China: progress, challenges and recommendations for improving implementation', *Energy Policy*, 51: 89–109.

SEEEX (2019) *Shanghai carbon market report* 2018, The Report of Shanghai environmental energy exchange, April.

Soutar, R. (2015) *A "Less Defensive" China can help Spur global climate deal*, https://www.chinadialogue.net/article/show/single/en/8248-A-less-defensive-China-can-help-spur-global-climate-deal (Accessed 19 December 2018).

Wang, T. (2012) 'China's carbon market challenge', *China Dialogue*, http://www.chinadialogue.net/article/show/single/en/4936-China-s-carbon-market-challenge (Accessed 19 April 2019).

Xinhuanet (2015) 'Spotlight: China Makes Active Contribution for Breakthrough at Paris Climate Talks', available at http://news.xinhuanet.com/english/2015-12/13/c_134912237.htm (accessed on 16.03.2019).

Xinhuanet (2017) *Xi Jinping's report at 19th CPC National Congress*, http://www.xinhuanet.com/english/special/2017-11/03/c_136725942.htm (Accessed 26 December 2019).

Yan, S. and Xiao, L. (2010) 'Evolution of China's position in international climate talks', *Journal of Contemporary Asia-Pacific Studies*, 1: 80–90. (in Chinese).

Zhang, Z. (2010) 'Copenhagen and beyond: reflections on China's stance and responses', In E. Cerda and X. Labandeira (eds.) *Climate change policies: global challenges and future prospects*, Cheltenham: Edward Elgar.

3   India

From climate laggard to global solar energy leader

Kirsten Jörgensen

Introduction

In international climate negotiations, India has for a long time been perceived as a nay-sayer, unwilling to commit to climate mitigation goals and persuading fellow developing countries to do the same (Michaelowa and Michaelowa, 2012). India argued that because industrialised countries had generated the problem of climate change, they should solve it, by reducing greenhouse gas emissions (GHGE) at home and providing funding for emission reductions in developing countries. Accordingly, one would not expect India to be an interesting case study for comparative research on international climate leaders and pioneers. However, this chapter will show that at least in recent years India has actually offered considerable climate leadership and pioneership in promoting solar energy. India's large market and population makes it an important player, especially in the G77 (Group of 77 at the United Nations is a coalition of 134 developing countries) but also in the G20 (Group of Twenty is an international forum for the governments and central bank governors from 19 countries and the European Union) and the BRICS (the association of five major emerging national economies: Brazil, Russia, India, China and South Africa).

The literature on climate leaders and pioneers focuses primarily on examples of countries which stand out due to their impressive record in climate mitigation, ambitious energy efficiency efforts and renewable energy policies. It is primarily interested in countries that strive to stimulate climate action in other countries and attract followers of its ambitious climate policies (e.g. Liefferink and Wurzel, 2017). The literature also explores examples of countries striving to convince other countries to form climate negotiation coalitions and to commit to joint goals in the context of the international climate negotiations (Agueda, Corneloup and Mol, 2014; Bäckstrand and Elgström, 2013; Torney and Mai'a, 2018). The largely Eurocentric climate leadership literature has rarely explored examples of climate leadership and pioneering actions in developing countries and the underlying conditions facilitating them. Very few studies examine India's role as a leader or pioneer in climate policy. For example, Jänicke discusses India's Solar Mission as part of a climate policy package and a systematic development of the country's own industry intended to make India a 'global leader in solar energy' (Jänicke, 2012: 52).

This chapter sets out to identify climate leaders and pioneers in India's climate policy landscape. It distinguishes between pioneers and leaders while drawing on Liefferink and Wurzel (2017) and Wurzel, Liefferink and Torney (2019) who have argued that environmental leaders usually try to attract followers which is normally not the case for pioneers (see also Chapters 1 and 14 in this volume). Moreover, this chapter differentiates between four different types of climate leadership. First, structural leadership/pioneership is mainly associated with economic power and military power although the latter does not play a significant role in climate governance. Secondly, entrepreneurial leadership/pioneership involves the use of negotiating and diplomatic skills with the aim of brokering integrative bargains and agreements. Third, cognitive leadership/pioneership is about defining/redefining problems and interests. Fourth, exemplary leadership/pioneership refers to the setting of examples for others. Climate leaders intentionally set examples for others to follow while pioneers only unintentionally provide exemplary pioneership (Liefferink and Wurzel, 2017; Wurzel, Liefferink and Torney, 2019; see also Chapter 1 in this volume).

India faces a complex 'climate trilemma' which makes the formation of consistent domestic climate policy preferences difficult and which has shaped India's domestic climate process and international negotiating strategy (Dubash, 2016: 4). Due to the size of its population and the rapid growth of its economy, India is one of the world's three largest global carbon dioxide (CO_2) emitters, even though its per capita emissions are among the lowest in world comparison (Dubash, 2019). High poverty rates in India are accompanied by low per capita electricity consumption because, amongst other factors, approximately 300 million people still lack access to electricity. In developed countries the per capita energy use is much higher than, for instance, in the US where per capita energy use is ten-fold higher compared to India (Jackson *et al.*, 2018).

At the national level and in India's states, climate change and environmental pollution are not high on the country's political agenda. In contrast, economic development takes priority. Poverty is a pressing issue that domestic and international decision-makers as well as economic and financing institutions and public opinion usually perceive as an issue that must be tackled through high growth rates regardless of the environmental costs involved. Thus, climate mitigation is often regarded as a threat, whereas securing energy for development is much more prevalent in the domestic debate (Dubash, 2016; Upadhyaya *et al.,* 2018). However, India, like many other developing countries, is extremely vulnerable to the effects of climate change, with the poorest sections of the population at greatest risk. Driven by international factors, the perceived increased vulnerability to climate impacts and new actor constellations, climate policy has been emerging as an important issue since 2007 (Fernandes *et al.,* 2020). In fact, India's contemporary climate policy can make use of policy innovations introduced in the past. Since the 1990s, regulatory frameworks and incentive systems for renewable energy and energy efficiency have been put into place, policies which were originally directed towards development and energy security concerns (Khosla *et al.*, 2017; Mathur, 2019). Despite the fact that climate protection is deemed to be a rather low politics issue in India, the country was, paradoxically, for the first time ranked among the top ten in the 2019 Germanwatch Climate Change Performance Index (Germanwatch, 2019). This high ranking is related to India's low per capita emissions and ambitious renewable energy targets discussed below. Although politically still contested despite the synergies between already present policies and institutions and new climate policy initiatives, the time has come for low-carbon development in India. The largest challenge ahead is the escape from India's carbon lock-in and, in particular, the phasing out of fossil fuel subsidies (Germanwatch, 2019).

This chapter sets out to examine climate leadership and pioneering action in India, a developing country with rapidly emerging markets and a growing population that is also a global climate power with significant influence on the future trajectories of the global system of climate governance. India was, up until 2015, when it launched highly ambitious solar energy goals, not considered a climate pioneer or leader at all. Instead, it was lumped together with the group of climate laggards, namely countries with low or no ambition in the field of climate change. However, a temporal examination of India's domestic climate policy processes and policy output as well as its moderately changing stances in international climate fora suggest a more significant role and a more differentiated perspective on both India's domestic as well as international climate governance.

From a developing country perspective, India was able to exert entrepreneurial and cognitive leadership early on, when it successfully pushed for the introduction of the 'common but differentiated responsibilities' (CBDR) principle that enshrined climate justice in the climate regime. This is an example of transformational leadership in the international climate regime (Liefferink and Wurzel, 2017; Wurzel, Connelly and Liefferink, 2017). Another example of India showing structural leadership is the country's international primacy in the expansion of a large market for solar energy. Leadership and pioneership are phenomena that are observable in international contexts. However, they are also identifiable and even more important in domestic policy structures. In India's domestic climate and energy policy process, the features of leadership and pioneership behaviour are already significant. There are examples of impressive national policy frameworks such as India's solar mission, which pursues the promotion of renewable energy while also contributing to climate mitigation. Other examples include initiatives taken in India's federal states and cities that are often helped by the vibrant NGO and think-tank sector which offers cognitive and entrepreneurial leadership stimulating climate action and pioneering new climate initiatives (Jörgensen *et al.*, 2015b; Jörgensen, 2020).

India's climate policy has evolved within its multi-tiered federal system. Most of its climate initiatives and policies therefore involve more than one governmental level as well as public and private actors. Successful implementation thus requires collaboration and negotiation between the various levels and actors involved. Considering the need for coordination, the multilevel governance (MLG) lens applied in this book is well suited to an analysis of India's climate policy. India's multilevel climate governance structures offers opportunity structures for climate-friendly innovations. In some cases, bottom-up approaches to renewable energy or other innovative green economy approaches developed in India's states and cities are uploaded to the national level or are diffused to other states.

While widening slightly this book's conceptualisation of climate policy, a broader definition of climate policy measures will be applied. Policy measures taken in neighbouring policy areas (such as agricultural policy), which have a positive spillover-effect for climate protection, will also be assessed briefly.

The first subsection introduces India's climate policy and the role it has played in the international climate regime giving illustrative examples of India's occasional and specific leadership role in the international context. The second subsection links the description of India's federal climate policy structures to the main theoretical concepts used in this book. It will be argued that the MLG lens is a helpful analytical tool for the analysis of pioneering actions and leadership in India. India's states, and increasingly also its cities, NGOs and think tanks, play an important role in shaping the country's climate and energy policy. The third subsection examines examples from the subnational state level. Finally, empirical examples and analytical insights will be discussed against the backdrop of theoretical concepts.

India's international climate policy

Since the advent of the international climate negotiations between 1990 and 2007, government officials and NGOs have shared an undisputed policy paradigm that has played a significant role in India's climate policy: the implication that developing countries like India should not be responsible for climate mitigation. Climate change was a problem created by industrialised countries which should therefore be held to account, accept responsibility for global warming and reduce their GHGE (Dubash, 2013; Isaksen and Stokke, 2014). India positioned itself as an opponent to any commitments to GHGE reductions in developing countries. Along with the historical responsibility of developed countries, another argument stated that developing countries still make a relatively small contribution to global GHGE and have very low per capita CO_2 emissions. Thus, they should not commit to GHGE reductions – instead, they should seek to catch up with industrialised countries economically. The related 'growth first' thinking holds that India's most important priority, as a former colony and developing country, should be development and poverty reduction which should not be hampered by emission reduction goals. 'Third Worldist diplomacy', which India had conducted since Nehru's time as Prime Minister, was applied in various international contexts, such as international trade negotiations India's climate justice narrative similarly builds on moral principles and distributive strategies (Michaelowa and Michaelowa, 2011: 3). It could therefore be argued that India has tried to offer cognitive climate leadership on climate justice issues in the international climate change negotiations. Climate justice implies that in contrast, industrialised countries should take responsibility and contribute to GHGE reduction efforts. This approach was also reflected in the Kigali agreement on hydrofluorocarbons (HFCs) in which India managed to postpone its phase-out until 2047 (International Energy Agency, 2019). India was quite influential in the international negotiation arena concerning the phasing down of HFCs in the context of the amendment of the Montreal Protocol on Substances that Deplete the Ozone Layer (Montreal Protocol). HFCs are chemicals used in air-conditioners and refrigerators and are extremely aggressive climate gases that are expected to contribute almost 20% of total global warming by 2050 (Gosh, 2019: 232). Driven by domestic industrial interests, the increasing consumer demand for air-conditioners and the projected costs of a transition, India was initially not interested in a reform of the Montreal Protocol and a ban on HFCs. Yet, the country shifted its position, motivated by domestic industrial stakeholder dialogues, which were initiated by Indian environmental non-governmental organisations (ENGOs) and think tanks, and improved expertise on the economic implications and climate impacts. India was able to 'make sense of complex technical issues on its own terms' and, switching from obstruction to cognitive leadership, submitted an influential proposal to amend the Montreal Protocol (Gosh, 2019: 243). India was able to show entrepreneurial leadership in the Kigali negotiations and brokered a less challenging HFC reduction schedule for a small group of the world's hottest countries.

The above-mentioned climate policy paradigm actually helped India to play a leading role at various stages of the global climate change negotiations (Dubash *et al.*, 2018). From the outset, India succeeded in placing the interests of the developing countries higher up the political agenda. India exhibited cognitive leadership by introducing the equity principle to the international climate negotiations, which was met with strong approval by fellow industrialising countries (Dutta *et al.*, 2016). This cognitive leadership can be traced back to India's NGO sector (Dubash *et al.*, 2018). In particular, Sunita Narain and the late Anil Agarwal – of the Centre for Science and Environment in Delhi, an ENGO founded by scientists, a journalist and environmental activists in 1980 – contributed expertise and data compilations, and insisted on per capita rather than total national emissions calculations. They also contributed to the framing of historical responsibility lying with the developed countries, whose emissions since the onset of the industrial era had triggered climate change (Raghunandan, 2020). India also played a role as global climate governance rule-maker by pushing successfully for the introduction of the CBDR concept into the 1992 United Nations Framework Convention on Climate Change (UNFCCC). India and China (see Chapter 2 in this volume) took the lead in the G77 negotiating group of 135 developing countries, in the run-up to the first Conference of the Parties (COP1) that took place in Berlin in 1995 (Gupta *et al.*, 2015). India and the other G77 countries insisted on the differentiated architecture of the treaties that took into account the CBDR concept, thus leaving developing countries free from obligations.

One side effect of the climate equity narrative and CBDR principle, which led to the creation of the Non-Annex 1 group of developing countries that did not commit itself to any obligations, was that it erected a firewall. This firewall sealed India's domestic political debate off from discussions about climate mitigation and the potential of low-carbon development in India, which regarded climate policy as a threat to development (Jörgensen, 2017).

Because of their rapidly-growing GHGE, India and other BRICS (Brazil, Russia, India, China and South Africa) countries came increasingly under pressure in the run-up to the 2009 Copenhagen Climate Change Conference (COP15). Shortly before the COP15, India joined the China-led alliance of BASIC countries to resist the mounting pressure from a US-led North to commit to mitigation obligations and dilute the CBDR (Sengupta, 2020). The BASIC countries drew up the Copenhagen Accord in direct negotiations with the US, sidelining the EU and the conference host Denmark (Andersen and Nielson, 2016). The Copenhagen Accord kept India's 'non-negotiables' and, in particular, the CBDR, equity principle, and the recognition for the 'overriding priorities' of poverty eradication and development (Sengupta, 2020: 124). Overall, it is not surprising that India did not emerge as a climate mitigation pioneer or leader during this phase. In the 2010s, India's government still made fewer commitments in the international climate negotiations than it carried out domestically (Betz, 2012).

Yet, India's climate policy paradigm shifted gradually, particularly in the years following the 2009 Copenhagen Climate Change Conference, setting the course more significantly towards climate mitigation (Dubash, 2019; Raghunandan, 2020). The post-Copenhagen position implies that 'despite not having been historically part of the problem, India was stepping forward to be part of the global effort towards a solution' (Raghunandan, 2020: 211). From 2008 onward, the domestic climate mitigation policy gradually emerged, spurred on by the international climate process and a newly developing domestic climate advocacy coalition, as well as the Environment Minister Jairam Ramesh (2009–2011), under the Congress-led United Progressive Alliance (UPA) government.

The international climate negotiations in the run-up to the 2015 Paris Climate Change Conference (COP21) stimulated a reformulation of climate policy. The polycentric architecture of the 2015 Paris Agreement, which emphasises voluntary pledges for emission reductions, learning, cooperation, trust and bottom–up initiatives rather than binding obligations was met with scepticism by India (Dubash, 2016). India's main concerns were related to climate justice and the lack of differentiation of responsibilities for GHGE reductions by industrialised and industrialising countries.

India nevertheless accepted the 1.5°C goal and ratified the Paris Agreement. Surprisingly, India's Intended Nationally Determined Contributions (INDCs) indicated in part a policy shift towards more ambitious goals. First, a rather moderate new goal for the reduction in the emissions intensity of India's gross domestic product (GDP) from 33% to 35% by 2030 (compared to 2005). Second, and even more importantly, India's INDCs include a highly ambitious goal for renewable energy according to which non-fossil fuels should make up a 40% share of the installed electricity mix by 2030. As large hydropower and nuclear energy are not expected to play a significant role in India's future energy mix, the transition from coal to renewable energy will take decades. Yet, the falling prices of renewable energy, concerns about local environmental impacts and the increasing costs of coal-based electricity 'are shifting the balance away from coal and towards renewable' energy (Sreenivas and Gambhir, 2019: 442). This means that fossil fuels will still play an important role for decades to come.

The third goal was additional carbon reductions from 2.5 billion to 3 billion tons through an increase in forest cover. The goal to enhance solar power capacity to 100 GW by 2022 particularly stands out. It is linked to a remarkable international initiative taken by India at the 2015 Paris Climate Change Conference, namely the creation of an International Solar Alliance involving more than 120 countries, including several African nations (Government of India, 2015), which can be interpreted as entrepreneurial leadership. Mohan and Wehnert (2019) argue more cautiously that India's NDC targets, including those for renewable energy and emission intensity, were formulated according to a business-as-usual scenario and also point to India's yet undecided expansion of coal capacities.

In 2018, India had a fully-fledged climate policy framework including mitigation and adaptation measures at the national level as well as climate policy plans by India's states (Jogesh and Dubash, 2015; Jörgensen *et al.*, 2015c; Mohan and Wehnert, 2019; Dubash *et al.*, 2018). Whereas India's domestic action surpassed its international commitments in 2008, the reverse scenario is likely in 2020: India has changed its role in international climate negotiations but its ambitious goals formulated within the global context are not fully reflected in India's domestic policy framework (Mohan and Wehnert, 2019). In the following section, the focus will be on India's domestic climate policy, which will be explored through a multilevel governance lens that will shed light on the interplay between the national and subnational state levels.

Multilevel climate governance in India

Located somewhere between polycentric and state-centric concepts (Wurzel, Liefferink and Torney, 2019) multilevel governance provides a suitable lens for the analysis of climate leaders and pioneers in India. India's federal system consists of a large number of government bodies, rural jurisdictions, urban local bodies in the cities, federal state and union territory governments, and government institutions at the national level. India comprises 29 states and eight union territories, more than 4,000 cities and 262,771 rural governments; all of these jurisdictions are challenged by the need to create urban and rural infrastructure and to adapt livelihoods to climate change. In large developing countries and transition economies, lower levels of government may play a more important role in the provision of public services and technological change (Bardhan, 2002). In India, policy-making takes place within a relatively centralised federal structure giving great authority to the Union Government. Despite continuous advocacy for greater decentralisation of India's political system, in climate policy, as in many other policy fields, the policy paradigm of the superiority of centralised top–down governing by the Union Government has been dominant since India's independence (Ciecierska-Holmes and Jörgensen, 2020). Perceived deficits of accountability to local citizens lead to various approaches to decentralisation and the empowerment of local governments (Bardhan and Mookherjee, 2006: 21). However, the implementation of devolution as performed by India's states varies and proceeds sluggishly with financial and decision-making powers staying centralised at the national and the subnational state level. Only in a few cases has decentralisation improved the fiscal and administrative capacity of local governments, and 'significantly affected patterns of representation and service delivery' (Bardhan and Mookherjee, 2006: 37).

India has a history of relatively centralised climate governance structures with a high degree of interdependence and need for cooperation between governmental levels. Major climate policy initiatives such as the National Action Plan on Climate Change (NAPCC) and sectoral policies are formulated and decided at the national level and need to be implemented at the lower levels of government involving overlaps in responsibilities which create the need for coordination and negotiation between different levels of government and the private sector. Climate initiatives taken in such contexts would fit better into MLG concepts, which assume a stronger role for governmental actors than polycentric concepts. Polycentric concepts deal with governance structures characterised by a high degree of self-coordination (Wurzel, Liefferink and Torney, 2019). In India, available cases of polycentric climate governance may play a more significant role in India's cities and rural areas where they are driven by, for example, specific needs for adaptation.

In large developmental states with rapidly growing economies, urbanisation, industrialisation processes, energy demands, and the upscaling and/or diffusion of best practice solutions can be powerful catalysts for the development of low carbon infrastructures and the adaptation to climate change. India is a rapidly-growing country that is challenged by persistent poverty, constantly growing energy demands and rapid urbanisation. There is a tremendous need for new infrastructure, including housing and transportation in India's urban areas, and access to electricity in rural areas (Singhal and Jain, 2020; Khosla and Bhardwaj, 2019). At the same time, there is a need for environmental leapfrogging, for clean technologies and green infrastructures to flatten the 'environmental Kuznets curve' (Dasgupta *et al.*, 2002). Kuznets postulated a causal relationship between economic growth and environmental pollution while assuming that indicators of environmental degradation rise before they reach a tipping point at which they start to fall while, at the same time, income per capita keeps on rising steadily. In other words, according to Kuznets a decoupling of environmental pollution and growth will take place once a certain level of affluence is reached by countries. It is a matter of urgency to decouple growth from resource consumption, reduce pollution and create healthy environments in India's urban and rural areas. Infrastructure development needs to consider the imperatives of climate change, which means adapting habitat and living space to global warming and, on the other hand, to mitigate climate change through low carbon infrastructure (Singhal and Jain, 2020). Climate-friendly infrastructure development and adaptation pose tremendous governance challenges, not least the difficulties in financing. This task requires horizontal and vertical coordination, input and cooperation from the public and private sector, civil society actors, think tanks and from other domestic actors as well as from international actors. It requires cooperation and coordination between political levels, and thus institutional structures and forms of multilevel governance which enable a shift in decision-making powers and leadership also beyond the central state.

Pioneers and leaders in India's states

India has a long history of centralised policy-making, and political leadership has traditionally originated from the national government. However, the second tier of government, the subnational states, has become more important since India's market liberalisation in 1991. An exploration of potential climate policy pioneers and leaders from this period onward is therefore important. The states in India have become more relevant, compensating for a lack of national initiatives, particularly in development policy, welfare and the regulation of industrial relations (Sinha, 2005). India's states can develop strategies for various economic sectors, which are of utmost importance for climate governance. Various climate-relevant legislative areas are subject to state regulation, including agriculture, water, waste and land use, or they are a responsibility shared between the states and the Union Government, such as electricity. Pioneering behaviour of the states emerges in the context of policy formulation, the development of state regulations and incentive systems. It can also be significant in the context of the implementation of national frameworks, as in regional planning processes and in the execution and further elaboration of national strategies.

India's first NAPCC (2008) was developed under the structural leadership of the national level and was not subject to a wide consultation process. Since then, in different phases with different levels of significance, more dynamic multilevel climate governance structures have emerged (Jörgensen *et al.*, 2015a). Domestic think tanks and academic organisations played a greater role offering cognitive leadership in agenda setting, policy formulation and implementation since 2009 (Dubash *et al.*, 2018; Fernandes *et al.*, 2020). Greater influence of the subnational levels is noticeable in the context of domestic and transnational networks involving NGOs, the corporate sector and donor organisations, all of which are influential at all levels of policymaking (Fisher, 2012). Subnational action plans, i.e. the State Action Plans on Climate Change (SAPCC), detail the objectives of regional climate action and the forms it should take (Jogesh and Dubash, 2015; Shukla *et al.*, 2015; Jörgensen *et al.*, 2015c). Thirty-two states and union territories had put in place SAPCCs before India's INDCs were submitted to the UNFCCC in 2015 (Government of India, 2015). Without the active role of India's states, union territories and cities, the mitigation and adaptation goals formulated in India's INDCs cannot be implemented.

Because of other pressing problems and political priorities, climate policy as a policy domain in its own right does receive rather low political attention in India's states. This is reflected in the institutionalisation of climate policy at India's state level. One state, Gujarat, was temporarily regarded a pioneer because it introduced a Department of Climate Change. Concerned by its vulnerability to the melting of Himalayan glaciers, the mountain state Sikkim institutionalised a State Council on Climate Change and a Glacier and Climate Change Commission. In the context of the climate action planning, a few states (Sikkim, Himachal Pradesh and Odisha) emphasised environmental issues (Dubash and Jogesh, 2014: 8) while other states gave high-level political support in the form of direct involvement of their political representatives and executive heads in the SAPCC (Gujarat, Maharashtra, Sikkim and Odisha) (Jörgensen *et al.*, 2015c).

Individual states, such as Madhya Pradesh, attached importance to regional consultations involving local stakeholders (Jogesh and Dubash, 2015: 257). Regarding policy change, the envisaged policy interventions in the states' climate action plans were considered more incremental (or transactional) than transformational. States' climate action plans were significantly shaped by the objectives and measures formulated in India's National Climate Action Plan 2008 (Jogesh and Dubash, 2015: 250). The majority of policy matters considered important in the states' climate action plans relate to specific regional concerns and in particular to vulnerability, adaptation and resilience building. This includes electricity transmission and distribution losses in the electricity sector (Odisha), water conservation (Sikkim) and the payment for ecosystem services (Himachal Pradesh) (Jogesh and Dubash, 2015: 250).

In the context of environmental policy, due to the scale of the pollution problem and their limited administrative-political capacities, India's states face difficulties simply acting, never mind pioneering environmental innovations (Jörgensen, 2020). In various neighbouring policy areas, such as agriculture and renewable energy, significant policy innovations can be observed. Agriculture is an important economic sector which provides employment to 48.9% of the workforce and contributes 17.4% of the GDP (TERI, 2016: 197). The sector requires improvements in resource efficiency, conservation and water use through more sustainable forms of agriculture. 17.81% of the electricity consumed in India goes to agriculture, in 17 states between 40% and 89.2% of the land is degraded, and India's states suffer from water pollution and scarcity (TERI, 2016: 197). The state of Sikkim introduced a policy in 2003 to pioneer the state's transition to organic farming (Government of Sikkim, 2019). Sikkim's 'Organic Mission' is an economic and environmental competition strategy, involving organic standards and regulations, market development for organic food products, the development of an organic farming sector and bio villages, and the development of new technologies. In 2016, Sikkim became the first fully organic state. Sikkim is regarded as a model which can be emulated by other states; however, Sikkim's farmers still face economic challenges due to the lack of local demand for organic products (*The Guardian*, 2017).

In a few areas, such as renewable energy and energy efficiency, initiatives in which India's states and union territories achieved incremental policy change or even pioneered new policy solutions have influenced national policies. As part of its mitigation strategy, India pledged to achieve 40% cumulative installed electric power capacity from non-fossil-fuel-based energy resources by 2030. India has a rapidly growing renewable energy sector and there is significant potential for a transition to renewable energy following China (see Chapter 2 in this volume) with 'perhaps a ten-year lag' (Mathews, 2015: 10). The national renewable energy policy framework is well established, and has successfully promoted low-carbon technologies in India's wind and solar sector (Jänicke, 2012). India's Union Government has pursued renewable energy policy since the 1970s and reinforced the institutionalisation of this policy field after India's liberalisation in 1991. In 2008, renewable energy policy became an integral part of India's national climate policy. Renewable energy goals are steadily increasing and present important components of India's NDCs to climate mitigation as submitted to the UNFCCC in the run-up to the 2015 Paris Climate Change Conference (COP21).

Because of its economic, social, environmental and climate co-benefits, renewable energy policy is an attractive policy field offering opportunities for pioneers and leadership. International comparative case studies show that subnational states pursue renewable energy policy for perceived economic and political advantages (Beermann and Tews, 2017; Jörgensen *et al.*, 2015a; Rabe, 2008; Schreurs, 2008). Green energy policy is, for example, an area where policy-makers at the state level can seek to decarbonise by reducing the use of fossil fuel and promoting instead wind and solar energy, thereby concurrently reducing local air pollution (Krause, 2011), generating jobs in the renewable energy sector and addressing the global climate problem (Rabe, 2008).

India's national government exhibited structural leadership during the first stage of renewable energy policy development in the 1980s while the states were only involved in demonstration projects (Chaudhary *et al.*, 2015). Since the 1990s, a few of India's states gained increased relevance in the design and implementation of a national renewable energy policy (Jörgensen *et al.*, 2015c). In 2011, Arunachal Pradesh, Odisha, Madhya Pradesh, Maharashtra and Uttrakhand were ranked highest among India's states with regard to climate mitigation. This ranking was made on the basis of State Action Plans for Climate Change, renewable energy growth rates and the electricity intensity of the states' GDP as part of an environmental performance index prepared for the Planning Commission (Chandrasekharan *et al.*, 2013). Renewable energy development and energy efficiency policies, in particular, create opportunities for economic co-benefits at the subnational level, such as achieving investment and employment through technology and business location strategies.

The coastal states of Tamil Nadu, Karnataka, Rajasthan, Maharashtra and Gujarat performed very well in the creation of renewable energy policies while exploiting locational advantages in the wind and the solar sector. In the solar policy sector, for example, there is indication of state leadership. Gujarat, located in north-west India with a large and fast-growing economy, introduced a solar energy framework and boosted solar power development via fixed preferential tariffs nearly a year before the national solar policy framework of the Jawaharlal Nehru National Solar Mission was officially released. Along with the national policy framework, individual state policies were relevant forces in boosting solar energy capacity from 18 MW in 2010 to 2,750 MW by July 2014 (Johnson,

2015). India's multilevel solar governance has stimulated India's solar energy market and yielded remarkable growth in capacity allocation across India's states. Achievements concern capacity, deployment rates, regulatory and policy support, industrial dynamics and the creation of knowledge (Jolly and Raven, 2016).

Energy conservation is another policy area where the states occasionally serve as pioneers (Khosla *et al.*, 2017). India pursues the goal of reducing the energy intensity of the economy. The national framework consists of: the Energy Conservation Act 2001; the National Mission of Enhanced Energy Efficiency, which focuses on industry; the Perform, Achieve and Trade (PAT) scheme; and the National Mission on Sustainable Habitat, which promotes energy efficiency in buildings. A few states pioneered policy measures for household energy efficiency, education and the influencing of behaviour. Andhra Pradesh introduced a mandatory building energy code which was developed in a consultative process involving public actors and stakeholders from the private sector after a severe power outage in 2012 (Khosla, 2016). Andhra Pradesh also stood out with its LED programme and is regarded as a pioneer in data-based governance. Agency for the state's initiatives came via the Chief Minister Chandrababu Naidu, who announced the vision for Andhra Pradesh to become one of the three best states in India by 2022 and the best state in terms of inclusive development by 2029.

The co-benefits approach introduced in India's National Action Plan on Climate Change to 'promote our development objectives while also yielding co-benefits for addressing climate change effectively' (Government of India, Prime Minister's Council on Climate Change, 2008) resonates in the states' climate action plan. A study of ten climate action plans found that renewable energy had been given emphasis in all of the plans (Jörgensen *et al.*, 2015c). The majority of the initiatives suggested in the state plans studied were incremental in nature and included some innovative initiatives. India's National Solar Mission was the policy driver behind the focus on solar energy across the board in all of the plans. Yet, objectives linked to non-solar energy resources varied across the states and were related to the advantages of co-benefits that are relevant to the subnational state context, for instance, renewable energy applications in agriculture, industry, urban development, transportation, energy, tourism and sustainable habitat. A few states, namely Madhya Pradesh, Karnataka and Kerala, have focused on green tariffs as an innovative, price-based way to promote renewable energy.

Conclusions

While assessing the central analytical themes for this volume, this chapter has analysed India's climate policy at both the domestic level and the international climate governance level. India has become a global climate power for the following three reasons. Firstly, it is the world's third largest emitter; secondly, it is a potential veto player in the international negotiations; and, thirdly, as a rapidly growing developing country it has wielded influence at different stages of the international process. India has provided cognitive and entrepreneurial leadership aligning the developing country coalition and has contributed significantly to anchoring the concept of 'common but differentiated responsibilities' in the international climate process. Another area is the leadership role that India plays in diffusing the uptake of renewable energy to other developing countries, as the International Solar Alliance launched by India and France at the 2015 Paris Climate Change Conference (COP21) suggests.

India's domestic climate policy is strongly shaped and led by the national government. As a large developing country split into many subnational states, cities and local rural governments, policy implementation requires negotiation and cooperation between different government levels. Responsibilities are not always neatly distributed and initiatives come from different governmental levels and non-governmental actors. Renewable energy policy, in particular, is spurred by policies developed both at the national and subnational levels. Some federal states pioneered energy and agricultural policy innovations, which spill over to climate policy and help India achieve its international obligations. Considering the interdependencies between different governmental levels, and the involvement of non-governmental actors and international factors, India's climate policy landscape can be examined as multilevel climate governance. National policy is often a necessary condition for the development of climate policy. However, it is not a sufficient condition. Bottom–up initiatives from the federal states are significant. Among other factors, India's multilevel solar governance has boosted solar development in India. India's National Solar Mission was the policy driver behind a focus on solar energy across the board in all the climate action plans developed by India's states.

Bibliography

Agueda Corneloup, I. de and Mol, A.P.J. (2014) 'Small island developing states and international climate change negotiations: the power of moral "leadership"', *International Environmental Agreements: Politics, Law and Economics*, 14(3): 281–297.

Andersen, M.S. and Nielson, H.Ø. (2016) 'Denmark: small state with big voice and bigger dilemmas', In R. Wurzel, J. Connelly and D. Liefferink (eds.) *The European Union in international climate change politics: still taking a lead?* London: Routledge, 83–97.

Bäckstrand, K. and Elgström, O. (2013) 'The EU's role in climate change negotiations: from leader to "leadiator"', *Journal of European Public Policy*, 20(10): 1369–1386.

Bardhan, P. (2002) 'Decentralization of governance and development', *Journal of Economic Perspectives*, 16(4): 185–205.

Bardhan, P.K. and Mookherjee, D. (2006) *Decentralization and local governance in developing countries. A comparative perspective*, Cambridge: MIT Press.

Beermann, J. and Tews, K. (2017) 'Decentralised laboratories in the German energy transition. Why local renewable energy initiatives must reinvent themselves', *Journal of Cleaner Production*, 169: 125–134.

Betz, J. (2012) 'India's turn in climate policy: assessing the interplay of domestic and international policy change', *GIGA*, http://www.econstor.eu/bitstream/10419/57188/1/689598971.pdf (Accessed 19 June 2020).

Chandrasekharan, I. *et al.* (2013) 'Construction of environmental performance index and ranking of states', *Current Science*, 104(4): 435–439.

Chaudhary, A., Krishna, C. and Sagar, A. (2015) 'Policy making for renewable energy in India: lessons from wind and solar power sectors', *Climate Policy*, 15(1): 58–87.

Ciecierska Holmes, N. and Jörgensen, K. (2020) 'Environmental politics in India. Institutions, actors and environmental governance', In N. Ciecierska-Holmes et al. (eds.) *Environmental policy in India*, Abingdon: Routledge, 241–258.

Dubash, N. (ed.) (2019) *India in a warming world. integrating climate change and development*, New Dehli: Oxford University Press India.

Dubash, N.K. (2013) 'The politics of climate change in India: narratives of equity and cobenefits', *Wiley Interdisciplinary Reviews: Climate Change*, 4(3): 191–201.

Dubash, N.K. (2016) 'Safeguarding development and limiting vulnerability. India's stakes in the Paris agreement', *Wiley Interdisciplinary Reviews: Climate Change*, 8(2): e444.

Dubash, N.K. et al. (2018) 'India and climate change: evolving ideas and increasing policy engagement', *Annual Review of Environment and Resources*, 43(1): 395–424.

Dubash, N.K. and Jogesh, A. (2014) *From margins to mainstream? State climate change planning in India as a "Door Opener" to a sustainable future*, New Delhi: Centre for Policy Research, Climate Initiative.

Dutta, V. et al. (2016) 'Evaluating expert opinion on India's climate policy: opportunities and barriers to low-carbon inclusive growth', *Climate and Development*, 8(4): 1–15.

Fernandes, D., Jörgensen, K. and Narayanan, N.C. (2020) 'Factors shaping the climate policy process in India', In N. Ciecierska-Holmes et al. (eds.) *Environmental policy in India*, Abingdon: Routledge, 158–173.

Fisher, S. (2012) 'Policy storylines in Indian climate politics: opening new political spaces?', *Environment and Planning C: Government and Policy*, 30(1): 109–127.

Germanwatch (2019) Climate *change performance index* 2020, https://www.climate-change-performance-index.org/country/india.

Government of India (GOI) (2015) *India's intended nationally determined contribution: working towards climate justice*, Government of India.

Government of India, Prime Minister's Council on Climate Change (2008) *National action plan on climate change*, http://www.nicra-icar.in/nicrarevised/images/Mission Documents/National-Action-Plan-on-Climate-Change.pdf.

Government of Sikkim (2019) Mission 2015 / Sikkim Organic Mission, http://www.sikkimorganicmission.gov.in/mission-2015/.

Gupta, H., Kohli, R.K. and Ahluwalia, A.S. (2015) 'Mapping "consistency" in India's climate change position: dynamics and dilemmas of science diplomacy', *Ambio*, 44(6): 592–599.

International Energy Agency (2019) Historic global deal to cut super-pollutant HFC gases - *environmental investigation agency website title: Eia-international.org*, https://eia-international.org/press-releases/historic-global-deal-cut-super-pollutant-hfc-gases/ (Accessed 20 December 2019).

Isaksen, K.-A. and Stokke, K. (2014) 'Changing climate discourse and politics in India. Climate change as challenge and opportunity for diplomacy and development', *Geoforum*, 57: 110–119.

Jackson, R.B. et al. (2018) 'Global energy growth is outpacing decarbonization', *Environmental Research Letters*, 13(12): 120401.

Jänicke, M. (2012) 'Dynamic governance of clean-energy markets: how technical innovation could accelerate climate policies', *Journal of Cleaner Production*, 22(1): 50–59.

Jogesh, A. and Dubash, N.K. (2015) 'State-led experimentation or centrally-motivated replication? A study of state action plans on climate change in India', *Journal of Integrative Environmental Sciences*, 12(4): 247–266.

Johnson, O. (2015) 'Promoting green industrial development through local content requirements: India's National Solar Mission', *Climate Policy*, 16(2): 178–195.

Jolly, S. and Raven, R.P.J.M. (2016) 'Field configuring events shaping sustainability transitions? The case of solar PV in India', *Technological Forecasting and Social Change*, 103: 324–333.

Jörgensen, K. (2017) 'India: the global climate power torn between "growth-first" and "green growth"', In R. Wurzel, J. Connelly and D. Liefferink (eds.) *The European Union in international climate change politics. still taking a lead?* London: Routledge, 270–283.

Jörgensen, K. (2020) 'The role India's states play in environmental policymaking', In N. Ciecierska-Holmes et al., (eds.) *Environmental policy in India*, Abingdon: Routledge, 39–59.

Jörgensen, K., Jogesh, A. and Mishra, A. (2015a) 'Multi-level climate governance and the role of the subnational level', *Journal of Integrative Environmental Sciences*, 12(4): 235–245.

Jörgensen, K., Mishra, A. and Sarangi, G.K. (2015b) 'Multi-level climate governance in India: the role of the states in climate action planning and renewable energies', *Journal of Integrative Environmental Sciences*, 12(4): 267–283.

Jörgensen, K., Mishra, A. and Sarangi, G.K. (2015c) 'Multi-level climate governance in India: the role of the states in climate action planning and renewable energies', *Journal of Integrative Environmental Sciences*, 12(4): 267–283.

Khosla, R. (2016) 'Building energy code lessons from Andhra Pradesh closing the policy gap', *Economic & Political Weekly*, 2: 66–73.

Khosla, R. and Bhardwaj, A. (2019) 'Urbanization in the time of climate change: examining the response of Indian cities', *Wiley Interdisciplinary Reviews: Climate Change*, 10(1): e560.

Khosla, R., Sagar, A. and Mathur, A.(2017) 'Deploying Low-carbon Technologies in Developing Countries. A view from India's buildings sector', *Environmental Policy and Governance*, 27(2): 149–162.

Krause, R.M. (2011) 'Policy innovation, intergovernmental relations, and the adoption of climate protection initiatives by U.S. Cities', *Journal of Urban Affairs*, 33(1): 45–60.

Liefferink, D. and Wurzel, R.K.W. (2017) 'Environmental leaders and pioneers: agents of change?', *Journal of European Public Policy*, 24(7): 951–968.

Mathews, J., (2015) *Greening of capitalism. How Asia is driving the next great transformation*, Stanford: Stanford University Press.

Mathur, A. (2019) 'India and Paris: a pragmatic way forward', In N. Dubash (ed.) *India in a warming world. Integrating climate change and development*, India: Oxford University Press, 222–229.

Michaelowa, K. and Michaelowa, A. (2011) *India in the international climate negotiations: from traditional nay-sayer to dynamic broker*, Zürich: Center for Comparative and International Studies (ETH Zurich and University of Zürich).

Michaelowa, K. and Michaelowa, A. (2012) 'India as an emerging power in international climate negotiations', *Climate Policy*, 12(5): 575–590.

Mohan, A. and Wehnert, T. (2019) 'Is India pulling its weight? India's nationally determined contribution and future energy plans in global climate policy', *Climate Policy*, 19(3): 275–282.

Rabe, B.G. (2008) 'States on steroids: the intergovernmental odyssey of American climate policy', *Review of Policy Research*, 28(2): 105–128.

Raghunandan, D. (2020)'Factors shaping India's international climate policy', In N. Ciecierska-Holmes et al. (eds.) *Environmental policy in India*, Abingdon: Routledge.

Schreurs, M.A. (2008) 'From the bottom up local and subnational climate change politics', *Journal of Environment & Development*, 17(4): 343–355.

Shukla, P.R., Garg, A. and Dholakia, H.H. (2015) Energy-*emissions trends and policy landscape for* India. New Delhi: Allied Publishers.

Singhal, S. and Jain, S. (2020) 'Smart sustainable cities', In N. Ciecierska-Holmes et al. (eds.) *Environmental policy in India*, Abingdon: Routledge, 174–200.

Sinha, A. (2005) *The regional roots of developmental politics in India*. Bloomington: Indiana University Press.

Sreenivas, A. and Gambhir, A. (2019) 'Aligning energy, development, and mitigation', In N. Dubash (ed.) *India in a warming world. Integrating climate change and development*, New Dehli: Oxford University Press India, 427–458.

TERI (2016) *TERI energy & environment data diary and yearbook 2015/16*, New Delhi, India: TERI Press.

The Guardian (2017) Sikkim's organic revolution at risk as local consumers fail to buy into project, Sikkim's organic revolution at risk as local consumers fail to buy into project. …((Can you please check this reference. In the text you state 31.01.2017 as the date of publication. Please add it to your reference here. Is there a URL which you could state?)).

Torney, D. and Mai'a, K. (2018) 'Environmental and climate diplomacy: building coalitions through persuasion', In C. Adelle, K. Biedenkopf and D. Torney (eds.) European Union *external environmental policy*, Berlin: Springer, 39–58.

Upadhyaya, P. et al. (2018) 'Comparing climate policy processes in India, Brazil, and South Africa: domestic engagements with international climate policy frameworks', *The Journal of Environment & Development*, 27(2): 186–209.

Wurzel, R.K.W., Liefferink, D. and Torney, D. (2019) 'Pioneers, leaders and followers in multilevel and polycentric climate governance', *Environmental Politics*, 28(1): 1–21.

Wurzel, R.K.W, Connelly, J. and Liefferink, D. (eds.) (2017) *The European Union in international climate change politics: still taking a lead?* London: Routledge.

4 Costa Rica and Vietnam

Pioneers in green transformations

Frauke Urban, Giuseppina Siciliano, Alonso Villalobos, Dang Nguyen Anh and Markus Lederer

Introduction

Green transformations are required worldwide to enable economies and societies to operate within the planetary boundaries (Rockström *et al.*, 2009a; 2009b; Steffen *et al.*, 2015). Transformations of the energy sector are especially important as approximately 70% of global greenhouse gas emissions (GHGE) come from energy-related activities, according to estimates by the Intergovernmental Panel on Climate Change (IPCC, 2018). This includes especially fossil fuel combustion from electricity generation, heating, cooling and industrial processes, as well as fossil fuel use in transport (IPCC, 2018).

Costa Rica and Vietnam are two countries that are considered as pioneers in driving forward green transformations in the Global South although both have very different political systems with Costa Rica being democratic and Vietnam being a socialist one-party state. Costa Rica's government pledged carbon neutrality by 2021, a pledge it later changed to decarbonisation by 2050. Already by the late 2010s, 98% of the country's electricity came from renewable energy, most importantly hydropower, but also wind, solar and geothermal energy (IEA, 2020). Vietnam has a National Green Growth Strategy that aims at reducing GHGE, promoting renewable energy, increasing energy efficiency and introducing carbon trading. Just under 40% of its electricity comes from hydropower, the remaining share is from fossil fuels (IEA, 2020). Both countries market themselves as pioneers in green transformations, mainly driven by domestic reasons, yet the actual implementation of these goals is challenging.

This chapter compares the strategies and motives of green transformations for energy-related industries in Vietnam and Costa Rica, analyses what role their different political systems have played on climate governance, what progress has been made and what the barriers are. Our main argument is that we witness state-led pioneership in both cases but of very different kinds. In both countries, multilevel governance arrangements provided important input into national energy policies but the main reason why both countries became ambitious pioneers had more to do with domestic politics and the role of public officials. The major difference is that in Costa Rica, the government has been much more open to bottom–up inputs as well as polycentric governance while taking up important initiatives from civil society actors of different kinds. In contrast, the government in Vietnam has adopted a much more top–down approach as can be seen from its energy policy which is almost completely dependent on government and party officials.

This chapter adopts a wide definition of the energy sector by including energy generation, use, and supply and demand in the power sector, transport sector, industry and through other economic activities. It discusses what other countries can learn from these pioneers in green transformations. The next section discusses the conceptual framework and the methodology before the results from Costa Rica and Vietnam are presented in the section which follows. The penultimate section analyses the results from a comparative perspective while the final section concludes the chapter.

Conceptual framework

This chapter combines the concepts of pioneers, leaders and followers in multilevel and polycentric climate governance by Wurzel, Andersen and Tobin in the Introduction (see also Liefferink and Wurzel, 2017; Wurzel, Liefferink and Torney, 2019) with the concept of green transformations by Scoones *et al.* (2015) to form a unique conceptual framework.

'Green' represents the environment, whereas 'transformations' are wide-ranging systemic changes across all sectors of an economy, affecting many groups in society. Transformations often involve 'challenging incumbent structures' through processes that tend to be driven by changes in knowledge and innovations (Stirling, 2015: 62). Green transformations can be defined as the reconfiguration of political, social, economic and technological systems to enable economies and societies to operate within the planetary boundaries (Scoones *et al.*, 2015). According to the Heinrich Böll Foundation (2013: 1) industrial societies should be transformed into 'climate compatible, resourceconserving and sustainable' systems. Achieving this requires farreaching and longterm changes in scientific, technological, social, economic, institutional and political systems and across global, regional, national and local levels (Lederer *et al.*, 2019; Urban *et al.*, 2018). Green transformations thus are multi-actor and multi-sectoral processes which pose the question: who are the agents providing most input? Scoones *et al.* (2015) have argued that drivers for green transformations stem either from technocentric or market-based innovations or are state- or citizen-led. We can differentiate *'material'-centred* types (technocentric and market-based transformations) and *'actor'-centred* types (citizen-led and state-led transformations). As our focus is on leaders and pioneers in this chapter, the actor-centred type seems to fit better. We focus more strongly on pioneership than leadership before justifying why the notion of state-led pioneership in a multilevel governance setting is the most appropriate analytical concept for a critical analysis of our two case countries.

We differentiate between pioneers, leaders and followers while largely following the definitions put forward in the introduction in this volume (see Chapter 1; see also Liefferink and Wurzel, 2017; Wurzel, Lifferink and Torney, 2019). Both

Costa Rica and Vietnam are therefore classified as pioneers with regard to green transformations in the energy-related industries. *Pioneers* tend to adopt ambitious climate and/or energy policies for domestic reasons. They may attract followers (e.g. other countries with similar policies), although they do so mostly unintentionally. *Leaders,* on the other hand, usually try to attract *followers*. For example, in case of the European Union (EU), one Member State may adopt an ambitious climate policy in the hope that another one will follow its example. Whether leaders and pioneers attract followers will need to be established empirically (Liefferink and Wurzel, 2017; Wurzel, Connelly and Liefferink, 2017; Wurzel, Liefferink and Torney, 2019).

In Central America, Costa Rica cannot be seen as a leader nor has it followed any leader in its neighbourhood. The same is true for Vietnam's role in Southeast Asia. As will be discussed below, this may provide empirical evidence for the leader concept not being applicable to Central America and Southeast Asia. There are, however, outside influences (i.e. exogenous factors) which we will assess in the empirical parts of this chapter. As these exogenous factors are too diverse, one cannot argue that either country is a clear follower of one other country, for example, the US in Costa Rica's case and South Korea for Vietnam. Focusing on these cases, we thus have to take into account the complex relationship between domestic politics and various inputs from the external environment. In order to be able to do so, we make use of the concept of multilevel governance.

Multilevel governance (MLG) focuses on the bi-directional dependency of governance between various actors at various levels, most importantly between supranational and subnational governmental players (Hooghe and Marks, 2001). MLG is, for example, frequently observed in the EU, where the European Commission plays a major role as well as national ministries. For climate governance, most scholars have suggested that there is no one central steering point. Instead, governance takes place at various governance levels by numerous actors (Jordan *et al.*, 2015). MLG also stresses the importance of coordinated action between environmental leaders/pioneers and the reinforcement of leadership/pioneership at multiple levels especially by supranational actors (Schreurs and Tiberghien, 2007). This concept is well suited also for our two case country studies, as no clear-cut leader-follower relationship can be identified, and we do not witness simple diffusion mechanisms without any involvement of domestic actors.

Polycentric governance by contrast focuses more on the role of societal actors such as businesses, NGOs and civil society. Self-coordination of these actors leads to multiple, decentralised decision-making units at various levels. Each of these units can function relatively independently, adopting its own norms and rules (Ostrom, 2010). This polycentric leadership/pioneership of societal actors supports the functioning of global climate governance. Important roles are attributed to self-organisation, experimentation and learning (Ostrom, 2014; Jordan *et al.*, 2015). We will show that Costa Rica and Vietnam became pioneers by incorporating external stimuli within their governance arrangements and that the way in which this happened was strongly influenced by the ideas, interests and institutions of the respective country. For both case countries, we will focus on how this has played out particularly at the government level as most analyses of multilevel governance have done. We therefore neglect some of the influence that societal actors have had and pay less attention to processes of polycentric governance although particularly in Costa Rica, civil society has had a tremendous influence on environmental and energy policies. And even in Vietnam non-state actors should not be completely neglected as we will show below. Nevertheless, in both cases we do not witness the degree of self-organisation or decentralised decision-making that one would expect in polycentric governance arrangements.

Methodology

This research draws on insights from in-depths qualitative fieldwork in Costa Rica and Vietnam during the period 2016–2019. The project team conducted 20 interviews in Vietnam and 27 in Costa Rica. The interviews were semi-structured and open ended. Four types of groups were interviewed: (1) policy-makers from government and bureaucracy; (2) representatives from firms and entrepreneurs; (3) experts from civil society and academia; and (4) representatives from multilateral organisations and donors. The interviews were recorded and stored on digital media, where possible. The interview data was analysed according to the conceptual framework, using narrative analysis to understand and interpret the findings. The primary data was supplemented with secondary data. This included qualitative data (e.g. policy documents) and quantitative data (e.g. energy and emission data from the International Energy Agency (2019) and the World Bank (2019)).

Costa Rica's green transformations in energy-related industries

Costa Rica, a small country in Central America, has a longstanding reputation of being a pioneer in green transformations. It has even been labelled a 'green republic' (Evans, 1999). It is a democracy in which civil society plays an important role and the country has no own military since 1948. From an environmental perspective, the country has hardly any extractive industries (e.g. mining). It has a high forest cover and a low rate of deforestation. Already in 1961, the government decided to dedicate a substantial part of Costa Rica's land to conservation by introducing a protected area system, particularly for forested areas. Over decades the forest cover has indeed increased. In 2020, about one-quarter of the country was under some form of protection and 55% of the country is once again covered by forests (World Bank, 2019). A Sustainable Development Strategy for national development was introduced in 1988, followed by the introduction of a Payments for Ecosystem Services (PES) scheme under the Forestry Law in 1996. In 2008, Costa Rica's government decided on a Carbon Neutrality Goal to be achieved by 2021, a pledge it later changed to decarbonisation by 2050. Costa Rica introduced the National Climate Change Strategy (ENCC) in 2009. It launched the National Climate Development Plan in 2015, which was followed by the National Decarbonisation Plan for the period 2018–2050.

In terms of energy, about 98% of the country's electricity comes from renewable energy, mostly hydropower (about 74%), but geothermal energy (about 12%), wind (about 11%), biofuels and solar play an important role (IEA, 2019). Hence, for about 70 years, Costa Rica has been a pioneer in green transformations, particularly in relation to forestry and energy issues for the following three main reasons. First, its green transformations are mainly domestically motivated. This is the case for protecting valuable natural resources such as forests where strong international actors (e.g. the World Bank) and also domestic actors (e.g. scientists but also indigenous communities) pushed for conservation schemes already in the 1980s. Similarly, when it comes to energy dependency, costly fossil fuel imports were reduced. This was motivated by economic reasons and allowed the provision of relatively cheap electricity in the whole country with almost all communities being connected to the grid system. Secondly, while Costa Rica has built a reputation of being green due to ambitious climate and energy goals, it has relatively few followers. When, however, it has attracted followers, it has happened unintentionally and much more in the field of PES and forestry than in the energy sector. Politically, Costa Rica is rather isolated in Central America in many policy fields (e.g. defence, environmental, energy and other policies) and no other country in the region ever officially labelled Costa Rica as an example that it wants to follow. Thus, with regard to climate and energy governance, Costa Rica is more progressive than its neighbouring countries, but this has not resulted in the country attracting followers. Finally, domestic pioneership has been taken up by different groups in Costa Rica, mainly from civil society, scientists and some government officials. Domestic leaders have faced strong opposition but also received some external support including donor agencies. For example, the development of the PES scheme was built on ideas that originated from the World Bank and that were taken up by scientists at the Universidad de Costa Rica and domestic NGOs. However, it was the central government that set up the scheme, regulated/re-regulated it and made sure that financial means, through the taxation of gasoline, were made available (Porras *et al.*, 2013). Overall, the PES scheme is thus a successful example of multilevel governance with elements of polycentricity.

Policy objectives, progress and current status

Already in 2008, Costa Rica pledged to become carbon neutral by 2021 and in 2015, the government also suggested that the country would be limiting its emissions to a maximum of 9.37 Mt CO_2 equivalent annually by 2030 (compared to about 12 Mt in 2012), as part of its Nationally Determined Contribution (NDC) for the Paris Agreement under the United Nations Framework Convention on Climate Change (UNFCCC, 2015a). Costa Rica's government also suggested ambitious goals for reducing per capita emissions, namely from more than 3.5 t CO_2 per person in 2015 to about 1.73 t by 2030, 1.19 t by 2050 and -0.27 t by 2100 (UNFCCC, 2015a). However, in 2018, the carbon neutrality goal for 2021 had to be replaced with the goal to decarbonise completely only by 2050 as it became evident that the earlier date was simply not achievable. In the energy field, Costa

Rica is already well advanced in the use of renewable energy sources, especially for electricity production. In 2017, the country set a new record when its electricity needs were completely covered by renewables over a period of 300 days. Most of the electricity generation comes from renewables, especially hydropower, as well as geothermal energy, wind, biofuels and solar (IEA, 2019). Solar energy is expected to grow rapidly in the future. The actual installed capacity of solar energy amounts to approximately 8.4 MW compared to the identified potential which amounts to 120 MW (Acuna *et al.*, 2018). From 2010–2017, Costa Rica attracted $1.9 billion for clean energy investments: over a third was directed to small hydro plants, followed by geothermal and wind (Rapid Transition Alliance, 2019).

Strategies and motives

Costa Rica's decarbonisation strategy has the highest political backing. For example, President Carlos Alvarado Quesada was quoted as saying:

> *Costa Rica knows that decarbonisation is the great task of our generation, and we want to be the first country in the world to achieve it. We are putting decarbonisation at the heart of our national development, public investment and long-term strategic plan. Our nation has understood that responding to climate change requires transformational – not incremental – shifts, and that the government has a key role to play in charting the path for such transformations.*
>
> (2050Pathways, 2019: 1)

Similarly, the Minister of Environment and Energy, Carlos Manuel Rodriguez, argued:

> *Decarbonisation is a commitment of Costa Rica with current and future generations, it means transforming the development model to a sustainable one, free of fossil fuels, that improves the country's competitiveness and the quality of life of people. The goal is to be a country with net zero emissions by 2050. These transformations are not new to the country, we have done it before. We are a tropical country that stopped deforestation and tripled our per capita income, which generates 99% of our electricity from renewable sources, we abolished the army and instead decided to invest in education. Costa Rica is ready for the challenge of decarbonisation.*
>
> (2050Pathways, 2019: 1)

Thus, the government identifies itself as a pioneer that is clearly 'moving ahead of the troops' (Liefferink and Wurzel, 2017: 2–3) arguing that decarbonisation is central to any policy-making and will go beyond incremental change, hence embarking on transformational pioneership driven forward at the top policy level. Furthermore, the green transformation is framed as being part of a tradition in environmental policy providing justification for radical change. Finally, as the National Decarbonisation Plan is labelled 'No one left behind – Decarbonisation and resilience are based on the principles of inclusion, respect for human rights, and gender equality' (Costa Rican Government, 2018: 3) and a rights-based legitimation strategy is being employed. The motives for decarbonisation in Costa Rica are therefore not only environmental and economic, but – at least on the official level – also related to quality of life for its people, intergenerational equity and social fairness.

Strategies for decarbonising Costa Rica's economy, as part of the National Decarbonisation Plan (2018–2050), target public and private transport, energy, industry, buildings, waste management, agriculture and land use management. The goal is that 100% of all electricity should come from renewable energy by 2050, which is a very realistic target to achieve, as about 98% of the electricity comes from renewables already today. However, more demanding is the transport sector. Most proposed emission abatement measures therefore rely on a greater use of electric transportation such as for buses, taxis and private vehicles. Other strategies include improving energy efficiency, encouraging energy conservation and fuel switching to reduce emissions in the built environment (e.g. housing, and residential energy use) and in industrial processes (Costa Rican Government, 2018). Several programmes are in place to reduce the impacts on emissions of the agricultural sector including Nationally Appropriate Mitigation Actions (NAMAs) to reduce GHGE from coffee production and processing, the livestock sector and the biomass sector.

Challenges and barriers

Despite these ambitious, aspirational goals for green transformations and decarbonisation, there are various challenges and barriers. First of all, it needs to be discussed why and how the goalposts have been shifted from the 2021 carbon neutrality goal to the 2050 decarbonisation goal. One interviewee argued that '[d]ecarbonisation is a step forward in comparison to carbon neutrality. Decarbonisation is more than having zero emissions' (Interview, climate and energy expert, 2019). Another interviewee pointed out that '[d]ecarbonisation is a transformation process, it is not just emissions' compensation. Decarbonising the economy is much better than carbon neutrality since you can involve several sectors of the Costa Rican economy' (Interview, environmental lawyer, 2019). Hence, expert interviewees seem to consider the change in policy a positive move that shifts from a strong reliance on forests as carbon sinks to wider emission reductions across every economic sector. Yet, changing the time frame of the goal from 2021 to 2050 also means that actions are postponed by several decades, hence buying time at a moment when the carbon neutrality goal might be difficult to achieve. Second, progress is quite uneven across different sectors and the transport sector in particular is lacking far behind as it is still heavily reliant on oil-based combustion vehicles. Transport makes up 50% of Costa Rica's energy demand and the transport sector accounts for more than 80% of oil product demand (IEA, 2019). Figure 4.1 shows the growing CO_2 emissions from transport as a percentage of total fuel combustion in Costa Rica. A sharp increase in emissions from transport has been found, amounting to nearly 70% of national emissions in Costa Rica (World Bank, 2019). The contribution of the transport sector to CO_2 emissions is predicted to grow in future, as Costa Rica's car market is growing between 3.5% and 6% per year (RECOPE, 2018). The abatement of transport-related emissions will thus require an ambitious investment portfolio in sustainable transportation systems (e.g. in electric vehicles and infrastructure) over the coming decades. In short, decarbonisation will only be possible if the Costa Rican transport sector will be transformed as emissions from other sectors were comparatively low: industries and construction emit about 15% of national CO_2 emissions, the remaining 15% of emissions are from households, services and agriculture (World Bank, 2019).

State-led pioneership?

Costa Rica's motives and strategies for achieving carbon neutrality and complete decarbonisation are mainly internally-driven which can partially be explained through the high vulnerability of the country. Climate change is considered a real threat, for example, due to increasing droughts, water stress and extreme weather events although this is also the case for Costa Rica's neighbours. We therefore argue that we also have to focus on the political system and ask how strong government officials pushed for state-led transformative processes. In all of our interviews, it became apparent that both the political elite as well as street-level

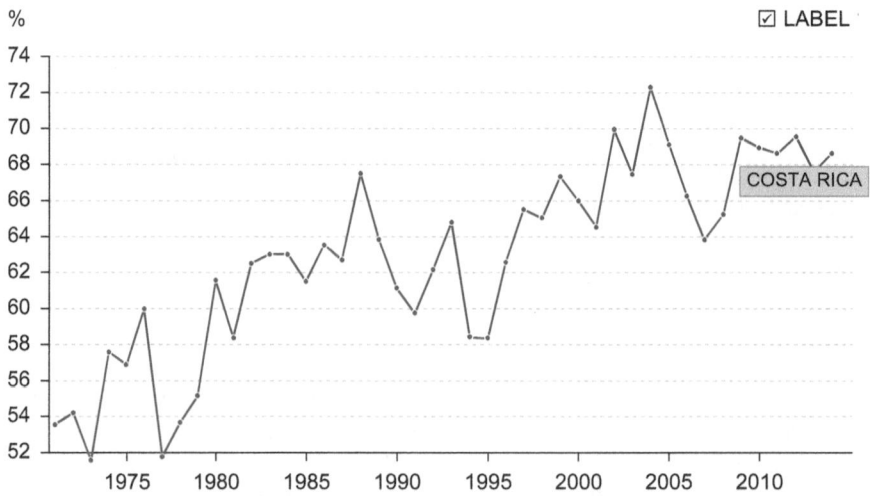

Figure 4.1 CO_2 emissions from transport as a percentage of total fuel combustion in Costa Rica. *Source:* World Bank (2019); Based on IEA Statistics © OECD/IEA 2014, //www.iea.org/statistics, All rights reserved.

bureaucrats consider green transformations as an opportunity to improve resilience, competitiveness, quality of life, inclusion and gender equality. The notion of Costa Rica becoming a pioneer in green transformations has therefore evolved into a national project. Historically, the country has a rather strong public service (e.g. in the field of health and education) and has been labelled a 'social democratic developmental state' (Sandbrook *et al.*, 2007) in which the bureaucracy not only enjoys some autonomy but also has strong capacities and a rather high degree of legitimacy (Lederer *et al.*, 2019). There is a strong correlation between the energy sector's actions proposed in the National Development Plan by the Ministry of National Planning and Economic Policy (MIDEPLAN) for the last two administrations (2014–2018 and 2018–2022) and the legal reforms approved by Parliament, executive orders from the Presidency and directives from Ministries in the last six years. This shows the strong influence of government in the energy-related industries, particularly for renewables, electric cars and trains, and the reduction of energy consumption from state institutions.

Yet, there is nevertheless some discernible external influence (e.g. from international donors) that is supporting green transformations. Interviewees reported the following:

> *International donor agencies have taken advantage of the experience of Costa Rica. Costa Rica is a sort of laboratory for new projects. They push certain kind of projects. For example GIZ is promoting that airport taxis become electric. GIZ thinks that Costa Rica can be an example of a green transformation.*
> (Interview, Ministry of Environment and Energy representative, 2018)

Similarly, the former UNFCCC Executive Secretary Christiana Figueres has argued that 'Costa Rica's decarbonisation plan is an excellent example for the rest of the world to follow' and the former US Vice President Al Gore claims, 'I am so excited to see that Costa Rica continues its role as world leader to help solve the climate crisis with the rapid deployment of the strategic plan to completely decarbonise its economy' (2050pathways, 2019). We can therefore state that while Costa Rica has domestic motives for a green transformation, the country has, on the one hand, been very good at promoting a 'green' image to the rest of the world, partly to create revenue from eco-tourism and to attract donors. On the other hand, the domestically-motivated pioneer is being pushed to take on the role of a leader in the future, both by internal and external actors.

Vietnam's green transformations in energy-related industries

Vietnam, a socialist country in Southeast Asia, has also become a pioneer in green transformations since about 2010. Vietnam is an autocratic country with a strong state apparatus where party and state as well as private and public issues are closely intertwined. Vietnam can therefore be characterised as a 'market economy with socialist orientation' (London, 2014: 2; Hansen, 2015: 92). It also features extractive industries (e.g. mining, oil and gas). From an environmental perspective, forest cover is high, and the deforestation rate is low although forest degradation is a major issue and many primary forests are turned into plantations. In fact, nearly 50% of Vietnam's land area is covered by tropical and sub-tropical forests, with increasing forest area over the last few decades (World Bank, 2019). In 2010, Vietnam introduced a user-led PES programme, under its Biodiversity Law, being the first in Southeast Asia. In 2011, the government adopted the National Climate Change Strategy, followed by the National Green Growth Strategy in 2012 and the Sustainable Development Strategy for the period 2011–2020. In 2014, the Vietnamese government issued the Green Growth Action Plan – again being ahead of all its neighbours. The Vietnam Green Growth Strategy (VGGS) aims to promote economic restructuring, increase economic competitiveness and achieve poverty reduction while at the same time using natural resources more efficiently, reducing GHGE and adapting to climate change (Urban *et al.*, 2018). It has quantifiable targets for GHGE reductions, namely, unconditionally to reduce GHGE by 8% compared to business as usual (BAU) by 2030, and a reduction of 25% if access to finance and climate-relevant technology is granted (UNFCCC, 2015b).

In terms of energy, total primary energy supply is still dominated by fossil fuels, mainly coal and oil. However, a growing part of Vietnam's electricity is generated from hydropower (currently about 39%), coal (about 33%), gas (about 28%) and a very small share of oil and wind energy (IEA, 2019). There are increasing investments in non-hydro renewable energy, most importantly wind and solar, but the generation capacities are still negligible compared to the contribution large-scale hydropower, coal and gas make (World Bank, 2019). Nevertheless, Vietnam can be characterised as a pioneer in green transformations for the following reasons. Firstly, its green transformations goals are ambitious and they are driven by domestic motives that are underpinning the Green Growth Strategy. Increasing poverty reduction and achieving economic growth through sustainable and resource-efficient economic restructuring are at the heart of the Green Growth Strategy. Secondly, in the past Vietnam could have been described as a follower that first learned from South Korea's 2008 Green Growth Strategy before it adopted its own VGSS in 2012. Ever since, the country has given its strategy a clearly Vietnamese perspective within which domestic interests in the bureaucracy but also with state-owned enterprises have dominated energy policies. Domestic leaders have also received some external support including donor agencies like GIZ or the World Bank. We can again identify a multilevel governance setting, highlighting the bi-directional dependency of governance between various actors at various levels. Thus, on the one hand, the Vietnamese government is deliberately aiming for international donor money in the field of climate and energy as other funds are often no longer available for a middle-income country like Vietnam. On the other hand, donors are actively trying to push the country into a green direction but, as stated above, the Vietnamese government – like many others in the Global South – has become quite self-confident in formulating its own priorities.

Similar to Costa Rica, Vietnam has no clear followers in the region although the country is well integrated in the Association of Southeast Asian Nations

(ASEAN) and more active than other ASEAN countries with regard to climate and energy policies.

Policy objectives, progress and current status

The main policy framework for green transformations in Vietnam is the Green Growth Strategy (GIZ, 2012). It has the following quantifiable targets: a GHGE intensity reduction of 8–10% compared to 2010; a decrease of energy intensity by 1–1.5% per year and GHGE from energy-related industries by 10–20% compared to 2010. The emissions could further be reduced to 20% with international financial support under the 2015 Paris Agreement or they could be limited to 10% domestic efforts without external support. The Strategy was also informative for Vietnam's NDC in the aftermath of the 2015 Paris Climate Change Conference (see below for details).

Vietnam's geography is highly favourable to renewable energy installations because renewable energy resources, such as for hydropower and solar energy, are abundant throughout the country. The current energy planning places greater emphasis on Vietnam's domestic potential for renewable energies for industrialisation and domestic demand of consumers. Politically, some experts argue that the country needs to become more independent from the favoured coal import market, which is mainly dependent on China. Public campaigns are in place to raise awareness and to help attract both domestic and foreign private sector investments and green technology in this growing field. The Vietnamese government also aims to increase the share of renewable energy among electricity generation (excluding hydropower) to 5–8% by 2020 compared with 3.5% in 2010. By 2025, the installed capacity of renewable energy should be 4,050 MW (Dang, 2016).

In terms of efforts and achievements, Vietnam increased its share of hydropower among the total energy supply, as well as wind and solar energy generation in recent years (IEA, 2019). About five wind farms are currently operating in Vietnam, nearly 30 further wind farms are currently under construction or in the pipeline. About 12 solar parks are in operation and about 30 are under construction or in the pipeline (DEVI Renewable Energy, 2017). Finally, old coal-fired power stations are being replaced by more modern, less polluting gas turbines and renewable energy sources (DEVI Renewable Energy, 2017). Vietnam has also been one of the major beneficiaries of the Clean Development Mechanism (CDM), achieving total GHGE reductions of about 137.4 million CO_2 emissions equivalent by 2015, and nearly 90% of the over 250 registered CDM projects were in the energy sector. There has also been an investment of about $150 million for renewable energy development and network expansion. This also included efforts to achieve rural electrification through grid extension. In addition, there were five to ten mini-hydro subprojects in off-grid mountainous areas where grid extensions were not feasible. These efforts aimed to provide electricity to about 100,000 households in rural areas. By 2015, Vietnam had therefore managed to achieve an electrification rate of 100%, even in the rural areas (World Bank, 2019).

Strategies and motives

The Green Growth Strategy as well as policies for climate change mitigation and adaptation are initiated and driven forward at the highest political level, as they are considered strategic national planning instruments to achieve economic growth, competitiveness, economic restructuring and poverty reduction. The Green Growth Strategy is also a way of tapping into new financing options and proving access to new technologies that would otherwise have been unavailable (Urban *et al.*, 2018). The Green Growth Strategy as well as policies for climate change mitigation and adaptation are embedded in Vietnam's NDC that the government submitted to the UNFCCC under the Paris Agreement. In line with national commitments, Vietnam aims to reduce its total GHGE by 8% by 2030 compared with BAU. Emission intensity reductions of 20% should be achieved by 2030 compared to 2010 levels. The NDC states that these targets will be increased to 25% GHGE reductions and 30% emission intensity reduction by 2030 compared to 2010 levels, if bilateral and multilateral financial and technical support is made available to Vietnam (Urban *et al.*, 2018). The unconditional goals can probably be met with some policy efforts, but meeting the conditional goals will depend on access to new technologies, new investments and much larger scale industrial restructuring.

One interviewee reported that 'Vietnam is viewed as a high-risk country for climate change' (Interview, representative from ADB, 2017). This is driving government's policies for green transformations and tackling climate change. Another interviewee argued that the 'government plans until 2030 offer new opportunities and advantages at wider economy level. This includes industrial policies for a green future ... this creates opportunities for economic growth, and it is driven at high-level' (Interview, representative from ADB, 2017). Diversifying the energy sector also increased energy security by avoiding load shedding and under-capacity in the power sector. While renewable energy only contributes to a small share of electricity at present, '...high-level policy-makers see the green growth strategy and the renewable energy expansion as important for national development and important for the energy security of supply' (Interview, representative from ADB, 2017). Based on our fieldwork we conclude that Vietnam is a pioneer in climate governance. Its reasons for driving forward green transformations are internally motivated. Some inspiration has been taken from South Korea's Green Growth Strategy or China's restructuring of its energy sector (see also Chapter 2 in this volume) and one might argue that Vietnam was a follower first before it became a pioneer. This initiative is driven primarily by the state bureaucracy, it is hardly influenced by civil society and NGOs, and only marginally by donor agencies.

Challenges and barriers

The biggest challenge for Vietnam is how to transform both the power sector and the industrial sector, given that the industrial sector accounts for about 35% of national CO_2 emissions (IEA, 2019). This is indicated in Figure 4.2.

One major challenge for Vietnam is that the energy sector is increasingly coal-dominated, despite investments in renewable energy as a way to achieve access

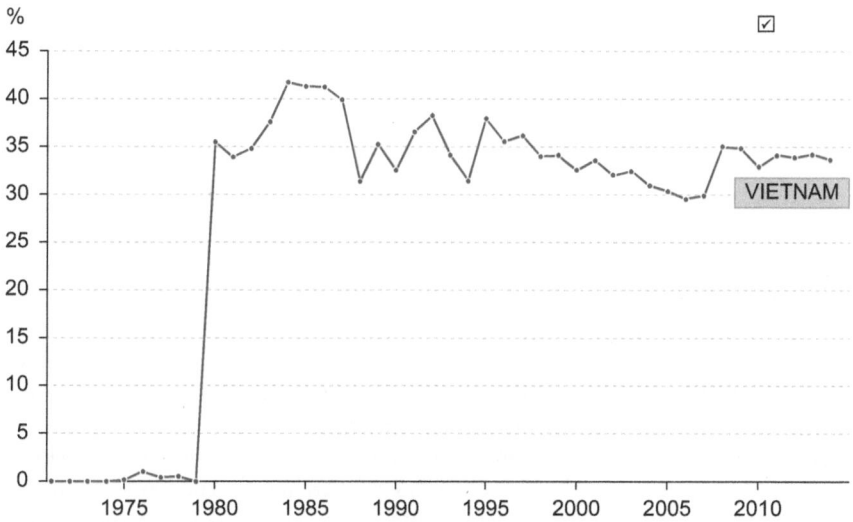

Figure 4.2 CO_2 emissions from industries as a percentage of total fuel combustion in Vietnam. *Source:* World Bank (2019); Based on IEA Statistics © OECD/IEA 2014, //www.iea.org/statistics, All rights reserved.

to electricity (Figure 4.3). According to data from the IEA (2019), in 2017 only 323 GWh of electricity was generated from wind energy and none from solar PV. According to expert opinion, Vietnam currently ranks globally as the 20th largest user of coal-fired power plants. This situation could worsen in future, unless the share of renewable energy will be increased significantly. A challenge is policy-making and the existing restrictive legislative framework for renewables. For example, power pricing rules currently favour fossil fuels due to price distortions by unequal government subsidies for fossil fuels. This keeps the price of fossil fuels artificially low and reduces the cost-competitiveness of renewables. To increase the uptake of renewable energy, the Vietnamese government issued a Made in Vietnam Energy Plan in October 2016 to encourage the private sector to invest in domestic renewable energy resources as an alternative to more imported coal.

Vietnamese interviewees mentioned that the country is keen to follow a more sustainable development trajectory to avoid some environmental problems China experienced recently (e.g. air pollution and fossil fuel resource depletion). They also stated that the challenge is to achieve a socially just green transformation to make sure that poor people are not being left behind. This relates again to power pricing, energy access and the impacts of energy development. For example, one interviewee suggested the following:

> *Key concerns regarding the energy sector are greenhouse gas emissions, impacts of energy infrastructure projects on local people and impacts on the*

environment, like hydropower projects flooding the natural environment and impacts on local people. These serious problems are being recognised by government and government think tanks. Some hydro projects—big dams— were cancelled due to environmental and social concerns, for example two large dam projects in central Vietnam.

(Interview, Institute of Energy Science representative, 2017)

State-led pioneership?

Vietnam can be considered a typical Asian developmental state (Beeson and Pham, 2012). However, the absence of a coherent industrial policy is a major problem for the country (Lederer *et al.*, 2019). Focusing on green issues and renewable energy has thus been perceived as a way out of stagnation and reviving modernisation. As stated above, our research found evidence for multilevel governance in relation to climate governance and green transformations. Overall, the central government has been the major driver. It has pushed state-owned enterprises to modernise and local or regional governments to issue green development plans. Furthermore, think tanks and research institutes are often linked to the state-elite and cannot be understood as independent actors. Thus, although some evidence is discernible for polycentric governance (especially through a partial opening up towards civil society), the case of Vietnam is much more state-centric than the one of Costa Rica.

Comparative analysis

Costa Rica and Vietnam are two middle-income countries with similar regional importance that are both ambitious pioneers in green transformations. In Costa Rica the green transformation is being driven by the government's Sustainable Development Strategy and in Vietnam it is embedded within the Green Growth Strategy. Both countries have increasing income levels, measured in gross domestic product (GDP) and gross national income (GNI) per capita. Both countries also have had several decades of increasing CO_2 emissions in total terms (although they have levelled off in Costa Rica since 2007) as well as increasing CO_2 emissions per capita. Per capita emissions for both Costa Rica and Vietnam are around 2 Mt per person, which is less than half of the world average (World Bank, 2019). Both countries also have rising energy demand and supply and growing electricity use. However, despite similar economic and environmental trends, there are striking differences between these two countries.

In terms of energy generation, the main differences are that in Costa Rica about 98% of the country's electricity comes from renewable energy, mostly hydropower, and Vietnam's electricity is predominantly generated from a mix of hydropower, coal and gas (IEA, 2019) (see Figure 4.3). In terms of economic structure, Vietnam depends more on energy- and carbon-intensive industries than Costa Rica. Energy use and emissions from the residential sector are also higher

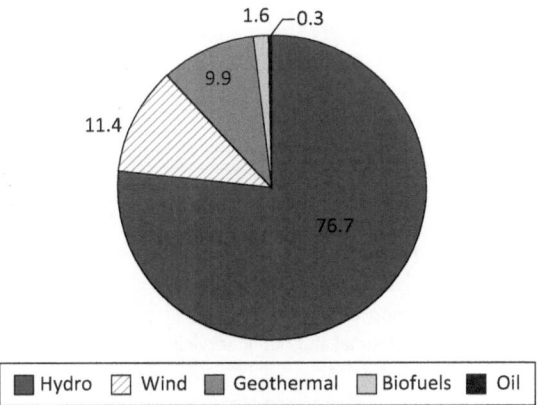

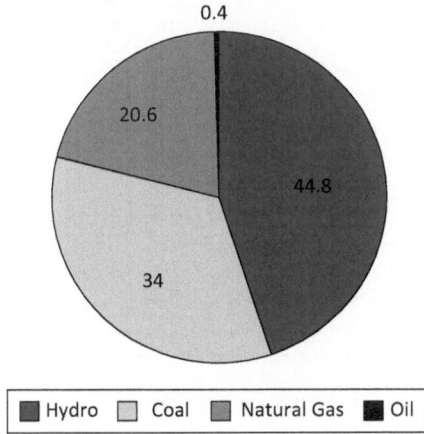

Figure 4.3 Comparative perspective of Costa Rica's and Vietnam's share of electricity generation by fuel. *Source:* IEA (2019).

compared to Costa Rica. This may be due to lower energy efficiency in buildings and potentially due to higher heating and cooling demand. The large majority of Costa Rica's CO_2 emissions are from the transport sector which amounted to about 70% in the late 2010s (World Bank, 2019). Figure 4.4 indicates the energy consumptions and CO_2 emissions by sectors.

In terms of opportunities, both Vietnam and Costa Rica are actively driving forward policies, strategies and actions to increase electricity production from hydropower, wind, solar and modern biomass to raise energy efficiency, conserve energy, and encourage fuel switching and economic restructuring. Both countries

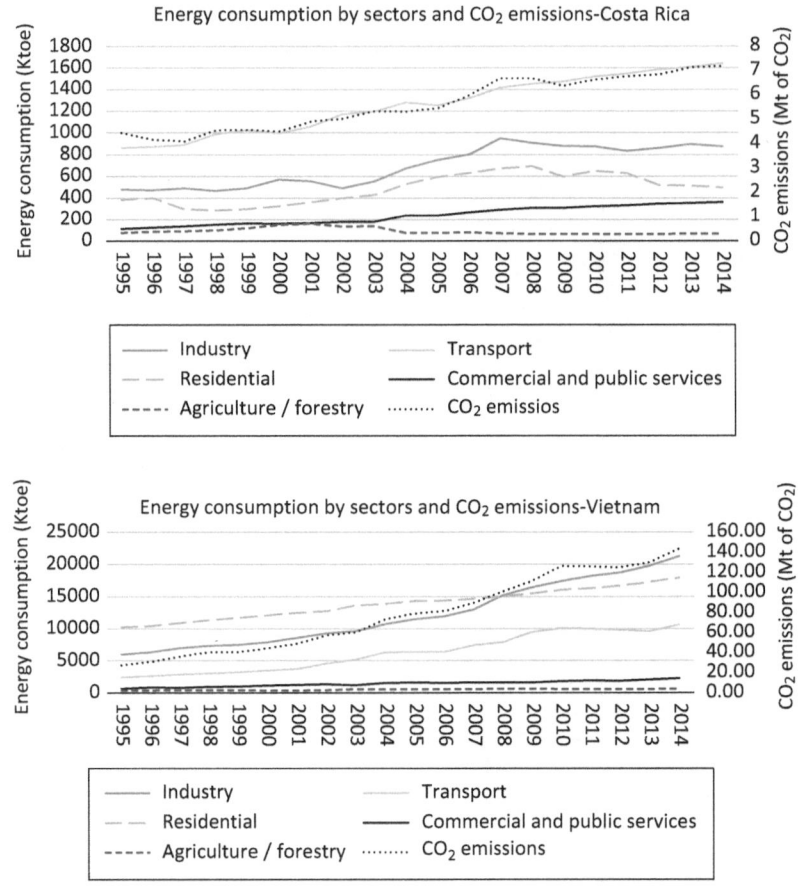

Figure 4.4 Comparison of Costa Rica and Vietnam's energy consumption and CO_2 emissions by sector. *Source:* IEA (2019).

are also attempting to integrate economic growth, sustainable development and climate change issues. However, looking at what has been achieved, we have been able to witness that Costa Rica has a far higher share of renewables (including hydro) than Vietnam, namely 98% compared to 39%, respectively. Furthermore, future goals of the Costa Rican government also appear more ambitious due to its target to achieve the complete decarbonisation of the economy by 2050. The challenges in Vietnam are greater due to the required economic restructuring that is needed regarding the industrial sector and the need to introduce renewable energy at larger scales.

In terms of challenges and barriers, over 70% of Vietnam's primary energy supply comes from fossil fuels, mainly coal and oil (IEA, 2019). Thus, in Vietnam the goals for introducing renewable energy is very modest and many of our interviewees claimed that it could be more ambitious. In the late 2010s, the aim has been to increase the share of renewable energy among electricity generation (excluding hydropower) to 5–8% by 2020, compared with 3.5% in 2010 (Dang, 2016). However, there are financial restrictions towards the large-scale uptake of renewables (such as fossil fuel subsidies that distort market prices), high upfront costs for renewable energy technology and limited frameworks for how to attract investments in renewables. At the same time there is a rapidly increasing energy demand and a growth in electricity generated from coal.

In Costa Rica the main barrier to decarbonisation is the transport sector which accounts for almost 70% of national CO_2 emissions. The car market is growing rapidly each year and there is strong oil dependency (IEA, 2019) which is rapidly driving up energy demand and thereby increasing emissions. In terms of similar challenges, interviewees in both Costa Rica and Vietnam showed a level of awareness for social justice in relation to energy issues including energy pricing which they argued needs to be fair and affordable for poorer people, as well as in relation to the siting for energy projects (in Vietnam with reference to hydropower and in Costa Rica with reference to geothermal energy).

Politically, both countries are very different. Costa Rica is a democracy and Vietnam is a socialist autocracy. Due to these political differences, the role of the state and the interplay with civil society also differ in relation to green transformations and climate governance. Although the bureaucracy is a central player in both polities, it works much more top–down in Vietnam and access for civil society is much more restricted. This should, however, not be perceived as a strict form of environmental authoritarianism, as our interviewees stressed that, for example, in comparison to China (see also Chapter 2 in this volume) or to other countries from Southeast Asia, Vietnamese civil society can make a difference and often does as became apparent in local protests against hydropower installations. Also, business interests have a chance of being heard and Schmitz and colleagues (2015) have argued that the Vietnamese government is working closely together with the private sector (including both national and foreign enterprises) to achieve economic development. Regarding Costa Rica, we can conclude that although the classification as an open democratic state with lots of access for civil society actors is generally accurate, not all civil society representatives are equally heard and particularly those from indigenous groups have a hard time being taken seriously. In both cases, we thus have been able to witness a much more nuanced picture when it comes to decision-making in the field of energy. However, in both states it was primarily governmental actors who made strategic choices. In Vietnam, the government decided to move from climate change adaptation to a focus on mitigation issues as a way of achieving economic restructuring, enable energy security, leverage of new finance and access to technology at a time when traditional development aid is coming to an end. Costa Rica's proactive democratic government has at the very top chosen to pursue climate neutrality and environmental protection as key government targets although the interlinkages with business, NGOs and scientists are very close so that frequently individuals not only know each other but also move in and out of the state apparatus. Over decades, hydropower has been pushed by the government as the main source of electricity and the Costa Rican government has supported renewable energy development through import tax exemptions for imported and local equipment and materials used in the renewable energy industry.

Conclusion

This chapter compared and evaluated the strategies and motives of green transformations for the energy sector in Vietnam and Costa Rica, analysed what role their different political systems play in relation to climate governance and pioneership, what progress has been made through different forms of central leadership within the two countries and what the barriers are for full implementation.

Overall, we can classify both countries as pioneers, a status which both achieved due to rather strong governments and capable bureaucracies. Opening the black box of the state as Wurzel, Liefferink and Torney (2019) have recommended is thus a worthwhile undertaking and our cases show many elements of vertical exemplary leadership/pioneership from the government. Through its top–down character the Vietnamese government could make more use of structural leadership, as it not only formulates policy agendas like the Green Growth Strategy but also largely monopolises all financial means to actually implement them. In Costa Rica, the government is also in the driving seat when it comes to finance, but it is also leaning towards cognitive leadership regarding societal actors by providing ideas and expertise. In both case countries, elements of an entrepreneurial pioneership are also visible with Vietnam's government strongly pushing for solar and wind installations and Costa Rica's focus primarily on hydropower. Interestingly, the role of external leadership from other countries or donors only plays a marginal role in both case countries. This might prove that the 2015 Paris Agreement's strong focus on domestic governmental action has been a smart choice as indeed neither civil society nor business nor subnational actors have played much of a role in formulating the respective NDC. Although the former actors might push for more ambitious action in the near future through the ratchet-up mechanism, true and lasting pioneership and leadership and thus the eventual implementation of progressive climate action will depend on the government's ownership and leadership.

What could other countries and also the research community learn from these two specific pioneers in green transformations? First, we are in a state of flux with changes occurring fast. Only a couple of years ago, one would not have thought of Vietnam as being a potential green pioneer. As we have shown in this chapter, the barriers to get there are still high and although progress in the field of energy might be visible it is slow. Nevertheless, some progress has been made which might lead to new leader-follower relationships. In the early 2020s, Southeast Asian countries are more concerned with China's Belt and Road Initiative than with green transformations although this might change soon, particularly when the Chinese government realises that it cannot externalise the costs of attempting to green its own economy by installing coal fired power plants in Vietnam or other countries on its periphery. We would therefore predict that countries in the region will try to emancipate their energy systems from Chinese influence and Vietnam might evolve as a leader. Similarly, Costa Rica's pathway towards decarbonisation might well evolve as a blueprint for its neighbours once countries within Central America start pushing for energy transitions. It will be interesting to see whether the ratchet-up mechanism of the Paris Agreement will lead to countries looking for solutions in our case countries' neighbourhoods. In both instances, these positive instances of leadership are not yet happening. Neither the Vietnamese nor the Costa Rican governments are yet intentionally leading, but there clearly is a high potential for intentionally setting an example for others to follow.

A second take-home message is that green transformations do not have to look alike and thus a large variety of energy transitions is possible taking into account very different contexts regarding size, economic structure, instrument choice and political systems. Similarly, processes of green transformation are varied and, even if successful, do not work in linear fashion as Costa Rica's ups and downs have shown. Finally, green transformations are in the end highly political projects and political alliances will have to be generated domestically to either transform Costa Rica's transport sector or lessen Vietnam's dependence on coal. This will not be easy as the low hanging fruits (e.g. regarding hydro) have already been picked. Multilevel governance can be of help, particularly through providing expertise and resources, but the influence should not be overestimated, as success eventually depends on political processes within these second-tier countries. And here, questions of coordination, fragmentation and institutional innovation are of importance. This is where more research on leaders and pioneers has to be undertaken.

Acknowledgements

We would like to thank Linda Wallbott (TU Darmstadt) for her contributions and Myro Athanassiou for help with the literature. We are grateful for funding from the Volkswagen Foundation, the Wellcome Trust and the Svenska Riksbankens Jubileumsfond. Thanks to all interviewees and project participants.

Bibliography

2050Pathways (2019) *Costa Rica launches decarbonization plan*, https://www.2050path ways.org/costa-rica-launches-decarbonisation-plan/, accessed 18 December 2019.

Acuna, E. *et al.* (2018) Energia Solar en Costa Rica (2014–2017), ACESOLAR, http://www.acesolar.org/datos-y-estadisticas/.

Costa Rican government (2018) Decarbonization *plan commitment of the bicentennial government*, https://www.2050pathways.org/wp-content/uploads/2019/02/Decarboni zationPlan-Costa-Rica.pdf, accessed 18 December 2019.

Beeson, M. and Pham, H.H. (2012) 'Developmentalism with vietnamese characteristics: the persistence of state-led development in East Asia', *Journal of Contemporary Asia*, 42(4): 539–559.

Dang, A. (2016) Renewable energy in Vietnam: *towards* green transformations. Presentation given at the GreeTS workshop, CATIE, Turrialba, Costa Rica.

DEVI Renewable Energy (2017) *Maps of solar and wind farms in Vietnam*, https://www.google.com/maps/d/viewer?t=m&oe=UTF8&ctz=-420&vpsrc=0&msa=0&ie=UTF8&mid=1NHY5MwPrzOgHvKXpWl3D9aDulas&ll=15.345920119078658%2C109.59857940000006&z=6.

Evans, S. (1999) *The green republic: a conservation history of Costa Rica*, Austin: University of Texas Press.

GIZ (2012) Vietnam *national green growth strategy*, https://www.giz.de/en/downloads/VietNam-GreenGrowth-Strategy.pdf, accessed 18 December 2019.

Hansen, A. (2015) 'The best of both worlds? The power and pitfalls fo Vietnam's development model', In A. Hansen and U. Wethal (eds.) *Emerging economies and challenges to sustainability. theories, strategies, local realities*, London: Routledge, 92–105.

Heinrich-Böll-Stiftung (2013) *Research for and on the great transformation*, Berlin: Heinrich-Böll-Stiftung.

Hooghe, L. and Marks, G. (2001) *Multi-Level governance and European integration, governance in Europe*, Lanham: Rowman & Littlefield.

IEA (2019) *Statistics*, https://www.iea.org/statistics/?country=VNM&isISO=true, accessed 18 December 2019.

IEA (2020), Data and statistics. https://www.iea.org/data-and-statistics?country=COSTARICA&fuel=Energy%20supply&indicator=ElecGenByFuel

IPCC (2018) *Special report. Global warming of 1.5 °C. summary for policy-makers*, https://www.ipcc.ch/sr15/chapter/spm/.

Jordan, A. *et al.* (2015) 'Emergence of polycentric climate governance and its future prospects', *Nature Climate Change*, 5(11): 977–982.

Lederer, M., Wallbott, L. et al. (2019) 'Green transformations and state bureaucracy', In R. Fouquet (ed.) *Handbook on green growth*, Cheltenham: Edward Elgar, 404–424.

Liefferink, D. and Wurzel, R. (2017) 'Environmental leaders and pioneers: agents of change?', *Journal of European Public Policy*, 24(7): 651–668.

London, J.D. (2014) 'Politics in contemporary Vietnam', In J.D. London (ed.) *Politics in contemporary vietnam: party, state and authority relations*, Basingstoke: Palgrave Macmillan, 1–20.

Ostrom, E. (2010) 'Polycentric systems for coping with collective action and global environmental change', *Global Environmental Change*, 20(4): 550–557.

Ostrom, E. (2014) 'A polycentric approach for coping with climate change', *Annals of Economics and Finance*, 15(1): 71–108.

Porras, I.T. *et al.* (2013) *Learning from 20 years of payments for ecosystem services in Costa Rica*, London: International Institute for Environment and Development.

Rapid Transition Alliance (2019) Can Costa Rica's path to carbon neutrality be replicated by other countries? https://www.rapidtransition.org/stories/can-costa-ricas-path-tocarbon-neutrality-be-replicated-by-other-countries/, accessed 18 December 2019.

RECOPE (2018) *Plan de Descarbonización del Sector Transporte Terrestre*, San José: Refinadora Costarricense de Petróleo (RECOPE).

Rockström, J. *et al.* (2009a) 'Planetary boundaries: exploring the safe operating space for humanity', *Ecology and Society*, 14(2): article 32.

Rockström, J. *et al.* (2009b) 'A safe operating space for humanity', *Nature*, 461: 472–475.

Sandbrook, R. *et al.* (2007) *Social democracy in the global periphery*, Cambridge: Cambridge University Press.

Schmitz, H. et al. (2015) 'Drivers of economic reform in Vietnam's provinces', *Development Policy Review*, 33(2): 175–193.

Schreurs, M. and Tiberghien, Y. (2007) 'Multi-level reinforcement: explaining European Union leadership in climate change mitigation', *Global Environmental Politics*, 7(4): 19–46.

Scoones, I. *et al.* (eds) (2015) *The politics of green transformations: pathways to sustainability*, Abingdon: Routledge.

Steffen, W. *et al.* (2015) 'Planetary boundaries: guiding human development on a changing planet', *Science*, 347(6223): 736–748.

Stirling, A. (2015) 'Emancipating transformations: from controlling 'the transition' to culturing plural radical progress', In I. Scoones, M. Leach and P. Newell (eds.) *The politics of green transformations*, London: Earthscan, 68–95.

UNFCCC (2015a) *Costa Rica's nationally determined contribution*, Bonn: UNFCCC.

UNFCCC (2015b) *Vietnam's nationally determined contribution*, Bonn: UNFCCC.

Urban, F. *et al.* (2018) 'Green transformations in Vietnam's energy sector', *Asia & the Pacific Policy Studies*, 5(3): 558–582.

World Bank (2019) *World Bank open data*, https://data.worldbank.org/.

Wurzel, R.K.W., Connelly, J. and Liefferink, D. (eds.) (2017) *The European Union in international climate change politics: still taking a lead*, Abingdon: Routledge.

Wurzel, R.K.W., Liefferink, D. and Torney, D. (2019) 'Pioneers, leaders and followers in multilevel and polycentric climate governance', *Environmental Politics*, 28(1): 1–21.

5 Rhetoric and reality in New Zealand's climate leadership

'My generation's nuclear-free moment'

David Hall[1]

Introduction

It was seven weeks prior to the 2017 general election that Jacinda Ardern became Leader of the New Zealand (NZ) Labour Party. Her predecessor resigned amidst the election campaign, unable to arrest the decline in his polling. Ardern's ascension dynamised the election, aided by her personal warmth and exemplary communication skills. In her first major speech as leader, she captured the media's attention by describing the climate crisis as 'my generation's nuclear-free moment' (Levine, 2018).

This phrase warrants some explication. NZ was the first developed nation to establish a nuclear-free zone, a policy pioneered by the NZ Labour Party. The Third Labour Government opposed nuclear testing in the Pacific in the early 1970s, and the Fourth Labour Government went further by barring nuclear-powered and nuclear-armed ships from entering NZ territorial waters in 1984, which eventually provoked the US into suspending its Australia, NZ, United States Security (ANZUS) treaty obligations to NZ (Clements, 1988). The moral case was brilliantly articulated by Prime Minister David Lange who, in a 1985 Oxford Union debate with American televangelist Reverend Jerry Falwell, famously quipped to his interlocuter, 'Hold your breath just for a moment. I can smell the uranium on it as you lean forward'. NZ's nuclear-free stance became a cornerstone of national identity, crystallising a self-image of a small island nation that, through its exercise of moral leadership, could hold its own against the world's superpowers.

With this backdrop, Ardern's 'nuclear-free' phrase staked multiple claims at once. It was a proud gesture to the Labour Party's past and its progressive capacity for nation-changing moments. It was a tacit critique of the incumbent National Government which failed to act substantively on climate change. It was a statement of generational change, a pledge by a 37-year-old political leader to embody new priorities. And most importantly for this collection, it was a future-focused declaration that NZ would be a leader on climate change, as it had been on the nuclear issue.

Has Ardern lived up to this promise? More generally, has NZ, through its climate change policy, fulfilled its own self-image of global leadership? This chapter will explore this question by reference to the literature on leadership and followership in polycentric and multilevel climate governance (Liefferink and Wurzel, 2017; Wurzel, Liefferink and Torney, 2019; Torney, 2019). It will analyse key policy achievements – and non-achievements – and the political justifications for them. Although the focus is the Ardern Government (2017 onwards), the chapter begins by reviewing the Fifth Labour Government (1999–2008) and Fifth National Government (2008–2017) for historical context. Finally, the chapter introduces the concept of emotional leadership – that is, the capacity to express and induce emotions in others – to explain the esteem that Ardern is globally held in, even when her immediate impact on emissions is modest.

Fifth Labour Government: looking to lead

The idea of leadership is a persistent feature of the NZ political imaginary. Jon Johansson, a political scientist who specialises in political leadership, notes: 'One of our celebrated myths … is our frequent claim to be "first in the world" to achieve some notable feat or another' (Johansson, 2009: 43). This is personified by well-known New Zealanders, such as Sir Edmund Hilary and his pioneering ascent of Mount Everest. It is also represented in pioneering social innovation, such as universal suffrage in 1893 and the aforementioned anti-nuclear stance. Coupled with NZ's geographical isolation and relatively small size – a population of nearly 5 million on a land area 22% larger than Great Britain – there is a zeal for achievements that, as the local idiom goes, 'put New Zealand on the map'.

But Johansson is quick to qualify this national myth. He notes that: 'New Zealand's periodic fits of policy progressiveness are overshadowed by an innate pragmatism that is an even stronger hallmark of our political culture'. While NZ celebrates its great leaps forward, these are temporary interruptions in 'long periods of piecemeal reform and consolidation [that] have tended to dominate our politics' (Johansson, 2009: 44).

This pattern of progressive moments and conservative interludes is also characteristic of NZ's climate change policy. In their comprehensive review, Hopkins *et al.* (2015: 568) remark: 'It is leadership in [issues like universal suffrage and anti-nuclear] that has contributed to a self-image of New Zealand as a "trail-blazing social laboratory", and led to domestic calls for New Zealand to take a leading role in global climate change governance. However, while New Zealand has participated in global climate regimes … New Zealand's record on climate change has been modest'.

NZ contributes only 0.17% of global emissions (MfE and Stats NZ, 2017). However, its emissions intensity – that is, its total emissions per million dollars of GDP – was 11th highest of 31 OECD countries in 2015 (Stats NZ, 2019). Since 1990, gross emissions have risen 24% up to 2018 (MfE, 2020: 2). Most of this growth occurred during the 1990s and early 2000s, with emissions roughly plateauing since the mid-2000s and not yet emphatically in decline. NZ's position is also at risk of reversals of the negative emissions generated by the Land Use, Land-Use Change and Forestry (LULUCF) sector, specifically by changes to plantation forestry. Between 1990 and 2017, carbon removals from LULUCF reduced by 23.1% due to declining rates of afforestation and large-scale land-use conversions from forestry to dairy agriculture (MfE, 2018: 49). The prospect of large-scale harvesting over the next decade – as trees from a 1990s planting boom reach maturity – further complicates the challenge of reducing emissions.

Under the 2015 Paris Agreement, NZ pledged a 2030 target to reduce emissions to 30% below 2005 levels (equivalent to 11% below 1990 levels). Notably, the Nationally Determined Contribution (NDC) is indeterminate about LULUCF accounting, stating that: 'We reserve the right to adjust our selection of methodologies, without reducing ambition' (NZ Government, 2015). Regardless of whether the Government chooses to register its gross or net emissions, however, official emissions projections anticipate that NZ will substantially overshoot its target. If gross emissions, NZ in 2030 is expected to be 3.7% below 2015 levels (or 19.6% above 1990 levels). If net emissions, then due to predicted forest harvesting, NZ in 2030 is expected to be 29.7% above 2015 levels (or 112.5% above 1990 levels) (MfE, 2017: 108). NZ's overshoot is expected to be addressed by purchasing international units, which increases the country's exposure to uncertainties over future availability, supply and price.

This reflects a failure of policy – to be discussed shortly – but also NZ's unusual circumstances. Investment into hydropower schemes from the mid-1940s to late 1980s (Kelly, 2011) meant that, in 1990, the baseline year for the Kyoto Protocol, NZ's renewable share of electricity generation was 78.5%, only slightly below its current rate of 82.3% (MBIE, 2019). This leaves NZ in an enviable position, ranked second after Norway among IEA countries for its renewable electricity share (IEA, 2017: 15). However it also means that NZ has diminished opportunities to reduce national emissions by converting electricity generation to renewable sources (OECD, 2017; Productivity Commission, 2018); and that the land sector features far larger in NZ's emissions profile than most industrialised nations (with the notable exception of Ireland; see Chapter 12 in this volume). Agricultural emissions make up nearly half (47.8% in 2018) of gross emissions under current accounting metrics (MfE, 2020: 60). Meanwhile, the negative emissions from Kyoto-eligible forests offset nearly one-third (29.7% in 2018) of NZ's gross emissions (ibid.: 69). This is why land use – as we shall see – plays an oversized role in NZ's climate politics.

Climate change became an explicit policy concern for the NZ Government around the time the Intergovernmental Panel on Climate Change (IPCC) was established in 1988. Various targets and policy packages were subsequently announced by left- and right-wing governments (for review, see Rive, 2011), reflecting a bipartisan acceptance of climate science. But the issue of carbon pricing, whether by carbon tax or emissions trading, came to occupy centre stage. Arguably, this preference for market mechanisms reflects the ideological predilections of government institutions, transformed by the New Public Management paradigm in the mid-1980s. These reforms predisposed government agencies to see climate action as a cost, and instilled a reluctance to intervene in market activity (Kelly, 2010).

After years of false starts, a carbon price was eventually implemented by the Fifth Labour Government (1999–2008), led by Prime Minister Helen Clark.

Originally, her Government proposed a carbon tax, to be introduced in April 2007 at NZD$15 per tonne, accompanied by a levy on agricultural methane to fund research on agricultural emissions. But the proposals met strong resistance from interest groups and opposition parties. Memorably, the agricultural emissions research levy was dubbed a 'fart tax' by lobby groups, a framing which was biologically inaccurate (methane is emitted through ruminant belching) but politically effective. The Government abandoned the levy in 2004, instead co-funding the Pastoral Greenhouse Gas Research Consortium with the agricultural sector, which gave the latter greater influence over funding and research strategy on agricultural emissions. The Government also abandoned the carbon tax proposal the following year, as a precondition for forming government with the NZ First Party after the general election in September 2005.

By this stage, NZ's climate change policy was characterised by 'lengthy and inconclusive debates, a lack of consensus amongst key stakeholders, governmental indecision and prevarication and a series of significant policy reversals' (Chapman and Boston, 2007: 113). Given the lack of substantive progress, the Fifth Labour Government refocused its efforts on implementing an emissions trading scheme (ETS). Details were introduced in September 2007. The ETS would cover six major greenhouse gases and all economic sectors, albeit with a phase-in period. Forestry would have full obligations at the beginning of 2008, stationary energy and industrial processes from 2010, liquid fossil fuels in 2011, then agriculture, waste and all other emissions in 2013. A free allocation of an unspecified volume of units was announced as a transitional measure. The ETS would allow international linking to provide flexibility, such that units could be purchased elsewhere. Finally, the ETS would operate within the limits prescribed by the first commitment period of the Kyoto Protocol 2008–2012), rather than set a domestic cap on emissions (Cameron, 2011: chaps. 6–8). It was passed through Parliament in September 2008, just prior to the general election on 19 November, which the Labour Party lost.

NZ was not the first to implement an ETS – the EU introduced its ETS in January 2005. However, NZ's ETS had innovative features, including its broad sectoral coverage and absence of a hard limit on emissions (Leining, Kerr and Bruce-Brand, 2020). Moreover, by implementing one of the world's first schemes, NZ exercised leadership ambitions. In introducing the ETS, Clark (2007) used uncharacteristically aspirational language: 'I have set out the challenge to our nation to become the first truly sustainable nation on earth – and to dare to aspire to be carbon neutral'. More pragmatically, by inspiring other countries to establish complementary mechanisms, NZ would facilitate the international trading of emissions reductions, which would give NZ greater flexibility to pursue least-cost mitigation.

On the downside, the Government's use of parliamentary urgency to rush the ETS through the legislative process, prior to the 2008 election, reinforced the instrument's 'post-political' status, shaped more by managerialist and technocratic approaches to governance than by values of transparency and democratic legitimacy (Driver *et al.*, 2018). Without public buy-in, or a broad-based understanding of how it works, the ETS was left vulnerable to manipulation by political actors (see next section), especially through sectoral exemptions and exploiting the lack of a cap.

The heavy focus on the ETS as a national-level response also meant that NZ became a 'lost opportunity' for a multilevel governance approach to climate change (Harker *et al.*, 2017). Such an approach was initially advanced through the 1991 Resource Management Act (RMA), a genuinely pioneering legislative framework for sustainable development (OECD, 2017: 25–29) introduced by the Fourth Labour Government. It devolved power through a three-tier system that required central government to provide statutory guidance, regional government to regulate environmental resources and local councils to regulate land-use activities through planning. However, a 2004 amendment by the Fifth Labour Government 'removed the ability of regional councils to consider the adverse effects of greenhouse gas emissions from proposed activities, such as thermal power plants, when granting air discharge permits' (Harker, Taylor and Knight-Lenihan, 2017: 491). Moreover, proposed linkages between a national-level response to climate change and local-level guidance for urban development never eventuated. As a result, councils had responsibilities to reduce exposure to risks from 'natural hazards' including climate-related impacts, yet lacked the mandate to reduce emissions-intensive activities through the consenting process (Rive, 2011). This amendment was repealed in June 2020 (see discussion below), but, along with limited resourcing for councils to effectively discharge their responsibilities for climate adaptation, its legacy has been to hinder NZ's capacity for a multilevel approach to the climate challenge.

Fifth National Government: fast followership

John Key became the Leader of the National Party in November 2006. Although his predecessor had embraced reactionary politics, including climate contrarianism, Key reorientated the National Party towards the political centre, which included accepting the reality of climate change (Johansson, 2009: 140). This preserved the broad consensus on climate change among NZ's political parties (Hopkins *et al.*, 2015: 562).

However, Key's positioning on climate action was overtly informed by a NZ Institute report titled, 'We're Right Behind You', published in October 2007. It recommended that 'New Zealand adopt a "fast follower" approach', defined thus: 'New Zealand ought to begin to act now so that if the world does change in a substantial manner with respect to climate change, it is able to move quickly and efficiently, but in a way that *avoids investing unnecessarily in leading the way*' (Skilling and Boven, 2007: 47; emphasis added). It continued: 'The trajectory of the pathway should be set to follow the actions of the relevant group of comparator countries … [which] include the major developed countries, New Zealand's major trading partners, as well as countries with whom New Zealand is competing to be the location of choice for economic activity' (ibid.: 42). A key consideration was NZ's trade-exposed economy, which raised the prospect of 'carbon flight' or leakage, where 'economic activity leaves New Zealand for other countries with a less demanding approach to reducing emissions' (ibid.: 38-9). The report explicitly contrasted its approach to 'lofty rhetoric about saving the planet or being a world leader' (ibid.: 48), such as that deployed by the Fifth Labour Government.

Key, then-Leader of the Opposition, declared that the report was 'on the right track … you need to balance your economic opportunities with your environmental responsibilities' (NZPA, 2007). Indeed, 'fast follower' was a phrase that Key returned to again and again throughout his tenure as Prime Minister, from November 2008 to December 2016. Even in his last month as Prime Minister, he described his climate legacy thus: 'it is consistent with best practice in the world, except that the Government has always said that it wants to be *a fast follower, not a leader*' (Key, 2016; emphasis added).

This self-characterisation frequently came under criticism. It was criticised, primarily, for its lack of ambition. But Key was also criticised for failing to live up to even this humble goal, and indeed for undermining institutions that might have enabled substantive emissions reductions. As Jacinda Ardern (2008) argued in her maiden speech as a newly elected MP, 'National told us we should be fast followers, but now all I see are the many, many losers – the future generations whom some people in this House do not yet believe they have a responsibility to'.

When Key came to power, only months after the nadir of the global financial crisis, his Government launched an immediate review of the ETS, with the intent of 'moderating' it (Office of the Minister for Climate Change Issues, 2009). New transitional measures included the exclusion of biological emissions from agriculture until 2015; a price cap of NZ$25 per tonne; and a one-for-two deal on surrender obligations for non-forestry sectors, whereby the surrendering of one New Zealand Unit (NZU) would cancel two obligations (Bertram and Terry, 2010). These measures succeeded in diluting the ETS's impact (Luth Richter and Chambers, 2014; Diaz-Rainey and Tulloch, 2016), yet when a second ETS review recommended these transitional measures be phased out (ETSRP, 2011), the Government retained the status quo, even choosing to defer indefinitely any decision on agriculture's inclusion. Moreover, in 2012, when the first Kyoto commitment period finished, the Government refrained from committing to a second period (which stood in contrast to the EU (see Chapter 8 in this volume)). Because the ETS had relied upon Kyoto commitments as a proxy for an emissions cap, this made the ETS an uncapped trading system by default, not a cap-and-trade system at all. In the meantime, the ETS also faced a dramatic collapse in price in mid-2011. The cause was well known: NZ emitters were importing cheap credits of dubious integrity for their surrender obligations, while stockpiling NZUs for whenever prices rose again. However, the Government acted slowly to respond, only prohibiting the use of international units in 2015 (Leining and Kerr, 2016). Consequently, an international comparative analysis concluded that NZ's ETS performs 'relatively poorly', with low rankings on all evaluative criteria (Narassimham *et al.*, 2018: 983).

This demonstrates how vulnerable the ETS is to political whim. The instrument is so complex that manipulations are hard to comprehend for the general public

(Palmer, 2015). Moreover, official regard for the ETS as 'the primary tool under-pinning NZ's domestic action to reduce emissions' (MfE, 2017: 67) has tended to come at the expense of complementary measures (Macey, 2014). The National Government weakened measures that did exist, such as home insulation subsidies, and also supported a range of climate-misaligned activities, such as irrigation subsidies and roadworks projects. Accordingly, while gross emissions declined slightly during the first few years of the Key Government, this occurred in the aftermath of the global financial crisis which produced a dip in emissions among Annex 1 countries (Peters *et al.*, 2012). NZ's emissions began increasing again in 2012.

Consequently, NZ came to be described as a 'laggard' on climate change (Barrett *et al.*, 2015). To put it formally, the Fifth National Government had low internal and low external ambitions (Wurzel, Liefferink and Torney, 2019: 8), by falling short of followership in its domestic policy and minimising its international responsibilities. While the National Government did oversee the signing of the Paris Agreement, its pledge was classified by Climate Action Tracker as 'inadequate', which added that there are 'virtually no policies in place in New Zealand to address the fastest-growing sources of emissions ... including transport and industrial sources' (Climate Action Tracker, 2015: 1–3). The OECD (2017: 3) recommended that a 'transition towards a low-carbon, greener economy would help New Zealand defend the "green" reputation it has acquired at international level'.

From a multilevel governance perspective, the Key Government also further entrenched the trend towards centralisation, introducing a series of 'major amendments to the RMA ... [to] enable nationally significant plan changes and development applications to circumvent local government decision-making processes and limit public stakeholder participation' (Harker, Taylor and Knight-Lenihan, 2017: 491). In its final year in office, ministers even obstructed central government's capacity to provide statutory guidance, by impeding the release of the Ministry for the Environment's updated guidance to local councils on planning for sea-level rise (Gibson, 2017). In sum, the Fifth National Government took 'a minimalist approach' to environmental governance which resulted in a 'fragmented and poorly performing environmental policy sector, reflecting an overriding ideological commitment to neoliberal economic thinking and policy capture by economic elites who were seemingly indifferent to environmental concerns' (Kurian and Smith, 2018: 251–252).

The upshot is that non-state actors took increasingly proactive roles to fill the leadership vacuum, which resulted in the spontaneous emergence of a less centralised, more polycentric governance regime (Ostrom, 2014; Wurzel, Liefferink and Torney, 2019). Four examples will suffice. The first is Generation Zero, a youth activist group established in 2012 whose policy advocacy sowed the seeds of the so-called Zero Carbon Act (see next section). The second is GLOBE-NZ, a cross-party collaboration of parliamentarians who created favourable political conditions for passing the aforementioned Act (Graham, 2018). Notably, GLOBE-NZ was partially funded by three foreign embassies, an expression of international soft power (ibid., 38).

The third example is the high-profile resistance by Māori tribal organisations and activist groups to oil and gas exploration and seismic testing in NZ waters. One high-profile protest led to Petrobras's early withdrawal from a five-year exploration permit in 2012. These interventions are *pioneering* (as defined by Liefferink and Wurzel, 2017), because their principal concern was to fulfil cultural duties such as *kaitiakitanga*, rather than provide example to others (Selby, Moore and Mulholland, 2010; Salmond, 2017: chap. 11). But Māori resistance also mobilised support from environmental NGOs, Greenpeace in particular, as well as mutual engagement with indigenous organisations engaged in similar struggles elsewhere in the world.

The fourth example is the emergence of business representative groups (Macey 2014: 54), such as the Sustainable Business Network, Sustainable Business Council and Pure Advantage, which offered a counter-narrative to the Fifth National Government's prioritisation of economic growth over environmental protection. This created space for more sustainable visions of economic prosperity – albeit via contested concepts such as 'green growth' – and reiterated the importance of NZ's 'clean, green' reputation for market access and export premiums (e.g. Vivid Economics, 2012). Similarly, NZ Super Fund, the country's sovereign wealth fund, began implementing a climate risk mitigation strategy in 2016, joining the One Planet Sovereign Wealth Fund Working Group the following year as a founding member. As such, climate leadership emerged throughout the NZ economy in spite of, or even as a reaction to, the shortcomings of central government.

Ardern's Coalition Government: between ambition and delivery

Jacinda Ardern's ascent to Prime Minister was enabled by growing frustration with the Fifth National Government's failure to grapple with multiple systemic crises. Public polling showed that poverty, inequality, housing, immigration and the environment were increasingly perceived as the most important problems facing NZ society (Mills, 2018). With his personal popularity ratings gradually declining, John Key resigned in December 2016, creating an opportunity for the National Party to refresh itself under new leadership. The strategy appeared to be working, with Labour lagging in the polls in the run-up to the election.

Yet with an unexpected change of leadership, the Labour Party's fortunes rapidly changed. Ardern campaigned vigorously on those wicked problems that the Government had prevaricated over, especially housing unaffordability, child poverty and climate change. Her campaign slogan, 'Let's do this', conveyed 'positivity, hope and inclusiveness in a call to action' (Ardern, 2018: 37). As Prime Minister, she pledged: 'This will be a government of transformation' (Ardern, 2017).

However, Ardern's capacity to govern is complicated by the fact that she leads a minority government. The Labour Party, which gained 46 of 120 seats, relies upon a coalition agreement with the NZ First Party (with nine seats), plus a confidence and supply agreement with the Green Party of Aotearoa New Zealand (eight seats). This has not always made governing easy. The cosmopolitanism and social liberalism of the Greens frequently brings it into conflict with the nationalism and social conservatism of NZ First. Although NZ First does not deny the reality of climate change, given its rural voter base, it is disposed to shield the agricultural sector from the disruptions of climate policy. However, these differences over-shadow some points of commonality. By supporting economic protectionism, NZ First shares with Labour and the Greens a proactive ideal of government, which is open to intervening in markets to advance the public good. Indeed, when NZ First Leader Winston Peters announced his decision to form a coalition with Labour, he argued 'that capitalism must regain its responsible, its human, face' (Peters, 2017). This remark illuminates the common ground upon which climate change policy is being grown.

The Coalition Government's first tranche of policy was a product of coali-tion negotiations, a cluster of ready-made manifesto promises that rushed into the policy pipeline. In regard to climate change, the most relevant are:

- Introduce a Zero Carbon Act to set 2050 emissions reductions targets and establish an independent Climate Change Commission.
- Initiate the One Billion Trees Programme to plant an annual average of 100 million trees over ten years.
- Establish a NZD$100 million Green Investment Finance to catalyse private investment through strategic public investments.
- Transition to 100% renewable electricity (in a normal hydrological year) by 2035.
- Cessation of subsidies for irrigation schemes, which incentivise agricultural intensification and hence biological emissions.
- Transition to an emissions-free government vehicle fleet by 2025/6.

Other climate-aligned initiatives enacted over the subsequent three years include major reforms to the ETS including the setting of a domestic cap, com-pletion of the first National Climate Change Risk Assessment, budget boosts for rail and sustainable land use, a regional fuel tax in Auckland to fund public transport, renewed powers for local authorities to consider climate mitigation when consenting under the Resource Management Act, enhanced climate-aligned investment signals in the research and innovation sector and manda-tory requirements for reporting of climate-related risks under the Task Force on Climate-related Financial Disclosures (TCFD) framework. Further initia-tives set in motion include proposals to exclude fossil fuel investments from default superannuation funds, develop a green hydrogen strategy, introduce fuel efficiency standards and establish vehicle feebates which would levy fees on high-emissions vehicles to grant rebates on low-emissions vehicles. Finally, a sustainable finance roadmap is being developed by the Aotearoa Circle, a part-nership between public- and private-sector chief executives, which – along with the newly formed Climate Leaders Coalition – expands business-led advocacy for climate action.

More concretely, the Government banned new exploration permits for off-shore oil and gas exploration (Ardern, 2018). This announcement was less decisive than it might have been: there are 22 existing offshore permits which remain unaffected, and the Government continues to offer permits for onshore acreage. Nevertheless, this is a significant supply-side intervention which transmits a clear signal for future prospects. Amidst criticism from regional leaders and oil and gas sector representatives, the Government also initiated a regional engagement process under a 'just transitions' framework (Huggard, 2019), including a NZD$27 million public investment into the National New Energy Development centre in Taranaki to create new 'green jobs' for oil and gas workers.

At the international level, the Coalition Government has announced a 'Pacific reset' with its small island neighbours to explore regional issues such as climate-induced migration, as well as a refresh of NZ's trade strategy, the Trade for All agenda, which emphasises the importance of environmental goods and services (Trade for All Advisory Board, 2019). NZ has also – along with Costa Rica (see also Chapter 4 in this volume), Fiji, Iceland and Norway (see also Chapter 11 in this volume) – launched the Agreement on Climate Change, Trade and Sustainability (ACCTS), a multilateral commitment to explore the use of trade rules to discipline fossil fuel subsidies, and to remove tariffs and barriers to environmental goods and services (Ministry of Foreign Affairs and Trade, 2019). Finally, within United Nations (UN) negotiations, NZ was a founding member of the Carbon Neutrality Coalition in 2017 and invited to join the High Ambition Coalition in 2018 (see also Chapter 8 in this volume), an affirmation of its new-found integrity on climate action.

The overlapping challenges of climate change and multilateralism is a theme that Ardern frequently raises on the international stage. Her positive reception in certain quarters is partly situational, given the contrast she strikes with US president Donald Trump. As individuals, the differences are stark, she being a women and feminist in her late-thirties who gave birth during her first term in office, him being a man in his early-seventies whose political life is dogged by accusations of infidelity and sexual harassment. As political leaders, moreover, they embody the open/closed distinction that is posed (overly simplistically) as the dividing line of contemporary politics (*Economist*, 2016). In 2018, Ardern warned the UN General Assembly that: 'Any disintegration of multilateralism – any undermining of climate related targets and agreements … are catastrophic' for Pacific island states (Ardern, 2018). No one in attendance could miss the allusion to Trump who, one year earlier, announced his intention to withdraw the US from the Paris Agreement. In his speech to the Assembly a few days earlier, Trump failed to mention climate change at all (Trump, 2018; see also Chapter 7 in this volume).

In this context, Ardern has acquired, and NZ by proxy has renewed, its reputation for moral leadership. But to what extent is this reputation for leadership complemented by measurable gains? Is NZ merely a 'symbolic leader' (Liefferink and Wurzel, 2017) that is buoyed by Ardern's manifest political skills on the international stage, but uncorroborated by domestic mitigation and adaptation activities?

Given that – at the time of writing – it is less than three years into Ardern's tenure, any evaluation must necessarily be tentative. Yet a few observations can be made.

Firstly, with her language of transformation, Ardern leaves herself vulnerable to under-delivering on policy. This is most acute in other policy areas, especially housing, where the Labour Party's manifesto target for 100,000 new homes was exposed as unviable and subsequently abandoned. But so too for low-emissions transport. The Government backtracked on its commitment to an emissions-free vehicle fleet by mid-2025, only to later backtrack on its backtrack – and still without a clear path to implementation (Daalder, 2019; ibid., 2020). The feebate policy, meanwhile, was mothballed due to resistance from NZ First (Coughlan, 2020). Rail infrastructure faces delays while roading projects received additional funding before the 2020 election. In short, Ardern's aspirational spirit can easily outpace her capacity to shape government policy, leaving her struggling to catch up with her own reputation for leadership.

Secondly, much of NZ's deluge of climate policy involves followership rather than leadership. Torney (2019: 169) defines 'climate followership as the adoption of a policy, idea, institution, approach or technique for responding to climate change by one actor by subsequent reference to its previous adoption by another actor'. This is true of the so-called Zero Carbon Act, the centrepiece of the Government's climate strategy. It sets emissions targets for 2050 in legislation, creates a Climate Change Commission that sets five-yearly carbon budgets along with policy recommendations for achieving them and requires Government to provide a written response to the Commission's recommendations. Not only is this legislative framework adapted from the UK's Climate Change Act 2008 (see also Chapter 10 in this volume), but its development and advocacy came from outside of government. The proposal first emerged via Generation Zero, the youth climate activist group mentioned earlier, and subsequently endorsed by the Parliamentary Commissioner for the Environment (2017), an independent government agency that provides advice on environmental issues. Both cited the UK framework as an exemplar. The Climate Change Response (Zero Carbon) Amendment Bill passed its third reading on 7 November 2019, shepherded through Parliament by Minister for Climate Change James Shaw who built upon the collaborative GLOBE-NZ network to secure cross-party support from the National Party. It is a policy story that exemplifies polycentric and multilevel climate governance, involving leadership from civil society actors and policy diffusion along well-trodden pathways between Anglosphere countries (Legrand, 2015).

Other flagship policies also involve followership. Vehicle emissions standards and feebates (also known as bonus-malus schemes) are well-established in Europe. NZ Green Investment Finance takes it cues from other publicly funded green investment vehicles throughout the world. Similarly, the prohibition of offshore oil exploration permits follows from pioneering resistance by Māori tribes and organisations, as well as more comprehensive moratoriums by Costa Rica in 2011 (see also Chapter 4 in this volume) and France in 2017. Thus, rather than being a trailblazer or innovator, much of NZ's recent progress involves fulfilling the previous Key Government's promise of 'fast followership'. Ardern's (2019a) insistence that 'New Zealand will not be a slow follower' is increasingly assured. What is less assured is NZ's progression beyond fast followership to exemplary leadership that sets new models for others to follow.

One potential exception is agriculture. A significant innovation in the Zero Carbon Act is its 'two-baskets' approach, which excludes biogenic methane from the net-zero target that all other greenhouse gases face by 2050. Instead, biogenic methane must be reduced by >10% by 2030 and 24–47% by 2050 (against a 2017 baseline). This approach acknowledges the lower warming potential of a short-lived gas like methane, compared to carbon dioxide or nitrous oxide (Cain *et al.*, 2019) and the proposed reductions are consistent with mitigation pathways to 1.5°C. However, the Labour Party has not yet fulfilled its campaign commitment to include agriculture in the ETS. Instead, the Government reached a compromise in October 2019, whereby agriculture would work voluntarily towards delivering its Primary Sector Climate Change Commitment, *He Waka Eke Noa – Our Future in Our Hands*, which includes implementing an alternative emissions pricing mechanism. If progress is not adequate, Government has options to bring agriculture into the ETS in either 2022 or 2025. Ardern (2019b) described this as 'a world-first agreement … by reaching an historic consensus with our primary sector'. Whether it sets an example that others follow depends on at least three things. Firstly, whether the underlying science is incorporated into international accounting frameworks; for example, by substituting Global Warming Potential (GWP) metric GWP_{100} for GWP* which better reflects the warming impact of short-lived greenhouse gases (Lynch *et al.*, 2020). Secondly, whether the primary sector fulfils its commitments rather than prevaricates in anticipation of a change in government. And thirdly, whether the Commitment's bottom–up, sector-driven, farm-level approach delivers better mitigation and adaptation outcomes than a top–down, market-driven, national-level approach like the ETS. Time will tell.

To sum up, the Ardern Government is best evaluated across various levels of analysis. At the level of domestic politics, where a government is judged vis-à-vis the governments that preceded it, Ardern is entitled to declare (as she did when the Zero Carbon Act was passed) that: 'We have done more in 24 months than any Government in New Zealand has ever done on climate action' (Ardern, 2019a). However, at the international level, where NZ is judged against the actions of other countries and its international commitments, it is more a fast follower than a leader, defined by policy uptake and advocacy rather than innovation. And at the normative level, where NZ is judged against objectives such as the 1.5°C carbon budget, its actions remain inadequate. For example, Climate Action Tracker (2019), while commending NZ's 'leadership by having passed the world's second-ever Zero Carbon Act', nevertheless classifies its 2030 target as 'insufficient' and notes that 'it lacks the strong policies required to implement it'. Indeed, official projections have barely budged between 2017 and 2019, with new policies only reducing the projected gross emissions for 2030 by –2.6% (MfE, 2019: 76). Of course, this does not tell the whole story, not least because the Government's emphasis thus far on institutional reform is more likely to yield long-term rather than short-term impacts. But it does reflect a deficit of concrete, project-level initiatives to produce immediate emissions reductions.

Ardern: an exemplar of emotional leadership?

This leaves us with a conundrum. Why is Ardern treated as an international role model for climate leadership (e.g. Gore, 2017) when her achievements are mixed and her leadership is more domestic than international?

The four types of leadership described by Liefferenk and Wurzel (2017: 957) only help us so far. In regard to *exemplary leadership*, NZ has yet to set examples for others to follow, although NZ may emerge as an exemplar in agriculture and land use. As a small nation, *structural leadership*, which relates to 'an actor's hard power', is something that NZ mostly lacks (except in relation to its Pacific island neighbours). Instead, NZ must exercise 'soft power', such as *entrepreneurial leadership* which 'involves diplomatic, negotiating and bargaining skills in facilitating compromise, solutions and agreements'; and *cognitive leadership* which involves 'defining or redefining of interests through ideas, as embodied in concepts such as sustainable development' (ibid.). To be sure, the Ardern Government has refocused its diplomatic skills towards progress on climate policy, both domestically and internationally, rather than adroitly minimising its responsibilities as the Key Government tended to do. In regard to cognitive leadership, however, the Ardern Government has mostly been a taker of ideas, adopting existing frameworks rather than devising its own. Even its 'well-being approach' to national accounting, which incorporates social and environmental indicators alongside GDP, builds on a workstream initiated by the previous government and heavily informed by the OECD's *How's Life* framework (Hall, 2019).

Yet this partly misses what makes Ardern so successful as a politician: that is, her embodiment of NZ's longstanding ideal of *moral leadership*. This has an important domestic function because it shores up popular support by appealing to a '"first in the world" mantra ... that [has], over time, come to hold force with people' (Johansson, 2009: 39). But her moral leadership also has a strong international dimension. A recent *Time* profile opined that 'Ardern's real gift is her ability to articulate a form of leadership that embodies strength and sanity, while also pushing an agenda of compassion and community' (Luscombe, 2020). This involves cognitive leadership. For example, when speaking to the UN General Assembly about NZ's response to 'wicked problems' like climate change, Ardern said: 'If I could distil it down into one concept that we are pursuing in NZ it is simple and it is this: *kindness*' (Ardern, 2018; emphasis added). This is a novel appeal in the present milieu, a resonant contribution to the exchange of political ideas.

But it also demonstrates *emotional intelligence* – that is, 'the ability to perceive emotions, to access and generate emotions so as to assist thought, to understand emotions and emotional knowledge, and to reflectively regulate emotions so as to promote emotional and intellectual growth' (Mayer and Salovey, 1997: 5). Although the relationship between emotional intelligence and leadership is underexplored (George, 2000), it is hypothesised that leaders with emotional intelligence can reduce fear and anxiety among followers in times of crisis (Meisler, Vigoda-Gadot and Drory *et al.*, 2013). At a time when ordinary citizens are becoming increasingly conscious of the climate crisis, and increasingly experiencing ecological grief as a consequence (Lertzman, 2015; Ellis and Cunsolo, 2018), emotional leadership could play a role in managing the denial and paralysis that negative affects engender. Ardern's emotional leadership, widely celebrated in her responses to both the Christchurch mosque attacks in March 2019 and the COVID-19 pandemic in early 2020, is not only valued by New Zealand citizens, but also the citizens of other jurisdictions, particularly where emotional intelligence among state leaders is lacking. In this sense, Ardern could be an authentic source of inspiration, providing moral leadership on climate change even when her policy contribution is still emerging. What remains unclear is what effect her emotional leadership has on those for whom it resonates, whether it merely offers succour and mollification, or mobilises and motivates by changing people's beliefs, priorities and behaviours. The emotional and affective dimensions of climate leadership deserve further research.

Conclusion

Moral leadership is a powerful ideal in the NZ social imaginary. As a small nation that lacks hard power in the conventional military and economic sense, NZ has long cultivated its soft power to exert an international influence through diplomacy, innovation and moral example. In regard to climate change, however, NZ's small size and its correspondingly small contribution to global emissions has provided an excuse to eschew leadership. Under Key, leadership was explicitly neglected for followership – and even this was only episodically lived up to. Under Ardern, as under Clark, the rhetoric of leadership is explicit, but the policy reality is more incrementalist. The introduction of a proportional representation voting system in 1993 arguably hindered the capacity for radical reform, due to the dependence on minority parties, especially the centrist NZ First Party, to form government. But this explanation is not sufficient, given that climate leadership has been achieved by countries with proportional representation, such as Germany (Christoff and Eckersley, 2011). On the other hand, NZ's climate policy fits a longstanding historical pattern of progressive upheavals being tempered by a prevailing conventionalism (Johansson, 2009). This moderating effect clearly weighs on the Ardern Government, and whether she can further substantiate her moral leadership through domestic-level gains, either by delivering policy more effectively or forming a new government, remains to be seen.

In the context of multilevel and polycentric governance, the NZ case study reveals patterns of climate leadership and pioneership in wider society, partly in response to government's modest progress. Non-state actors such as environmental NGOs, and youth and business groups have provided vital momentum, and contributed to policy transfer and diffusion, when state action on climate change has lagged. Māori have pioneered domestic climate action, guided by Māori cultural values, and establishing precedents that non-Māori follow for alternative environmental, social or economic reasons. By contrast, between 2004 and 2020, local government had an attenuated role because climate mitigation was largely excluded as a consideration in resource consenting. In order to actualise the promise of international leadership, NZ's central government would do well to endorse the presence and potential of domestic climate leadership, not only by following its example, but also enhancing its capabilities through greater resourcing and empowerment. An explicitly polycentric and multilevel approach may assist NZ's would-be climate leaders to align reality to rhetoric.

Note

1 **Disclaimer**
   Dr David Hall is a member of the Technical Working Group for Aotearoa Circle's Sustainable Finance Forum, Co-Chair of the Independent Advisory Group for Auckland Council's Climate Action Framework and a former researcher for Pure Advantage.

Bibliography

Ardern, J. (2008) 'Maiden statements', *Hansard*, 651(753), 16 December 2008.

Ardern, J. (2017) 'Speech from the throne', *Delivered on the Occasion of the State Opening of Parliament*, 8.11.2017. https://www.beehive.govt.nz/speech/speech-throne-2017

Ardern, J. (2018) 'Labour 2017: the Prime Minister's perspective', In S. Levine (ed.) *Stardust and substance: the New Zealand general election of 2017*, Wellington: Victoria University Press, 33–44.

Ardern, J. (2019a) 'Parliamentary debates for climate change response (Zero Carbon) amendment bill – third reading', *Hansard*, 742, 7 November 2019.

Ardern, J. (2019b) 'World-first plan for farmers to reduce emissions', *Media Release*. 24.10.2019. https://www.beehive.govt.nz/release/world-first-plan-farmers-reduce-em issions

Barrett, P.N., Wright, J. and Kurian, P. (2015) 'Environmental security and the contradictory politics of New Zealand's climate change policies in the Pacific', In I. Watson, and C. Pandey (eds.) *Environmental security in the Asia-Pacific*, Basingstoke: Palgrave Macmillan, 157–178.

Bertram, G. and Terry, S. (2010) *The carbon challenge: New Zealand's emissions trading scheme*, Wellington: Bridget Williams Books.

Cain, M., Lynch, J., Allen, M.R., Fuglestvedt, J.S., Frame, D.J. and Macey, A.H. (2019) 'Improved calculation of warming-equivalent emissions for short-lived climate pollutants', NPJ Climate and Atmospheric Science, 2: 29.

Cameron, Alastair (ed.) (2011) *Climate change law and policy in NZ*, Wellington: Lexis Nexis, 166–213.

Chapman, R. and Boston, J. (2007) 'The social implications of decarbonising the New Zealand economy', *Social Policy Journal of New Zealand*, 31 July 2007.

Christoff, P. and EcKersley, R. (2011) 'Comparing state responses', In J.S. Dryzek, R.B. Norgaard and D. Schlosberg (eds.) *The Oxford handbook of climate change and society*, Oxford: Oxford University Press, 431–448.

Clark, H. (2007) 'Launch of emissions trading scheme', *Beehive Website*, https://www.bee hive.govt.nz/speech/launch-emissions-trading-scheme

Clements, K. (1988) *Back from the brink: the creation of a nuclear-free New Zealand*, Wellington: Bridget Williams Books.

Climate Action Tracker (2015) *New Zealand deploys creative accounting to allow emissions to* rise, Policy Brief, Climate Analytics, NewClimate Institute, PIK and Ecofys, https://climateactiontracker.org/documents/35/CAT_2015-06-15_NewZealandDeploysCreative_PolicyBrief.pdf

Climate Action Tracker (2019) *Country summary: New Zealand*, Climate Action Tracker, https://climateactiontracker.org/countries/new-zealand/ (retrieved on 30 November 2019).

Coughlan, T. (2020) 'NZ first axes government's "feebate" electric vehicle subsidy plan, while greens vow to take the policy to the election', *Stuff*, 21 February 2020, https://www.stuff.co.nz/national/119713361/nz-first-axe-governments-electric-vehicle-subsidy-plan-while-greens-vow-to-take-the-policy-to-the-election

Daalder, M. (2019) 'Govt quietly abandons electric vehicle target', Newsroom, 8.10.2019, https://www.newsroom.co.nz/2019/10/08/847665/government-quietly-abandons-electric-vehicle-target

Daalder, M. (2020) 'Government revives implausible EV target', Newsroom, 16.3.2020, https://www.newsroom.co.nz/2020/03/16/1079233/government-revives-implausible-ev-target

Diaz-Rainey, I. and Tulloch, D. (2016) *Carbon pricing in New Zealand's emissions trading scheme*, USAEE Working Paper No. 16–255.

Driver, E., Parsons, M. and Fisher, K. (2018) 'Technically political: the post-politics(?) of the NZ emissions trading scheme', *Geoforum*, 97: 253–267.

Economist (2016) 'Drawbridges up', *The Economist*, 30.7.2016, https://www.economist.com/briefing/2016/07/30/drawbridges-up

Ellis, N. and Cunsolo, A. (2018) 'Ecological grief as a mental health response to climate change-related loss', *Nature Climate Change*, 8(4): 275–281.

ETSRP (Emissions Trading Scheme Review Panel) (2011) *Doing New Zealand's fair share. Emissions trading scheme review 2011: final report*, Wellington: Ministry for the Environment.

George, J.M. (2000) 'Emotions and leadership: the role of emotional intelligence', *Human Relations*, 53: 1027–1055.

Gibson, E. (2017) 'Officials' long struggle to publish new sea level guidance', Newsroom, 21.12.2017, https://www.newsroom.co.nz/2017/12/21/70263/officials-long-struggle-to-publish-new-sea-level-guidance

Gore, A., (2017) 24 *hours of reality: truth in action*, https://www.24hoursofreality.org/

Graham, K. (2018) 'Cross-party collaboration on climate policy: the experience of GLOBE-NZ', *Policy Quarterly*, 14(1): 37–43.

Hall, D. (2019) 'New Zealand's living standards framework: what might Amartya Sen say?', *Policy Quarterly*, 15(1): 38–45.

Harker, J., Taylor, P. and Knight-Lenihan, S. (2017) 'Multi-level governance and climate change mitigation in New Zealand: lost opportunities', *Climate Policy*, 17(4): 485–500. doi:10.1080/14693062.2015.1122567

Hopkins, D., Campbell-Hunt, C., Carter, L., Higham, J. and Rosin, C. (2015) 'Climate change and Aotearoa New Zealand', *Wiley Interdiscip. Rev. Climate Change*, 6(6): 559–583.

Huggard, S. (2019) 'A just transition', In D. Hall (ed.) *A careful revolution: towards a low-emissions future*, Wellington: BWB Texts, 80–95.

IEA (2017) *New Zealand: 2017 review*, Paris: OECD/IEA.

Johansson, J. (2009) *The politics of possibility: leadership in changing times*, Wellington: Dunmore Publishing.

Kelly, G. (2010) 'Climate change policy: actions and barriers in New Zealand', *International Journal of Climate Change Impacts and Responses*, 2(1): 277–290.

Kelly, G. (2011) 'History and potential of renewable energy development in New Zealand', *Renewable and Sustainable Energy Reviews*, 15: 2501–2509.

Key, J. (2016) 'Climate change policy—leadership and Donald Trump', *Hansard Debates: Oral Questions*, https://www.parliament.nz/en/pb/hansard-debates/rhr/document/Hans S_20161116_051900000/5-climate-change-policy-leadership-and-donald-trump

Kurian, P. and Smith, M. (2018) 'New Zealand environmental policy in the Key era: escalating crises in a time of neo-liberal economic dominance', In S. Levine (ed.) *Stardust and substance: the New Zealand general election of 2017*, Wellington: Victoria University Press, 251–264.

Legrand, T. (2015) 'Transgovernmental policy networks in the anglosphere', *Public Administration*, 93(4): 973–991. https://doi:10.1111/padm.12198

Leining, C. and Kerr, S. (2016) *Lessons learned from the New Zealand emissions trading scheme'*, *Motu economic and public policy research trust*, Motu Working Paper 16–06.

Leining, C., Kerr, S. and Bruce-Brand, B. (2020) 'The New Zealand Emissions Trading Scheme: critical review and future outlook for three design innovations', *Climate Policy*, 20(2), 246–264.

Lertzman, R. (2015) *Environmental Melancholia: psychoanalytic dimensions of engagement*, New York: Routledge.

Levine, S. (ed.) (2018) *Stardust and substance: the New Zealand general election of 2017*, Wellington: Victoria University Press.

Liefferink, D. and Wurzel, R.K.W. (2017) 'Environmental leaders and pioneers: agents of change?', *Journal of European Public Policy*, 24(7): 951–968, doi:10.1080/13501763 .2016.1161657

Luscombe, B. (2020) 'A year after Christchurch, Jacinda Ardern has the world's attention. How will she use it?', *Time*, 20.2.2020, https://time.com/5787443/jacinda-ardern-chris tchurch-new-zealand-anniversary/

Luth Richter, J. and Chambers, L. (2014) 'Reflections and outlook for the New Zealand ETS: must uncertain times mean uncertain measures?', *Policy Quarterly*, 10(2): 57–66.

Lynch, J.M., Cain, M., Pierrehumbert, R.T. and Allen, M. (2020) 'Demonstrating GWP*: a means of reporting warming-equivalent emissions that captures the contrasting impacts of short- and long-lived climate pollutants', *Environmental Research Letters*, https:// doi.org/10.1088/1748-9326/ab6d7e

Macey, A. (2014) 'Climate change: towards policy coherence', *Policy Quarterly*, 10(2), 49–56.

Mayer, J.D. and Salovey, P. (1997) 'What is emotional intelligence?', In P. Salovey and D. Sluyter (eds.), *Emotional development and emotional intelligence: Educational implications*, New York: Basic Books, 3–31.

Meisler, G., Vigoda-Gadot, E. and Drory, A. (2013) 'Leadership beyond rationality: emotional leadership in times of organizational crisis', In A.J. DuBrin (ed.) *The handbook of research on crisis leadership in organizations*, Cheltenham: Edward Elgar Publishing, 110–126.

Mills, S. (2018) 'Survey findings and the 2017 election', In *Stardust and substance: the New Zealand general election of 2017*, Wellington: Victoria University Press, 365–378.

Ministry for Business, Innovation and Employment (MBIE) (2019) *Electricity graph and data tables. Produced by markets team, evidence and insights branch, MBIE website*, https://www.mbie.govt.nz/building-and-energy/energy-and-natural-resources/energy-statistics-and-modelling/energy-statistics/electricity-statistics/

Ministry for the Environment (1997) The State of New Zealand's *environment* 1997, Wellington: NZ Government, https://www.mfe.govt.nz/publications/environmental-reporting/state-new-zealand%E2%80%99s-environment-1997

Ministry for the Environment (2017) *New Zealand's seventh national communication – fulfilling reporting requirements under the United Nations framework convention on climate change and the kyoto protocol*, Wellington: Ministry for the Environment.

Ministry for the Environment (2018) *Our land 2018: data to 2017*, Wellington: NZ Government.

Ministry for the Environment (2019) *New Zealand's fourth biennial report under the United Nations framework convention on climate change*, Wellington: NZ Government.

Ministry for the Environment (2020) *New Zealand's greenhouse gas inventory 1990–2018*, Wellington: NZ Government.

Ministry for the Environment & Stats NZ (2017). *New Zealand's environmental reporting series: our atmosphere and climate 2017*, Wellington: NZ Government.

Ministry of Foreign Affairs and Trade (2019) 'Agreement on climate change, trade and sustainability (ACCTS) negotiations', *MFAT Website*, https://www.mfat.govt.nz/en/trade/free-trade-agreements/climate/agreement-on-climate-change-trade-and-sustainability-accts-negotiations/

Narassimhan, E., Gallagher, K. S., Koester, S. and Rivera Alejo, J. (2018) 'Carbon pricing in practice: a review of existing emissions trading systems', *Climate Policy*, 18(8): 967–991.

NZ Government (2008) Major *design features of the emissions trading scheme, factsheet 16*, Wellington: NZ Government.

NZ Government (2015) 'Submission under the Paris agreement New Zealand's nationally determined contribution', *NDC Registry*, UNFCC Secretariat, https://www4.unfccc.int/sites/ndcstaging/PublishedDocuments/New%20Zealand%20First/New%20Zealand%20first%20NDC.pdf

NZPA (2007) 'Climate change report on right track – Nats', *Stuff*, http://www.stuff.co.nz/environment/24678/Climate-change-report-on-right-track-Nats

NZ Treasury (2018) *Living standards framework: background and future work*, Treasury Paper, https://treasury.govt.nz/publications/tp/living-standards-framework-background-and-future-work

OECD (2017) *New Zealand: OECD environmental performance reviews*, Paris: OECD.

Office of the Minister for Climate Change Issues (2009) *Moderated emissions trading scheme – proposed amendments to the climate change response act 2002*, Cabinet Paper, https://www.mfe.govt.nz/more/cabinet-papers-and-related-material-search/cabinet-papers/climate-change/climate-change-and-35

Ostrom, E. (2014) 'A polycentric approach for coping with climate change', *Annals of Economics and Finance*, 15(1): 97–134.

Palmer, G. (2015) New Zealand's Defective Law on Climate Change. Speech at Old Government Buildings, Victoria University of Wellington.

Parliamentary Commissioner for the Environment (2017) *Stepping stones to Paris and beyond: climate change, progress, and predictability*, Wellington: NZ Government.

Peters, G., Marland, G., Le Quéré, C. *et al.* (2012) 'Rapid growth in CO_2 emissions after the 2008–2009 global financial crisis', *Nature Climate Change*, 2: 2–4. doi:10.1038/nclimate1332

Peters, W. (2017) 'Winston Peters on why he chose a Labour-led government', *New ZealandZ Herald*, 20.10.2017, https://www.nzherald.co.nz/nz/news/article.cfm?c_id =1&objectid=11934973

Productivity Commission (2018) *Low-emissions economy*, Wellington: NZ Government.

Rive, V.J.C. (2011) 'New Zealand climate change regulation', In *Climate change law and policy in New Zealand*, Wellington: Lexis Nexis, 166–213.

Salmond, A. (2017) *Tears of Rangi: experiments across worlds*, Auckland: Auckland University Press.

Selby, R., Moore, P. and Mulholland, M. (eds.) (2010) *Māori and the environment: Kaitiakitanga*, Wellington: Huia. Skilling, D. and Boven, D. (2007) *We're right behind you: a proposed New Zealand approach to emissions reduction*, Auckland: The New Zealand Institute.

Stats NZ (2019). *Environmental-economic accounts: 2019 (data to 2017)*. Wellington: NZ Government.

Torney, D. (2019) 'Follow the leader? Conceptualising the relationship between leaders and followers in polycentric climate governance', *Environmental Politics*, 28(1): 167–86, https://doi.org/10.1080/09644016.2019.1522029.

Trade for All Advisory Board (2019) *Report for the trade for all advisory board*, https:// www.tradeforalladvisoryboard.org.nz/

Trump, D. (2018) *Remarks by President Trump to the 73rd session of the United Nations general assembly | New York, NY*, https://www.whitehouse.gov/briefings-statements/ remarks-president-trump-73rd-session-united-nations-general-assembly-new-york-ny/

Vivid Economics and Energy Centre, University of Auckland Business School (2012) *Green growth: opportunities for New Zealand*, Auckland: New Zealand Green Growth Research Trust.

Wurzel, R., Liefferink, D. and Torney, D. (2019) 'Pioneers, leaders and followers in multilevel and polycentric climate governance', *Environmental Politics*, 28(1): 1–21, https://doi.org/10.1080/09644016.2019.1522033

6 Multilevel climate governance in Brazil and Indonesia

Domestic pioneership and leadership in the Global South

Markus Lederer, Chris Höhne, Fee Stehle,
Thomas Hickmann and Harald Fuhr

Introduction

While more than two decades of international negotiations and carbon govern-ance have delivered only limited positive results, one issue is receiving increasing importance in international climate policy-making: the Global South, in particular the growing middle classes of emerging economies, is rapidly increasing its share of greenhouse gas emissions (GHGE). With some 58% of GHGE in 2017, the rise of the South has been firmly acknowledged in the Paris Agreement adopted in 2015. Governments of the Global South have also committed themselves to issue nationally determined contributions and thus agreed to deviate from a busi-ness-as-usual scenario and unsustainable growth trajectories. Interestingly, many countries in the Global South have been very active players in global climate politics for some time prior to the adoption of the 2015 Paris Agreement, and the question arises whether we can observe leadership or pioneership also in the 'developing world'?

Our chapter aims to identify domestic pioneership and leadership that – com-pared to other sectors, governmental levels or jurisdictions within the same nation-state – move 'ahead of the troops' (Liefferink and Wurzel, 2017: 2–3). Moreover, we analyse what role multilevel governance has played in bringing about domes-tic pioneership and leadership, characterise different types, and scrutinise whether these pioneers and leaders trigger other domestic actors to follow and replicate climate change mitigation actions. Focusing on processes of multilevel govern-ance, we are interested in exploring who constitutes pioneers and leaders as well as how and with what consequences such processes take place. By doing so, we build upon earlier calls to open up the black box of the nation-state for the study of climate governance (Hickmann *et al.*, 2017; Höhne, 2018; Lederer, 2015) and for the analysis of leadership (Wurzel, Liefferink and Torney, 2019) in order to understand how external, national and subnational actors interact.

In our case studies, we focus on these interactions in two emerging economies, namely Brazil and Indonesia. Both countries and their governments are impor-tant political and economic actors that are responsible for a significant share of global GHGE. Moreover, the two countries are governed democratically and have decentralised structures of government. Analysing large decentralised democracies in the Global South allows us to understand processes of pioneership and leadership that might emerge at different levels of government, but that can also be influenced by endogenous processes and players. Empirically, we will analyse two policy fields that are highly relevant for climate governance and deeply enmeshed in processes of global multilevel governance: (1) urban climate policy and (2) forest policy. We will thus investigate two specific carbon governance arrangements which address those sectors and at the same time operate at different horizontal and vertical levels (Hickmann *et al.*, 2017). The first arrangement is labelled Reducing Emissions from Deforestation and Forest Degradation (REDD+) and works mostly in a top–down fashion from a vertical multilevel perspective. Our second governance arrangement constitutes Transnational City Networks (TCNs), which can be defined as non-hierarchical and horizontal institutions that aim to facilitate polycentric cooperation between city governments around the world (Bulkeley and Betsill, 2013). TCNs therefore mainly work bottom–up from a vertical multilevel governance perspective.

Our main argument to be developed in the remainder of this chapter is that international and transnational processes, incentives and ideas often trigger the development of domestic pioneership and leadership. Such processes, however, cannot be understood properly if domestic politics and dynamics across governmental levels within the nation-state are not taken into account. Similar to previous discussions relating to innovation and performance (Blum *et al.*, 2019; Campbell and Fuhr, 2004; Levy, 2011), many instances of domestic pioneership and leadership are facing serious problems of sustained impact in the short run, and sustainability and 'survival' in the medium-long run. Many of them are simply dissolved after a couple of years due to a change in the domestic political economy, such as increasing political opposition by vested interests, changes in political leadership, economic problems, increasing financial constraints or a combination thereof. Despite such fragility, however, domestic pioneers or leaders often leave their traces and might serve as foundations for progressive climate policies in the future. To advance this argument, our chapter proceeds as follows. In the next section, we conceptualise domestic climate pioneership and leadership. The third section presents the findings from our case studies on Brazil and Indonesia based on expert interviews and qualitative analysis of primary and secondary documents. In the conclusion, we reflect upon our results and point to aspects that should merit attention in future research.

Conceptualising domestic climate pioneership and leadership in multilevel governance arrangements

In our chapter we define domestic pioneership as instances in which specific jurisdictions, such as cities, provinces or the federal level have started with progressive or innovative climate policies moving 'ahead of the troops'. Such instances cover more than a single project or climate governance experiment, but encompass a whole sector or a complete jurisdiction although they neither necessarily lead to more progressive change nor are they always enduring, stable or even long-lasting.

Slightly diverging from Liefferink and Wurzel (2017: 954), who argue that pioneers have a high internal and low external environmental ambition and thereby focus primarily on domestic environmental politics, we contend that domestic pioneership can both be set up with an internal and external audience in mind (e.g. following a globally set logic of appropriateness in the latter case). We call them pioneers, because they do not simply emulate practices of other countries, as the actual policies and organisational changes that have been implemented also reflect domestic interests, ideas, coalitions and institutions. We agree with Liefferink and Wurzel's claim that pioneers can only indirectly and unintentionally provide an example to others leading to processes of diffusion that they call exemplary leadership (Liefferink and Wurzel, 2017: 958; see also Wurzel *et al.*, 2019: 8). They further differentiate between different forms of horizontal leadership, namely structural (i.e. military power and economic strength), entrepreneurial (i.e. negotiation skills) and cognitive leadership (i.e. ideas and knowledge). We, however, extend their typology by arguing that these different leadership types (i.e. structural, entrepreneurial and cognitive) can also be used to characterise different types of pioneers in the form of structural, entrepreneurial and cognitive pioneership. Thus, pioneers can become frontrunners because of their structural capabilities (such as economic or military power), which can be termed *structural pioneership*, while *entrepreneurial pioneership* exists when a pioneer has put a new topic on the agenda and builds coalitions around it, and finally *cognitive pioneership* is enacted when a pioneer promotes new ideas.

We also focus on domestic leadership taking up Wurzel, Liefferink and Torney's (2019) call to open up the black box of the nation-state for the study of leadership. Wurzel, Liefferink and Torney (2019), however, focus on non-state actors, whereas we stress the role of vertical interactions among different governmental levels within nation-states. We highlight the following four types of vertical leadership: First, *vertical structural leadership* can be understood in terms of central governmental power to enforce or to push subnational governments to follow its lead. This can often be witnessed in policy fields where the central and subnational governmental levels share powers or where subnational governments are responsible for implementing central government policies. Second, in settings of shared power between governmental levels, we may also find stronger bargaining between these levels, but also the provision of financial resources, which can both be captured by the term *vertical entrepreneurial leadership*. Leaders obviously need bargaining skills and must establish coalitions, especially when they belong to a lower governmental level. Third, with regard to *vertical cognitive leadership*, both higher and lower governmental levels can use ideas and knowledge to persuade other governmental levels to follow their lead. Fourth, pioneers who unilaterally move ahead and do not aim to attract followers among higher or lower governmental levels may be characterised by *exemplary leadership*.

As our contribution focuses more on governmental and less on societal action, we rely on the concept of multilevel rather than polycentric governance. While both terms are stressing arrangements in which various agents can become author-itative (Wurzel, Liefferink and Torney, 2019: 10), the notion of polycentricity puts much more emphasis on self-regulation and conceives of the various actors as independent entities (Ostrom, 2010). Although multilevel governance has always stressed more strongly the role of governments, it is now also used to conceptual-ise the various processes that link international institutions, national governments, and to some extent sub- and non-state actors, in global climate policy-making (Fuhr *et al.*, 2018; Hickmann and Stehle, 2019; Höhne, 2018). This is of relevance for two reasons: First, the policies we will focus on are developed within a global setting and receive support from international donors, NGOs, etc. We, therefore, scrutinise whether this second-image reversed mechanism (Gourevitch, 1978), where the global level influences the national one, plays a major role. Second, the different governmental levels comprise varying degrees of political, admin-istrative, and financial powers in a respective policy field, and public actors can engage in uploading or downloading new policy initiatives or innovations to other governmental levels for policy formulation and implementation (Höhne *et al.,* 2018). This will also allow us to review the differences between the national and subnational levels within our cases and to analyse how domestic leadership and followership play out between the different levels. We will concentrate on devel-opments after 2005 as around this time, countries in the Global South started to discuss voluntary climate change mitigation actions.

Domestic pioneer and leadership in action

In the following, we will first provide a brief overview of the national climate policy in the countries and then investigate their respective forest and urban cli-mate policy.

Brazil – domestic pioneers retreating

With more than 206 million inhabitants, Brazil is the fifth largest country of the world, both in terms of size of area and population. Although 85% of Brazil's population live in urban areas, the main share of the country's total emissions of 2 gigatonnes (Gt) carbon dioxide equivalent (CO_2eq) in 2005 did not result from energy consumption, but from the enormous deforestation in the Amazon which was mostly carried out to enable cattle and soy production activities (WRI, 2016). Brazil consists of one federal district and 26 federal states, which are divided into 5,561 municipalities. The Constitution of 1988 allocates competencies and legis-lative authority between all three levels of government. They have the common duty to preserve an 'ecologically balanced environment', which includes forest and natural resource management (Gebara *et al.*, 2014). Brazil comes closest to be a potential leader in global climate politics among the countries of the global South, as, for example, it was one of the first countries worldwide to issue a highly ambitious national climate action plan already in 2007. In its NDC the country has pledged to reduce its emissions by 37 % by 2025 (compared to 2005). With significant REDD+ activity and strong involvement in the C40 Cities Climate Leadership Group (C40), Brazil has been considered as an important player in these two policy fields and thus a true pioneer – at least until 2018 when the government of President Bolsonaro reversed the country's course.

Brazilian forest governance

Until the end of the military regime in 1984, forest management in Brazil was highly centralised (Banerjee *et al.*, 2009). With the establishment of the democratic government and the adoption of the constitution in 1988, state governments gained profound autonomy, mostly through fiscal decentralisation. But while resources and tax raising authorities were transferred to the state governments, functions were not sufficiently clarified or remained at the central level (Gregersen *et al.*, 2004). Within the forestry sector, weak law enforcement and missing transparency led to exploding rates of deforestation, timber exploitation and the distribution of land titles were subject to criminality and corruption (Rajão *et al.*, 2012). Increasing domestic pressure and international attention forced the government to take action, leading to the adoption of the national Plan to Prevent and Control Deforestation in the Brazilian Amazon (PPCDAm) in 2004 (Caviglia-Harris *et al.*, 2016; Di Gregorio *et al.*, 2016).

The new regulations and institutional changes significantly contributed to the remarkable decrease in deforestation from 2005 onwards (Assunção *et al.*, 2012). This happened at the same time as REDD+ became an important topic on the international climate agenda. A number of domestic factors such as a soy moratorium that banished producers growing soy on land cleared after 2006 and the governmental satellite monitoring system PRODES contributed to Brazil becoming a cognitive pioneer in reducing deforestation (Nepstadt *et al.*, 2014). The structural and cognitive vertical leadership that Brazil's government demonstrated with its progressive and effective deforestation control had the potential to provide a major contribution to national emission reductions and thus attracted the interest of international donors. In 2008, the government established the Amazon Fund that received its largest share of US$ 1 billion from the Government of Norway (see also Chapter 11 in this volume), and US $28 million from Germany (DI-18052017; Tollefson, 2009).

Following the cognitive und structural vertical leadership of the national government, state governments, which are entitled to apply for funds through the Amazon Fund, started to create legal and organisational structures and to implement projects to reduce deforestation. This was further supported by a multilevel approach which both of the main donor countries pursued in their development cooperation, demonstrating both horizontal cognitive leadership by infusing new ideas into the system as well as entrepreneurial leadership by providing the financial means needed to implement them. Norway supported and stipulated actions at the subnational level, for instance by requesting state governments to establish own jurisdictional REDD+ policies, funding a range of state and non-state actors, as well as the transnational exchange between state governors (DI-18052017,

DI-12122017; see also Chapter 11 in this volume). For the German involvement, two prominent ways of engaging with subnational actors stand out (for Germany's domestic energy transition see Chapter 9 in this volume). On the one hand, German REDD+ support is focused on technical cooperation and capacity building by seconding experts and advisors, which became embedded in national ministries and state secretariats, as well as by funding civil society organisations (DI-16052017). In the case of Acre, for instance, development workers have supported forest management for more than 20 years (NI-GI-29112017). On the other hand, Germany's REDD+ for Early Movers programme (REM) figures as development assistance that distributes funds based on the results of reduced deforestation (DI-29112017). The REM project aims at building up capacity and projects at the state level through the provision of results-based donations (DI-29112017). As of 2018, the state government of Acre was compensated with 25 million Euro for avoided deforestation, equaling a reduction of CO_2 emissions by 6.47 $MtCO_2$ (NI-GI-29112017; GI-01122017a).

At the subnational level, Acre thus can be considered to be a cognitive pioneer and exemplary leader amongst the states with forest cover in its efforts to reduce deforestation. In addition, the state government also formulated and implemented policies to reduce deforestation since the early 2000s, well ahead of Brazil's national government (NI-GI-29112017; GI-01122017a). Here, a REDD+ unit was established in the form of fairly independent agencies and a cross-sectoral body organises the coordination between sectoral ministries, public and private actors. Civil society and indigenous organisations are included in governance processes and a jurisdictional payment for ecosystem services has been established, which demonstrates Acre's entrepreneurial pioneership considering its ability to negotiate between groups of stakeholders (GI-06112017; GI-30112017). Due to Acre's moving ahead in reducing deforestation and including safeguards, the state was the first to receive compensations for reducing deforestation from the German government under the REDD+ Early Movers programme REM (DI-29112017). Later on, the state also served as a role model for setting up REM in Mato Grosso showing exemplary leadership, as delegations of public officials visited Acre to learn from their experience (IMC, 2017; GI-01122017b).

But while the instruments and policy enforcements described above had an overtly positive effect on the reduction of deforestation rates until 2016, political processes slowed down at the national level and tensions have arisen since the beginning of Brazil's economic crisis in 2014 (RI-17052017). Nevertheless, in October 2015, seven years after the creation of the Amazon Fund, Brazil's National REDD+ strategy was released at the 21st Conference of the Parties (COP21) to the United Nations Framework Convention on Climate Change (UNFCCC). The strategy defines a number of objectives regarding governance structures and the allocation of functions between the three levels of government as well as the integration and coordination in respect to existing policies (Fatorelli *et al.,* 2015). However, the federal states, which had started to prepare for establishing separate REDD+ funding schemes, were largely sidelined by the 2015 national policy. The policy instituted the centralised control over payments for reductions and established a scheme to share all revenue between the central and state governments, thereby restricting the states' access to international carbon markets to receive payments for their deforestation reductions (NI-09052017, NI-01112017, RI-17052017). The central government thus provided structural leadership, but of the wrong kind.

After President Dilma Roussef's impeachment in 2016, the situation aggravated further (NI-09112017). When Michel Temer became acting president, the agricultural lobby group at the national level, the bancada ruralista, regained power. The legal amendments, which had improved the decentralisation of forest management and community participation and made Brazil a pioneer in this policy field, were substantially weakened by changes of legislation and budget cuts in services related to controlling deforestation (NI-09112017). With the inauguration of the presidency of Bolsonaro in 2018, this development further worsened.

Brazil's Urban Climate Governance

The two largest cities in Brazil are São Paulo with 12 million inhabitants and Rio de Janeiro with 6.5 million inhabitants (IBGE, 2015). The urbanisation rate is expected to grow in the years to come, a fact that makes cities an important player in national politics while it forces them to adapt infrastructure, energy supply, and transportation systems, and respond to environmental risks caused by, amongst other things, the coastal location of Brazil's major cities (Fernandes, 2007). Similar to the forest sector, the Constitution of 1988 included provisions on urban policy which enabled urban reform and municipality autonomy (NCA, 1988). But while Brazil gained international recognition for participatory elements for its urban policy-making model, most urban reform processes have slowed down due to a lack of federal legislation and the opposition of certain interest groups (Fernandes, 2007).

Most major Brazilian cities have implemented municipal plans to reduce emissions and set up councils for urban climate action while some of them actively participate in transnational fora (Kahn and Brandão, 2015). Transnational city networks (TCNs) supported the initial uptake of climate actions across cities in Brazil, but their involvement largely depends on local leadership, whose absence or ceasing can disrupt the institutionalisation of climate actions. ICLEI – Local Governments for Sustainability, which is a global network with more than 1,700 local and regional governments, has been instrumental in creating awareness on climate issues among many city officials. For instance, the network has supported peer-to-peer learning by facilitating the CB27, which is an annual conference of the environmental secretaries of the 27 largest Brazilian cities (NI-19052017). Yet, political changes, the domestic economic crisis of 2014 and a number of structural obstacles seem to have trapped local action on climate change in a state of inertia (Stehle *et al.*, 2019). One of the biggest obstacles is the absence of national leadership in aspects of urban climate politics. This manifests in a lack of national guidance and weak to non-existent channels of coordination between the national and local levels (DI-17052017; GI-12122017). Similarly, the national government does not provide cities with sufficient budgets to transform their infrastructure to become less carbon intensive and does not grant them with access to external funding (RI-20102017; GI-22102017; NI-16112017). In short, the instances of pioneership that exist are not supported, scaled-up or even opposed and consequently slowly fade away.

São Paulo joined ICLEI in 1994 and was a founding member of the C40 group in 2005 (Johnson *et al.*, 2015). In 2003, the city became a cognitive pioneer when it established its first GHGE inventory with estimated annual emissions of 15 million tonnes CO_2 equivalent and initiated the policy-making process for a climate change policy in 2005. One of the first steps was the establishment of a Municipal Committee on Climate Change and Sustainable Economy that represented actors from civil society, academia, municipal and state governments, and environmental organisations, thus also demonstrating entrepreneurial pioneership based on its ability to convene a multi-stakeholder negotiation process (Barbi and da Costa Ferreira, 2013). The climate policy that São Paulo adopted thereupon in 2009 is an example of exemplary leadership. By setting the specific goal for a GHGE reduction of 30% compared to emission levels from 2005 to 2010, it was highly proactive and inspired both the state and national climate change policies (de Macedo *et al.*, 2016; São Paulo, 2009). The policy encompasses concrete action plans for mitigation and adaptation to climate change and the implementation follows a cross-cutting, multi-sectoral approach (Barbi and da Costa Ferreira, 2013). But after a change of mayor and the implementation of a new agenda in 2012, most activities relating to the implementation of the climate policy as well as the city's transnational involvement were stopped.

Similar developments have happened in Rio de Janeiro, which is the second largest city of Brazil and to some extent has been a follower. Rio de Janeiro established a municipal Forum on Climate Change in 2009 which led to the adaption of the city's climate change policy in 2011 and a pledge to reduce its GHGE by 8% until 2012, 16% until 2016 and 20% until 2020 (Rio de Janeiro, 2011). The city joined the C40 network in 2006. It has implemented projects in collaboration with C40 in four out of the seven C40 categories on solid waste, clean transportation and sustainable urban development (Cohen, 2010). However, after the last election in 2016, the city's new mayor has so far not been a strong promoter of climate policy (GI-06122017).

In Brazil, we can thus see an interesting mix of multilevel pioneering with elements of domestic leadership in the forestry sector and some municipal pioneership. The national government, pushed by pioneer states and international and non-state actors, became a pioneer in promoting forest policy. This was further supported by the horizontal cognitive and entrepreneurial leadership of the donors Norway and Germany. The central government exerted vertical cognitive leadership by rolling out a national plan that had a substantial impact on reducing deforestation. The achievements in urban development in Brazil, which have mostly emerged due to cognitive pioneership at the local level (Fernandes, 2007), are more of an exemplary nature. For a period of time, cities were pioneers of climate change adaptation and mitigation that overtook the nation-state in climate change regulation and initiated the diffusion of good examples that have inspired other cities in Brazil. However, many forest and urban policies have been slowed down, reversed or stopped. This suggests that pioneering and leadership in Brazil are either temporarily restrained phenomena or, on a more hopeful note, institutions that might reemerge like buoys in a stormy sea.

Indonesia – weakening domestic leadership

Indonesia is a presidential republic with a three-tier political system (including provinces and districts) that was democratised in 1998 and decentralised through the Regional Governance Law of 1999 (Bünte, 2008: 38). The country is the world's fifth largest GHG emitter (Climate Watch, 2019) and its historical GHGE per capita emissions of 9.69 t CO_2 equivalent are already higher than in the European Union (7.19 tonnes CO_2 equivalent). The land use sector is responsible for 68% of these GHGE (World Resources Institute, 2018). But cities are growing and the energy sector is expected to surpass GHGE from the land use sector by 2027 (Chrysolite *et al.*, 2017).

Indonesia's forest governance

The country hosts the third largest tropical forests in the world, but its high rates of deforestation resulted in the decline of forest cover from 65% to 50% from 1990 to 2015 (World Bank, 2016). Indonesia has a history of exploitation of the state-owned forests through informal networks of politicians, bureaucrats and business actors. This practice was decentralised in 1999 (Ribot *et al.*, 2006), as districts were granted the authority to manage their forest resources and continued the legal and illegal exploitation of forests on the local level (Ardiansyah *et al.*, 2015: 6–8). Subsequent regulations and laws in 2002, 2004 and 2014 strengthened the competencies of the provinces and the central government (Höhne *et al.*, 2018). Hence, the forestry sector is the worst sector when it comes to GHGE although the engagement with REDD+ created some instances of pioneership starting to influence how the overall forestry sector is governed with the adoption of the bilateral agreement between Norway and Indonesia in 2010 (Lederer and Höhne, 2019).

From a multilevel perspective, the national government was the key pioneer for that process, supported by the most important external partner: Norway. But this partnership did not emerge in a simple leader-follower relationship, as both Norway and President Yudhoyono together defined the political principles of the REDD+ partnership in the Letter of Intent in 2010 (Agung *et al.*, 2014). While Indonesian REDD+ proponents in the President's Office and the National Council on Climate Change (DNPI) regarded REDD+ as an opportunity to reform the forest governance and to receive similar international recognition as Brazil (GI-BI-31082017; GI-DI-21082017), Norway tried to support those countries who were willing to take climate mitigation actions in the forest with financial incentives and capacity building activities (i.e. through entrepreneurial and cognitive horizontal leadership). But Indonesia's central government showed strong cognitive and entrepreneurial types of pioneership itself through the initiation of a forest policy and governance reform process which cannot be captured through a simple followership. This resulted in several outputs and processes that were advanced by the REDD+ Task Force/Agency (operating from 2010 until 2015), such as the issuance of a moratorium on new forest concessions (GI-08082017; Anderson *et al.*, 2015: 269), the integration of several conflicting and overlapping maps (indicating conflicting and overlapping natural resource usage licenses) in one authoritative map in order to resolve conflicting land claims (DI-15082017; GI-08082017; Wibowo and Giessen, 2015: 135), and the establishment of a forest reference level against which future forest cover changes can be measured (GI-05082016; NI-09082017; GI-DI-21082017; DI-15082017). Furthermore, the REDD+ process facilitated the long-planned establishment of forest management units based on the German *Forstamt* model at the subnational level and even supported the recognition of indigenous peoples' rights to forests (Lederer and Höhne, 2019). Overall, those outputs have been developed slower than expected, included several loopholes and the overall outcome still remains to be seen. However, it provided some initial reform dynamic, which continued even after an initial slowdown after the change of presidency from Susilo Bambang Yudhoyono (2004–2014) to Joko Widodo (since 2014) (Anderson *et al.*, 2016: 33) with first results-based payments for reduced GHGE by Norway having been agreed on in February 2019 (Jong, 2019).

The bilateral agreement with Norway also stipulated the advancement of REDD+ in one pilot province (Kingdom of Norway and Republic of Indonesia, 2010). Being more ambitious, the REDD+ Task Force even selected ten priority provinces in Papua, Sulawesi, Sumatra and Kalimantan alongside the REDD+ pilot province Central Kalimantan to work on REDD+ and forestry issues in all forestry relevant areas in Indonesia (Ahmad, 2012). The central government requested them to set up provincial REDD+ task forces and to develop provincial REDD+ strategies (RI-10082016). It furthermore provided training of provincial REDD+ task force staff (GI-DI-08082017) and enacted vertical structural leadership based on orders as well as vertical cognitive leadership based on forest governance reform ideas towards those provincial governments, which followed the lead of the central government.

The central government further recentralised forestry and empowered provinces with the Regional Governance Law of 2014, which was not initiated by REDD+ directly, but was enacted to increase the control over natural resources in the light of misconduct of district governments (Höhne *et al.*, 2018; Lederer and Höhne, 2019). When the new government under President Joko Widodo in 2014/2015 slowed down in its efforts at the national level it thereby also ended its vertical structural (i.e. no further national regulations) and cognitive leadership (i.e. no further new ideas) for provincial governments. Many provinces subsequently remained in limbo and some even retreated from earlier efforts on REDD+ and reforming forest governance. This was mainly caused through abolishing the national REDD+ Agency and the integration of its tasks into the newly merged Ministry of Environment and Forestry (MOEF) (Anderson *et al.*, 2016: 33).

But it was also caused by the collapse of any form of subnational pioneership in cases such as the REDD+ pilot province Central Kalimantan, which used to be a REDD+ pioneer when election brought a governor with a resource exploitation agenda to power (NI-09082017; GI-BI-22082017; DI-15082017; SGI-13092017; GI-DI-08082017; GI-07082017).

However, some provinces such as East Kalimantan continued to provide exemplary leadership in advancing REDD+ and forest governance reform. They were supported by external actors such as the World Bank's Forest Carbon Partnership Facility and domestic as well as external non-state actors in the process (DI-14082017). Especially cases such as East Kalimantan show that subnational pioneers must provide some form of cognitive pioneership, as the governor at the time, Awang Faroek Ishak (in power 2008–2018), envisioned a greener development path for his province long before any national REDD+ involvement of the country (SGI-1-19092017). But there is only so much sub-national governments can do, as pioneers like East Kalimantan still have to wait for central government regulations on REDD+ issues such as safeguards and the benefit sharing of the results-based payments by external actors between the involved domestic actors (SGI-2-19092017; DI-14082017). Furthermore, when it comes to safeguarding high conservation value forests in land ear-marked as potential agricultural areas, provinces are dependent upon the coop-eration by districts that are responsible for handing out agricultural permits. East Kalimantan was able to convince them, through vertical cognitive and entrepre-neurial leadership, to protect high conservation value forests in those areas in 2017 (SGI-1-19092017).

In conclusion, external ideas and incentives chimed with domestic ideas and interests. Indonesia's central government enacted entrepreneurial and cognitive pioneership in the national forest governance sector. Except for the case of East Kalimantan, through central government's vertical structural and cognitive lead-ership, provincial government started to engage on climate change issues in the forestry sector. Some of the governors, e.g. in East Kalimantan, were already engaging in cognitive and entrepreneurial pioneership beforehand and have con-tinued to do so while others, e.g. Central Kalimantan, are retreating from earlier achievements.

Indonesia's urban climate governance

Before the country's decentralisation in 1999, Indonesia was characterised by centralistic planning while after decentralisation, the central and provincial gov-ernments lost significant powers to the cities (Moeliono, 2011: 135, 140–141). Subsequent regulations and laws in 2004, 2007 and 2014 re-empowered the prov-inces and the central government at the expense of cities (Hickmann *et al.*, 2017). In Indonesia, cities are growing massively, depend increasingly on higher energy use (e.g. for cooling) and are mostly being built for individual traffic rather than public transport. This contributes to increasing GHGE in urban sectors, as energy is highly subsidised and is produced from fossil sources (Chrysolite *et al.*, 2017).

In contrast to the forestry sector, the central government did not take any leadership role with regard to climate mitigation by cities. Climate change emerged on the national agenda due to the cognitive pioneership by President Yudhoyono, who established the National Council on Climate Change in 2008 and issued the National Action Plan for Greenhouse Gas Emissions Reductions (RAN-GRK) in 2011 (ROI, 2011). The RAN-GRK mandated the provinces to develop provincial climate action plans in line with specific guidelines published by the Ministry of National Planning (ROI, 2011). But the central government has not instructed cities to develop climate action plans themselves. For the development of the national climate action plan, the central government did not even consult subnational governments (Anggraini *et al.*, 2011: 17). Similarly, provinces have not involved city governments in the development of the provincial climate action plans (SGI-26092017). The central government hence provided structural and cognitive leadership with regard to provinces, but not with regard to cities. This lack of central government climate leadership did not result in a polycentric upspring of urban climate action plans by cities, even though in many policy areas (such as transport, waste and energy efficiency) they possess sufficient powers to take actions (CI-GI-02102017). However, cities are often dependent on financial support from higher governmental levels for implementing activities such as on transport.

Only in the very few instances in which cities engaged intensively with transnational city networks, some form of small-scale pioneership emerged. TCNs have a hard time to convince Indonesian cities, as climate action is voluntarily for those cities and most of them are not interested in this issue (DI-03082016; CI-18082016). It hence needs some local motivation, especially from the mayor, to participate in climate change initiatives (SGI-26092017; CI-03102017; SGI-03102017; SGI-1-04102017). Cities such as Bogor and Balikpapan have engaged on climate change issues as participants in EU-funded ICLEI projects from 2012–2015. But the vast majority of cities in Indonesia have not taken any action on climate change and in most cases they only know very little about climate change. When cities engage on climate change then they usually do so in cooperation with external partners such as ICLEI or Germany's development agency (*Deutsche Gesellschaft für Internationale Zusammenarbeit* – GIZ) which provide capacity building and ideas and thereby mostly engage in cognitive leadership (CI-18082016). Cities such as Bogor participate in those initiatives as they align with a domestic urban agenda. For example, the mayor had a green vision of Bogor and used the support of ICLEI to advance this agenda in his city (CI-03102017), thereby showing cognitive pioneership. However, for the implementation of sector reform, such as public transport, cities often rely upon some regulatory and financial support from higher governmental levels (DI-18082016).

Jakarta is a special case. When President Yudhoyono announced Indonesia's climate mitigation target of 26% GHGE reduction by 2020 compared to business as usual at the G20 meeting in Pittsburg in 2009, the announcement was followed by Jakarta's governor who pledged a similar target for his province, adopting a 30% GHGE reduction target by 2020 compared to business as usual

(Susanti, 2011: 24). Different accounts have been presented as to why he did so, but all point towards the role of cognitive pioneership by either domestic actors (in the case of the President) or cognitive leadership by external actors (in the case of C40 which motivated its member cities to present own GHGE targets before the Copenhagen COP) (SGI-2-04102017; CI-03102017). In any case, this must also be regarded from a domestic perspective as the Governor of Jakarta was engaged in some climate planning and knowledge exchange with the head of the planning ministry who was keen on advancing environmental issues. But the governor also saw it as a good opportunity for campaigning and to attract funding (CI-03102017). He therefore also showed some cognitive pioneership to join the TCNs and to engage in some climate change activities. However, after the change of governor, most activities came to a halt (CI-03102017). Since Jakarta joined the C40 in 2007 (Susanti, 2011: 24), it has mostly been engaged in participating in some workshops and peer to peer learning such as on transport (CI-16012018; C40, 2012, 2013b, 2013a). However, those lessons learnt are rarely implemented (CI-16012018; SGI-1-05102017; SGI-2-04102017). While Jakarta included some climate change issues in its development plan, and thereby benefited from support by C40's implementation partner, climate action is very limited when it comes to changing sectoral policies. However, in the late 2010s, Jakarta started to address energy efficiency through the change of the building code for the purpose of reducing GHGE (SGI-2-05102017; DKI, Jakarta n.d. Mahendriyani, 2016).

In conclusion, there is some small-scale cognitive pioneering at the city level in Jakarta, Bogor and Balikpapan, mostly supported by horizontal cognitive and entrepreneurial leadership by TCNs that need to align with the domestic interests of local governments to advance the climate change agenda. There is, however, no vertical leadership from the central government or the provincial governments at the city level, with the exception of the province of Jakarta which was influenced by vertical structural and cognitive leadership from central government. Furthermore, no vertical bottom–up leadership from cities to national government were observed.

Conclusions

Comparing our two Global South case study countries, we can conclude that while in the forestry sector in Indonesia the central government was a pioneer and enacted different leadership types with regard to the provincial government, in Brazil's forestry sector a movement started that resulted in a tug of war between the national and state governments for the ownership of REDD+. In the urban sector in Indonesia, the central government did not provide any leadership with regard to cities and only very few cities engaged in some small-scale pioneering with the support of TCNs. In Brazil, the national level also fell short in guiding urban actors on implementing climate policies. For some time, this gap was filled by endogenous and transnational actors, but their influence on urban climate politics has been contested and in some cases, came to an end with changes in the government.

Overall, we have been able to identify two patterns regarding domestic pioneership as well as leadership. First, we have shown that both can occur at national and subnational levels. Brazilian cities, for example, were much more engaging in climate pioneership than Indonesian cities. While Brazilian cities' pioneership moved upwards and motivated climate change engagement of the central government, Indonesia has not seen a similar dynamic. Neither Brazilian nor Indonesian cities were benefiting from higher governmental level vertical leadership. This was different in the forestry sector, where Indonesia's central government first engaged in climate pioneership at the national level and then even provided vertical leadership at the provincial level. However, in a few cases, some provincial pioneering also occurred in Indonesia's forestry sector. Brazil has both witnessed national- and state-level pioneering in the forestry sector.

Second, regarding horizontal climate leadership, we have identified only elements of cognitive and entrepreneurial leadership provided by external actors, such as Norway's or Germany's development agencies. Overall, domestic politics and dynamics across governmental levels within the nation-state mattered a great deal more for the actual uptake of climate actions and we recognised bottom–up processes in Brazil's urban sector and more domestic top–down processes in Indonesia's forestry sector. These processes neither led to wide-scale diffusion nor an upscaling within the two countries, although central governments were able to download their policies quite effectively to lower governmental levels if they wanted to, as events in the forestry sector in both countries showed. In addition, they were inherently fragile: whenever there was a change in government at central and subnational level, there was a risk that instances of pioneership and leadership quickly disappear, particularly when politicians came into power with little interest in the climate agenda.

Where does this leave us in terms of lessons learnt and future research? First, we now know that diffusion, policy transfer or leadership do not work directly nor do they amount to emulation in a one-way-street fashion. However, we do not know yet under which conditions domestic pioneership can survive and maybe even mature, leading to the lock-in of progressive policies and potentially turning into leadership. Second, and to some extent sobering for development cooperation or other attempts of external actors to set up new practices, international or transnational influence on initiating and encouraging pioneers in the Global South seems to be quite limited. Although REDD+ as an international mechanism has certainly left its traces, it has only in very few instances turned into a game changer stopping deforestation. Similarly, the impact of TCNs and the role of cities overall should not be overestimated. Despite the fact that cities in both countries have been involved in climate initiatives, we have found little evidence in our cases that the C40 group or the ICLEI network have significantly influenced the way in which climate policies are carried out. Apparently, domestic climate pioneership and leadership in the Global South is much harder to establish, to enlarge and to sustain than conventional wisdom has it.

List of Interviewees

Brazil: 20 interviews

NI-09052017: NGO Interviewee, 9 May 2017
DI-16052017: Donor Interviewee, 16 May 2017
RI-17052017: Research Interviewee, 17 May 2017
DI-17052017: Donor Interviewee, 17 May 2017
DI-18052017: Donor Interviewee, 18 May 2017
NI-19052017: NGO Interviewee, 19 May 2017
RI-20102017: Research Interviewee, 20 October 2017
GI-22102017: Government Interviewee, 22 October 2017
NI-01112017: NGO Interviewee, 1 November 2017
GI-06112017: Government Interviewee, 6 November 2017
NI-09112017: NGO Interviewee, 9 November 2017
NI-16112017: NGO Interviewee, 16 November 2017
NI-GI-29112017: NGO/ former Government Interviewee, 29 November 2017
DI-29112017: Donor Interviewee, 29 November 2017
GI-30112017: Government Interviewee, 30 November 2017
GI-01122017a: Government Interviewee, 1 December 2017
GI-01122017b: Government Interviewee, 1 December 2017
GI-06122017: Government Interviewee, 6 December 2017
DI-12122017: Donor Interviewee, 12 December 2017
GI-12122017: Government Interviewee, 12 December 2017

Indonesia: 26 interviews

DI-03082016: Donor Interviewee, 3 August 2016
GI-05082016: Government Interviewee, 5 August 2016
RI-10082016: Research Interviewee, 10 August 2016
CI-18082016: Consultancy Interviewee, 18 August 2016
DI-18082016: Donor Interviewee, 18 August 2016
GI-07082017: Government interviewee, 7 August 2017
GI-08082017: Government Interviewee, 8 August 2017
GI-DI-08082017: Ex-government, Donor Interviewee, 8 August 2017
NI-09082017: NGO Interviewee, 9 August 2017
DI-14082017: Donor Interviewee, 14 August 2017
DI-15082017: Donor Interviewee, 15 August 2017
GI-DI-21082017: Ex-government, Donor Interviewee, 21 August 2017
GI-BI-22082017: Ex-government, Business Interviewee, 22 August 2017
GI-BI-31082017: Ex-government, Business Interviewee, 31 August 2017
SGI-13092017: Subnational Government Interviewee, 13 September 2017
SGI-1-19092017: Subnational Government Interviewee, 19 September 2017
SGI-2-19092017: Subnational Government Interviewee, 19 September 2017
SGI-26092017: Subnational Government Interviewee, 26 September 2017
CI-GI-02102017: Consultancy, Ex-government Interviewee, 2 October 2017

CI-03102017: Consultancy Interviewee, 3 October 2017
SGI-03102017: Subnational Government Interviewee, 3 October 2017
SGI-1-04102017: Subnational Government Interviewee, 4 October 2017
SGI-2-04102017: Subnational Government Interviewee, 4 October 2017
SGI-1-05102017: Subnational Government Interviewee, 5 October 2017
SGI-2-05102017: Subnational Government Interviewee, 5 October 2017
CI-16012018: Consultancy Interviewee by telephone, 16 January 2018

Bibliography

Agung, P. et al. (2014) 'Reform or reversal: the impact of REDD+ readiness on forest governance in Indonesia', *Climate Policy*, 14(6): 748–768.

Ahmad, M. (2012) 'Nesting state-national alignment', *GCF task force annual meeting presentations*, http://www.gcftaskforce.org/documents/2012_annual_meeting_present ations/Nesting%20State-National%20Alignment/Mubariq_Ahmad_Nesting_Alig nment.pdf, accessed 1 September 2016.

Anderson, P., Firdaus, A. and Mahaningtyas, A. (2015) 'Big commitments, small results: environmental governance and climate change mitigation under Yudhoyono', In E. Aspinall, M. Mietzner and D. Tomsa (eds.) *The Yudhoyono Presidency. Indonesia's decade of stability and stagnation*, Singapore: Institute of Southeast Asian Studies, 258–278.

Anderson, Z.R. *et al.* (2016) 'Green growth rhetoric versus reality: insights from Indonesia', *Global Environmental Change*, 38: 30–40.

Anggraini, S.D., Boer, R. and Dewi, R.G. (2011) *Study on carbon governance at sub-national level in Indonesia. Case study: Jakarta province*, Institute for Global Environmental Strategies (IGES).

Ardiansyah, F., Marthen, A.A. and Amalia, N. (2015) *Forest and land-use governance in a decentralized Indonesia. A legal and policy review*, CIFOR Occasional Paper 132, Bogor, Indonesia: CIFOR.

Assunção, J., Gandour, C. and Rocha, R. (2012) 'Deforestation slowdown in the legal amazon: prices or policies?', In PUC-Rio (ed.), *CPI working paper climate policy initiative*, Rio de Janeiro, Brazil: Climate Policy Initiative/PUC-Rio.

Banerjee, O., Macpherson, A.J. and Alavalapati, J. (2009) 'Toward a policy of sustainable forest management in Brazil: a historical analysis', *The Journal of Environment & Development*, 18(2): 130–153.

Barbi, F. and da Costa Ferreira, L. (2013) 'Climate change in Brazilian cities: policy strategies and responses to global warming', *International Journal of Environmental Science and Development*, 4(1): 49–51.

Blum, J.R., Ferreiro-Rodriguez, M. and Srivastava, V. (2019) *Paths between peace and public service: a comparative analysis of public service reform trajectories in postconflict countries*, Washington: World Bank.

Bulkeley, H. and Betsill, M. (2013) 'Revisiting the urban politics of climate change', *Environmental Politics*, 22(1): 136–154.

Bünte, M. (2008) 'Dezentralisierung und Demokratie in Südostasien', *Zeitschrift für Politikwissenschaft*, 18(1): 25–50.

C40 (2012) 'Melbourne convenes C40 workshop on sustainable communities', http://www .c40.org/blog_posts/melbourne-convenes-c40-workshop-on-sustainable-communities, accessed 28 March 2016.

C40 (2013a) 'Delegates from London, Madrid and Stockholm share green building ideas at European conference', http://www.c40.org/blog_posts/delegates-from-london-mad rid-and-stockholm-share-green-building-ideas-at-european-conference, accessed 28 March 2016.

C40 (2013b) 'C40 bus rapid transit workshop', http://www.c40.org/events/c40-bus-rapid-transit-workshop, accessed 28 March 2016.

Campbell, T. and Fuhr, H. (eds.) (2004) *Leadership and innovation in subnational government: case studies from Latin America*, Washington: World Bank.

Caviglia-Harris, J. *et al.* (2016) 'Busting the Boom–Bust pattern of development in the Brazilian Amazon', *World Development*, 79: 82–96.

Chrysolite, H. et al. (2017) 'Evaluating Indonesia's progress on its climate commitments', https://www.wri.org/blog/2017/10/evaluating-indonesias-progress-its-climate-comm itments, accessed 10 October 2018 .

Climate Watch (2019) 'Data explorer. historical emissions', https://www.climatew atchdata.org/data-explorer/historical-emissions?historical-emissions-data-sources =31&historical-emissions-end_year=&historical-emissions-gases=131&historical -emissions-gwps=1&historical-emissions-regions=All%20Selected&historical-emissio ns-sectors=377&historical-emissions-start_year=&page=1, accessed 22 March 2019.

Cohen, N. (2010) Green *cities: an* A-to-Z *guide*, Los Angeles: SAGE Publications.

de Macedo, L.S.V., Setzer, J. and Rei, F. (2016) 'Transnational action fostering climate protection in the city of São Paulo and beyond', *disP - The Planning Review*, 52(2): 35–44.

Di Gregorio, M. et al. (2016) *Integrating mitigation and adaptation in climate and land use policies in Brazil: a policy document analysis*, Leeds/London: University of Leeds/CIFOR.

DKI Jakarta (n.d) 'Jakarta green building', https://greenbuilding.jakarta.go.id/index-en. html, accessed 16 September 2019.

Fatorelli, L. et al. (2015) *The REDD+ governance landscape and the challenge of coordination in Brazil*, CIFOR infobrief no. 115.

Fernandes, E. (2007) 'Implementing the urban reform agenda in Brazil', *Environment and Urbanization*, 19(1): 177–189.

Fuhr, H., Hickmann, T. and Kern, K. (2018) 'The role of cities in multi-level climate governance: local climate policies and the 1.5 °C target', *Current Opinion in Environmental Sustainability*, 30: 1–6.

Gebara, M.F. et al. (2014) 'REDD+ policy networks in Brazil: constraints and opportunities for successful policy making', *Ecology & Society*, 19(3): 53.

Gourevitch, P.A. (1978) 'The second image reversed: international sources of domestic politics', *International Organization*, 32(4): 881–911.

Gregersen, H. et al. (2004) *Forest governance in federal systems: an overview of experiences and implications for decentralization*, Interlaken workshop on decentralization in Forestry Interlaken, Switzerland: CIFOR.

Hickmann, T. and Stehle, F. (2019) 'The embeddedness of urban climate politics in multilevel governance: a case study of South Africa's major cities', *The Journal of Environment & Development*, 28(1): 54–77.

Hickmann, T., et al. (2017) 'Carbon governance arrangements and the nation-state: the reconfiguration of public authority in developing countries', *Public Administration and Development*, 37(5): 331–343.

Höhne, C. (2018) 'From "talking the talk" to "walking the walk"?: Multi-level global governance of the Anthropocene in Indonesia', In T. Hickmann et al. (eds.) *The Anthropocene debate and political science*, London: Routledge, 124–145.

Höhne, C., *et al.* (2018) 'REDD+ and the reconfiguration of public authority in the forest sector: a comparative case study of Indonesia and Brazil', In E.O. Nuesiri (ed), *Global forest governance and climate change: interrogating representation, participation, and decentralization*, London: PalgraveMacmillan, 203–241.

IBGE, Instituto Brasileiro de Geografia e Estatística (2015) 'Cidades', http://www.cidades.ibge.gov.br/xtras/home.php, accessed 12 January 2015.

Instituto de Mudanças Climáticas e Regulação de Serviços Ambientais (IMC) (2017) 'Acre orienta Mato Grosso na implementação de política de Redd+', http://imc.ac.gov.br/acre-orienta-mato-grosso-na-implementacao-de-politica-de-redd/

Johnson, C., Toly, N. and Schroeder, H. (eds.) (2015) *The urban climate challenge: rethinking the role of cities in the global climate regime*, New York: Routledge.

Jong, H.N. (2019) 'Indonesia to get first payment from Norway under $1b REDD+ scheme', *Mongabay,* 20 February 2019.

Kahn, S. and Brandão, I. (2015) *The contribution of low-carbon cities to Brazil's greenhouse gas emissions reduction goals*, Stockholm Environment Institute (SEI), U.S. Center – Seattle Office.

Kingdom of Norway and Republic of Indonesia (2010) 'Letter of intent on cooperation on reducing greenhouse gas emissions from deforestation and forest degradation', https://www.regjeringen.no/contentassets/78ef00f5b01148e2973dca203463caee/letter-of-intent-indonesia-norway.pdf, accessed 16 June 2020.

Lederer, M. (2015) 'Global governance and climate change', in Karin Bäckstrand and Eva Lövbrand (eds.) *Research handbook on climate governance*, Cheltenham: Edward Elgar, 3–13.

Lederer, M. and Höhne, C. (2019) 'Max Weber in the tropics: how global climate politics facilitates the bureaucratization of forestry in Indonesia', *Regulation & Governance*, online first. https://doi.org/10.1111/rego.12270

Levy, B. (2011) *Can Islands of effectiveness thrive in difficult governance settings? The political economy of local-level collaborative governance*, Washington: World Bank.

Liefferink, D. and Wurzel, R.K.W. (2017) 'Environmental leaders and pioneers: agents of change?', *Journal of European Public Policy*, 24(7): 651–668.

Mahendriyani, D. (2016) 'Indonesia: Jakarta signs the green building "30:30 commitment" regulation', http://www.asiagreenbuildings.com/14726/indonesia-signed-green-building-commitment/, accessed 16 September 2019.

Moeliono, T.P. (2011) *Spatial management in Indonesia: from planning to implementation. Cases from West Java and Bandung. A socio-legal study*, Leiden: University of Leiden.

NCA, National Constituent Assembly (1988) 'Constituição da República federativa do Brasil de 1988', In National Constituent Assembly (ed.), Brasilia: Government of Brazil.

Nepstadt, D. *et al.* (2014) 'Slowing Amazon deforestation through public policy and interventions in beef and soy supply chains', *Science*, 344(6188): 1118–1123.

Ostrom, E. (2010) 'Polycentric systems for coping with collective action and global environmental change', *Global Environmental Change*, 20(4): 550–557.

Rajão, R., Azevedo, A. and Stabile, M.C.C. (2012) 'Institutional subversion and deforestation: learning lessons from the system for the environmental licencing of rural properties in Mato Grosso', *Public Administration and Development*, 32(3): 229–244.

Ribot, J.C., Agrawal, A. and Larson, A.M. (2006 'Recentralizing while decentralizing: how national governments reappropriate forest resources', *World Development*, 34(11): 1864–1886.

Rio de Janeiro (2011) 'Law No. 5.248/2011 - municipal policy on climate change and sustainable development', In *Secretaria Municipal de Meio Ambiente (SMAC)*, 5.248/2011, Rio de Janeiro: City of Rio de Janeiro.

ROI, Republic of Indonesia (2011) *Presidential regulation of the republic of Indonesia No. 61 Year 2011 on the national action plan for greenhouse gas emissions reduction*, Jakarta: Government of Indonesia.

São Paulo (2009) *Lei Municipal no 14.933, de junho de 2009. Institui a Política de Mudança do Clima (PMMC) no município de São Paulo*, São Paulo: Câmara Municipal.

Stehle, F. *et al.* (2019) 'The effects of transnational municipal networks on urban climate politics in the global South', In J. van der Heijden, H. Bulkeley and C. Certomà (eds.) *Urban climate politics: agency and empowerment*, Cambridge: Cambridge University Press.

Susanti, P. (2011) 'Measurements in climate change in Jakarta', http://citynet-ap.org/wp-content/uploads/2011/12/CLIMATE_CHANGE_MEASURES-Dhaka_27_November_2011.pdf, accessed 29 March 2016.

Tollefson, J. (2009) 'Paying to save the rainforests', *Nature*, 460: 936–937.

Wibowo, A. and Giessen, L. (2015) 'Absolute and relative power gains among state agencies in forest-related land use politics: the ministry of forestry and its competitors in the REDD+ programme and the one map policy in Indonesia', *Land Use Policy*, 49: 131–141.

World Bank (2016) 'Forest area (% of Land Area)', http://data.worldbank.org/indicator/AG.LND.FRST.ZS, accessed 6 January 2016.

WRI (2016) 'CAIT climate data explorer', http://cait.wri.org/historical/Country%20GHG%20Emissions?indicator[]=Total%20GHG%20Emissions%20Excluding%20Land-Use%20Change%20and%20Forestry&indicator[]=Total%20GHG%20Emissions%20Including%20Land-Use%20Change%20and%20Forestry&year[]=2012&sortIdx=1&sortDir=desc&chartType=geo, accessed 5 April 2016.

World Resources Institute (2018) 'CAIT climate data explorer. historical emissions', http://cait.wri.org/historical/, accessed 29 January 2018.

Wurzel, R.K.W., Liefferink, D. and Torney, D. (2019) 'Pioneers, leaders and followers in multilevel and polycentric climate governance', *Environmental Politics*, 28(1): 1–21.

Part 3
United States and Europe

7 Climate change politics and policy in the United States

Forward, reverse and through the looking glass

Henrik Selin and Stacy D. VanDeveer

Introduction

How should climate change leadership be characterised when a state and its component parts are as erratic over time and as internally contradictory as those in the United States (US) over the last three decades? This chapter analyses the content of, and changes in, US foreign and domestic climate change policy from the 1990s to 2020, while highlighting distinct phases and important developments. US climate change policy has been highly inconsistent and contradictory over time, with changing directions – at home and abroad – numerous times. US states and large cities often compete for leadership status in part by repeatedly challenging their own national government rhetorically, in regulations and law making, and in court. This federalist contestation occurs among actors seeking to advances climate change laws and regulations and those opposing policies aimed at reducing greenhouse gas emissions (GHGE) or furthering climate change adaptation. US groups of citizens as well as various political and economic actors are deeply divided on climate change policies. In such an environment, climate change leadership and its impacts remain challenging to characterise and explain.

US climate change politics has evolved against a backdrop of fluxes in GHGE trends. National GHGE increased almost every year throughout the 1990s and most of the 2000s, peaking in 2007 at almost 20% higher than in 1990[1]. Preliminary estimates suggest that US GHGE by 2019 had declined slightly more that 12% since their 2007 peak (Mufson, 2020; Houser and Pitt, 2020) – this reduction falls well short of US pledges at United Nations (UN) climate change conferences in Copenhagen in 2009 and in Paris in 2015. Most GHGE reductions are due to a large-scale shift away from coal to natural gas and modest increases in renewable energy sources (Houser and Pitt, 2020). US per capita emissions are similar to those of Canada and Australia, but two or three times higher than those in many European countries (see Chapters 9–13 in this volume) and China (see Chapter 2 in this volume), and often 10–20 times higher than in many of the poorest, low emission states in the developing world.[2]

This chapter characterises US foreign and domestic climate change policy over time as a combination of moving forward, reversing course and leaping through the looking glass. The latter characterisation borrows from the title of Lewis

Carroll's sequel to *Alice's Adventures in Wonderland*; passing through the mirror into a world where everything, including logic and causation, is reversed. We show that US climate change politics and policy demonstrate that prolonged multilevel governance does not guarantee 'better' governance over time. Federal and subnational authority to make climate change and energy policy has frequently failed to translate into serious outcomes, simultaneously giving opponents opportunities to block or reverse policy initiatives designed to advance climate change action. At times, federal authorities sought to prevent subnational leadership. At subnational scales, very different emissions trajectories and political environments demonstrate that some states, cities and private sector actors exhibit leadership even as others use multilevel governance dynamics to limit and obstruct climate change policy efforts.

Federalism, multilevel governance and leadership

The dynamics of US politics and policy-making on climate change are shaped by the federal structure of the US political and legal system (Selin and VanDeveer, 2009, 2012, 2013; Rabe 2004, 2010; Scheberle 2013; Karapin 2016). Because of federalism's importance, US climate change politics and policy can be helpfully examined through the lens of multilevel governance, a perspective stressing the importance of actors and actions at and across subnational, national and supranational levels (Liefferink and Wurzel, 2017 Wurzel, Liefferink and Torney, 2019). Subnational leaders – especially in state governments – have reacted to federal level inaction or hostility to climate change policy-making by offering leadership and pioneership in their states and, at times, at national and global scales. At the same time, private sector and civil society organisations have a high degree of autonomy, as often stressed in polycentric perspectives. While comparative federalism and multilevel governance frameworks are the more common lenses through which to examine US climate change politics and policy-making, we argue that a 'polycentric turn' is emerging in US politics, as state and city governments and a host of other societal actors engage and sometimes seek to offer leadership (Jordan *et al.*, 2015; Dorsch and Flachsland, 2017; Sovavool and Van de Graaf, 2018; Jordan *et al.*, 2018).

US presidents head the executive branch and are the chief architects of US foreign policy, including on climate change. Policies requiring legal or budgetary change need Congressional action. The bicameral legislative branch, consisting of the Senate and the House Representatives, has limited supervision of executive branch foreign policy-making, and the executive branch is required to regularly inform congressional oversight committees. New treaties requiring ratification must be approved by a two-thirds majority of the Senate (e.g. 67 of 100 senators). This is necessary when new federal legislation involving Congress is needed, but US presidents have carved out a right to unilaterally accept agreements not requiring legislative change. Presidents issue executive orders on specific issues, but these must be based on existing law or constitutional authority. For example, a president can only issue a new emissions or energy efficiency standard if the legal authority to do so already exists. Any executive order can be challenged in court and/or reversed by a subsequent president. US states play important roles in implementation and enforcement of federal policy and they have direct authority over many issues impacting GHGE.

Some observers argue that states serve as important 'laboratories of federalism' – a common phrase made famous by US Supreme Court Justice Louis Brandeis in the 1930s. This idea suggests that federalism allows states to engage in policy experimentation, engendering competition to create solutions to public challenges while periodically contesting or renegotiating some areas of authority with the federal government. The hope is that jurisdictions at varying levels complement and support each other, as particular decision-making authorities are reserved for different levels. However, federalist dynamics related to US climate change politics and policy-making have been highly contentious between federal and subnational levels and among subnational units. Patterns of US climate change politics have resulted in dynamics often characterised more by 'bottom–up' than by 'top–down' policy-making within the federal structure. US federal policy largely lags behind more ambitious subnational initiatives involving states and cities – a trend Derthick (2010) described as 'compensatory federalism'. Also, some federal and state environmental and energy policy changes have been subject to litigation, sometimes all the way to the US Supreme Court.

Much climate change leadership and pioneership within the US federal system is exercised by states. In this volume, leadership is seen to involve explicit efforts to gain followers, while pioneers are defined as actors who are looking to expand their efforts to address climate change without necessarily looking to attract additional followers (Liefferink and Wurzel, 2017; Wurzel, Liefferink and Torney, 2019). Some US states have shown exemplary leadership, as they lead by example, as well as structural leadership, as they use economic power to influence markets, standards and national policy-making. Yet, subnational leadership has mostly failed to push the 'top' toward more stringent action on climate change. The development of subnational policies and programmes allowed for policy experimentation and diffusion (Rabe, 2004, 2009, 2010; Selin and VanDeveer, 2009, 2012, 2013; Hoffmann, 2011), but these struggle to compensate for the lack of comprehensive federal policy. The federal government has exhibited moments of structural leadership (using its economic power) and entrepreneurial leadership (looking for compromises and championing new policy ideas), but only rarely.

US federalist climate change politics is characterised by intense lobbying of decision-makers by organised industrial and commercial interests often opposed to climate change action (e.g. Kraft and Kamieniecki, 2007; Layzer, 2012; Kamieniecki and Kraft, 2013; Klein, 2014; Karapin, 2016; Brulle, 2018; Mildenberger, 2020; Stokes, 2020). Such lobbying – critics use terms like influence peddling – occurs in Congress in Washington DC, in state capitals all over the country, and in many other domestic fora. US fossil-fuel-based energy companies and some mainly right-wing and free-market oriented think tanks financed and disseminated climate change denial and scepticism for the purpose of influencing decision-makers, media and the public (Dunlap and McCright, 2011; Dunlap and

Jacques, 2013). US public opinion has been heavily influenced by self-identified political party affiliation, leading to substantially different views of the validity of climate change science and the need for policy-making (Dunlap, McCright and Yarosh, 2016; Funk and Hefferon, 2019). Self-identified Republican voters have been less likely to favour policy action on climate change than Democratic voters, with Independent voters often split, generally moving toward greater support for climate change policy over time.

Tangled webs: The United States and global climate change agreements

US foreign policy in the area of climate change has changed significantly over the five presidential administrations of George Herbert Walker Bush (1989–1993), Bill Clinton (1993–2001), George Walker Bush (2001–2009), Barack Obama (2009–2017) and Donald Trump (2017–) (Rabe, 2010; Downie, 2013; Kincaid and Roberts, 2013; Vig, 2013, 2019; Selin and VanDeveer, 2019). US presidents' wide latitude to articulate and pursue foreign policies extends to most areas of international climate change and energy cooperation. However, Congressional hostility to presidential promises in multilateral fora that go beyond existing US law tempers and shapes US global leadership, even when presidents are inclined to offer it. Substantial Congressional opposition has existed to all global climate change agreements since the Senate ratified the United Nations Framework Convention on Climate Change (UNFCCC) in 1992. Congress also has primary budgetary authority, meaning that it must allocate any funds committed to international climate change funds or domestic implementation efforts.

Over the past three decades, few (if any) countries displayed more erratic and contradictory approaches to international climate change cooperation than the US. The positions of five US presidential administrations shifted from moderately supportive of the UNFCCC and the 1997 Kyoto Protocol under the administrations of George H.W. Bush and Bill Clinton, to openly hostile to the Kyoto Protocol and substantive multilateral climate change cooperation under George W. Bush (Selin and VanDeveer, 2009; Downie, 2014). The arrival of the Obama administration initially brought back stronger rhetorical support for multilateral cooperation coupled with the appointment of executive branch officials with substantial climate change and energy experience (Selin and VanDeveer, 2010; VanDeveer and Selin, 2010). The Obama administration's second term (2013–2017) witnessed growing US structural and entrepreneurial leadership in international climate change policy-making (Kincaid and Roberts, 2013; Karapin, 2016). Such leadership was critical to the formulation and adoption of the 2015 Paris Agreement. The subsequent Trump administration reintroduced a combination of US disengagement and open hostility to global climate change action in multilateral fora.

Widely divergent US engagement in global climate change cooperation illustrates that US leadership aimed at moving climate change policy forward toward higher mitigation goals and improved adaptation, as well as increases in climate financing through the Green Climate Fund and other multilateral sources, is not guaranteed. US support for the UNFCCC in the early 1990s and the Paris Agreement in the mid-2010s has been interrupted by efforts to actively block or roll back efforts to curb GHGE under the international climate change agreements and help developing countries to adapt to a changing climate. While only a minority of other states protective of fossil fuel-based interests explicitly joined the international anti-climate change cooperation efforts of the George W. Bush and Donald Trump administrations, they served to 'successfully' slow down global multilateral cooperation, at least in the short term. While this volume takes the view that the 'climate leadership' concept applies only to efforts to advance some types of positive climate change policy (Wurzel, Liefferink and Torney 2019), this conceptualisation should not gloss over the fact that powerful actors regularly champion opposition to serious climate change policy development and oppose science-based assessments.

In the beginning...

The US was an early UNFCCC supporter and among the first countries to ratify it. Before and after 1992, the US offered cognitive leadership by funding a substantial share of climate change–related scientific research. As a major emitter of GHGE, the US was among those wealthier countries committing to lead in reducing emissions under the UNFCCC principle of common but differentiated responsibilities. However, the US failed to live up to this commitment over the subsequent decades. The Clinton administration signed the Kyoto Protocol under which the US had an obligation to reduce GHGE by 7% below 1990 levels by 2012. Yet, the Protocol was never submitted to the Senate for ratification. The Senate, before the Kyoto Protocol was adopted in 1997, had passed a 'Sense of the Senate' resolution with a vote of 95–0 opposing the draft treaty because excluding developing countries from GHGE reduction mandates 'could result in serious harm to the United States economy, including significant job loss, trade disadvantages, increased energy and consumer costs, or any combination thereof'? Senators voted in favour of this resolution for a variety of reasons (Mildenberger, 2020), but it signalled Congressional hostility to global climate change agreements.

President George W. Bush, shortly after taking office in January 2001, made rejection of the Kyoto Protocol official US foreign policy. This opposition lasted throughout his eight years in office. The administration took steps to undermine Kyoto Protocol implementation, but failed to prevent it from entering into force in 2005. Strong US opposition to international emissions reduction commitments in the absence of similar commitments by China (see Chapter 2 in this volume), India (see Chapter 3 in this volume) and other large developing countries had a major impact on global negotiations. These disagreements came to a head at the 2009 Copenhagen conference. That the Obama administration had by then replaced the Bush administration did little to change these dynamics. Obama administration officials insisted on more active developing country efforts to reduce emissions, and they faced strong opposition in the US Senate to any Kyoto-like treaty. That prospects for Senate ratification of any meaningful climate change agreement have remained very low since the 1990s is also related to larger dynamics of domestic political polarisation, which meant that few multilateral treaties on any issue have been ratified by the Senate since the early 1990s.

From constructive engagement and commitment, back to hostility...

The 2009 Copenhagen Accord was cobbled together at the last minute by a small number of large emitters behind closed doors, including input from President Obama. The Accord, widely criticised both for its absence of substantive commitments and the way that it was negotiated, ushered in a new approach to global climate change policy-making (Fisher, 2010, 2011; Dimitrov, 2010; VanDeveer and Selin, 2010; McGregor, 2011). It was also the beginning of renewed structural and entrepreneurial leadership by the US. The legacy of Copenhagen included the formulation of national, voluntary commitments without making a fundamental distinction between industrialised and developing countries – even as most developing countries still expect wealthier ones to lead by example. The US pledged to reduce GHGE in the range of 17% below 2005 levels by 2020, 42% below 2005 levels by 2030 and 83% below 2005 levels by 2050. That large developing countries – including China, India and Brazil – also submitted national targets was seen as positive by Obama era negotiators. This laid political groundwork for building more inter-state agreement in the run-up to the 2015 Paris conference.

The US and China exhibited shared structural and entrepreneurial leadership when they concluded an important bilateral climate change agreement in November of 2014 (see also Chapter 2 in this volume) (Gallagher and Xuan, 2019). Under this bilateral agreement between the globe's two largest GHG emitters, both countries affirmed their commitment to reach an ambitious agreement in Paris one year later that respected the principle of common but differentiated responsibilities. The US set a goal of a 26–28% GHGE reduction from 2005 levels by 2025, while China committed to achieve peak emissions by 2030. The bilateral agreement was explicitly framed as an attempt to build momentum on the road to Paris. It also served as the basis for the US' first target embedded in its National Determined Contributions (NDCs) under the Paris Agreement. The US became a Paris Agreement party in September of 2016. This was made possible by the Obama administration's classification of the Paris Agreement as an executive agreement requiring neither Senate approval nor changes to federal law. The administration determined that existing measures under the 1990 Clean Air Act combined with ongoing trends in energy markets would be sufficient to meet the US commitment.

The departure of the Obama administration and the arrival of the Trump administration resulted in a sharp reversal of official US attitudes toward the Paris Agreement and global climate change cooperation more broadly. The Obama era logic of committed structural and entrepreneurial leadership designed to make identifiable, if incremental, progress toward reducing global GHGE was quickly reversed under Trump's 'America First' banner. Given policy debates in US climate change politics since the 1990s, the switch at least partly represents continuity despite often being seen as a major disjuncture (MacNeil and Paterson, 2019). But many aspects of US climate change politics – both foreign and domestic – went through the looking glass. During his election campaign, Trump repeatedly stated his intention to withdraw the US from the Paris Agreement, calling it one of many 'bad deals' negotiated by earlier administrations. Trump administration engagement at global climate change meetings from 2017 to 2019 became notorious for advancing pro-fossil fuel rhetoric.

The Paris Agreement entered into force on 4 November 2016 with the US as a party. By early 2020, 188 countries and the EU were parties. Article 28 of the Paris Agreement stipulates that a party can begin a formal withdrawal process no earlier than three years after the treaty entered into force. The Trump administration used this option when it notified the Secretary-General of the UN on 4 November 2019 that it intended to leave the agreement. A formal notice of withdrawal becomes effective one year after notification – in this case, that means 4 November 2020. That is one day after the 2020 presidential election. US withdrawal will not spell an end to the Paris Agreement, just as George W. Bush's renunciation of the Kyoto Protocol did not prohibit its entry into force. But the withdrawal sends a strong signal to the rest of the world about US unilateralism. When the Trump administration announced its withdrawal decision, the only other UN members not party to the Paris Agreement were Angola, Eritrea, Iran, Iraq, Kyrgyzstan, Lebanon, Libya, South Sudan, Turkey and Yemen.

A future US president may try to restore US global structural leadership by bringing the country back into the Paris Agreement, which is made possible by the fact that the US is still a UNFCCC party. Article 21 of the Paris Agreement stipulates that a country can join 30 days after submitting the necessary legal documentation. This procedure is the same irrespective of whether the country joins for the first time, or if a former party rejoins after having previously withdrawn. A US president who supports the Paris Agreement would need to decide if Senate approval is necessary to rejoin, as it remains difficult to imagine any climate change agreement garnering the necessary support of 67 senators, at least in the near term. A president seeking to reverse – again – the direction and quality of US leadership will face a challenge no previous president has overcome; to achieve significant, national GHGE reductions the president must successfully work with Congress to pass a serious piece of federal climate change and energy legislation. Such legislation also must survive inevitable court challenges and be reasonably well-implemented across the states.

Leading whom, from below to where?

Subnational climate change action in the United States develops within the country's federalist structures, and thus with complex and contentious interactions with federal authorities and institutions (Rabe, 2004, 2010, 2018; Selin and VanDeveer, 2009, 2012, 2013, 2019). Congress held 175 hearings on climate change between 1975 and 2006, but passed no climate change specific legislation (Rabe, 2009). Few US states developed climate change policies during the

1990s, as many waited for the development of federal mandates and incentives. Despite strong Congressional opposition to the Kyoto Protocol, supporters of more climate change action hoped that the federal government would act domestically. The Clinton administration, however, elected not to move forward on carbon dioxide (CO_2) regulations under the Clean Air Act. In 1999, environmental non-governmental organisations petitioned the Environmental Protection Agency (EPA) to set standards for CO_2 emissions from vehicles, but the agency did not act on this request. Many environmental advocates supported then Vice President Al Gore in the 2000 presidential election, hoping he would offer more substantial climate change leadership.

On the presidential campaign trail, George W. Bush expressed support for regulating CO_2 emissions from large power plants, but he reversed this position during his first months in office in early 2001. His administration instead took a position that supported scientific research and technological development, but rejected the need for federal regulations of GHGE. As a result, much US climate change politics in the 2000s was characterised by bottom–up dynamics with a growing number of states and cities taking action in the policy vacuum left by the federal government and Congress – sometimes referred to as 'environmental federalism'. US states attempting to address climate change responded in two main ways: (1) seeking to use the federal court system to force federal-level action, and (2) moving ahead with state-level initiatives most often focused on increased energy efficiency, expansion of renewable energy production, and the formulation of state-level GHGE reductions targets. Some of these efforts bore fruit at the state level, but they did not force any consistent federal level action.

Out of the congress and into the courts

Many issues concerning environmental regulatory authority have been reviewed by US courts over the past 50 years, including more recently on climate change (Duane, 2013; O'Leary, 2019). Several states initiated legal processes against the federal government for its refusal to introduce mandatory regulations on CO_2 emissions under the Clean Air Act. In 2003, attorneys general from 12 states filed a legal suit in federal court. In March 2006, Massachusetts and 28 other parties requested that the Supreme Court review the case. The Supreme Court accepted this request and agreed to make a decision on the federal government's obligation to control CO_2 emissions from vehicles. In 2007, the Supreme Court ruled that the Clean Air Act gave the EPA the authority to set vehicles emission standards. In December 2007, Congress, with the support of the Bush administration, passed a bill increasing Corporate Average Fuel Economy (CAFE) standards for vehicles. Setting the target of 35 miles per gallon by 2020, it was the first increase in CAFE standards by Congress in over 30 years. President Bush signed these standards into law in the Energy Independence and Security Act of 2007.

Although Congress and the Bush administration introduced legislative measures that increased vehicle standards aimed at limiting CO_2 emissions, both the legislative and executive branches largely failed to regulate GHGE from power plants and other industrial point sources. The attorneys general of Massachusetts and 16 other states (together with the corporation counsel for New York City, the city solicitor of Baltimore and 13 environmental advocacy groups) responded to this lack of regulatory action by filing a Petition for Mandamus with the US Court of Appeals for the District of Columbia in 2008, requesting that the EPA be required to act on CO_2 emissions from all major sources under the Clean Air Act. In 2011, the Supreme Court blocked this lawsuit, stating that the authority to decide over the regulation of GHGE rested with the EPA and not federal judges. It was thus then up to the Obama administration and Congress to decide on any further federal level initiatives to address GHGE.

In 2009, the Obama administration EPA issued an 'endangerment finding', stating that contemporary and projected atmospheric GHG concentrations threatened the public health and welfare of current and future generations. This was an administratively and politically important procedural step to take further regulatory action under the Clean Air Act. Following this endangerment finding, the EPA began to explore options for developing additional GHGE controls through administrative and regulatory means. In parallel, the Obama administration called on Congress to pass more comprehensive climate change and energy legislation, including through the adoption of a national GHGE trading scheme. Such legislation passed the House of Representatives in 2009, but it failed to gain approval in the Senate. In fact, it was never even put to a vote in the face of strong opposition from mainly Republican senators. A subsequent decade of elections confirmed this opposition to climate change policy among a large majority of Republican members of Congress.

Litigation in courts beyond the previously mentioned rulings by the Supreme Court have long played very significant roles in US environmental politics, and climate change issues are no exception (Duane, 2013; O'Leary, 2019). Lawsuits have been filed by both supporters and opponents of climate change and energy policy changes, be these private sector actors, city and state governments or environmental advocacy organisations. By 2018, over 80 such climate change-related cases had been submitted to the courts.[3] Some lawsuits by climate change policy advocates target government agencies for a failure to address GHGE. Others focus on the private sector, including the responsibilities of large fossil fuel companies. For example, New York state sued ExxonMobil in 2018, arguing that the company had misled its investors about how regulations to address climate change may impact the company. Importantly, there have also been multiple, both successful and unsuccessful, lawsuits by firms and others looking to constrain the adoption of more stringent climate change action (Setzer and Byrnes, 2019). All of this indicates that legal battles are likely to remain common in US climate change politics, shaping policy throughout the federal system.

Busy and diverse laboratories of federalism

Behind the US national trend in GHGE there is substantial variation among the 50 states in emission levels (both total and per capita) and trends over time. Between

1990 and 2017, state-wide CO_2 emissions in Idaho increased by over 63% while Maryland's CO_2 emissions decreased by more than 26%, for example.[4] Many US states emit GHGs on par with larger and middle-sized countries. States' emission profiles are influenced by a range of factors, including varying combinations of fossil fuel-based and non-fossil sources of energy, economic growth rates, population changes, differences in renewable energy standards and environmental policies, and diverging transportation networks and needs. All of this means that individual US states face widely different situations and challenges in reducing GHGE. In addition, the threats from a changing climate such as the fuelling of wildfires, increased droughts, sea-level rise and melting glaciers play out very differently across the country, from California to Maine and from Florida to Hawaii and Alaska.

Throughout the 2000s, US states attempting to show leadership on climate change and cleaner energy launched several types of initiatives (Rabe, 2004; Selin and VanDeveer, 2009). California in particular – demonstrating structural leadership based on the size of its share of the US market and entrepreneurial and cognitive leadership by generating a plethora of policy experiments and calling for others to act – enacted GHGE reduction and renewable energy goals and set vehicle emission standards that go beyond federal mandates (Farrell and Hanemann, 2009; Houle *et al.,* 2015; Bang *et al.*, 2017; Vogel, 2018). California launched its own emissions trading scheme and led the development of a series of multi-state emissions measurement and reduction initiatives, some of which included Mexican states and Canadian provinces. The California case illustrates the difficulty in some empirical cases of attempting to make a clear distinction between climate leaders who seek to attract followers and pioneers who are not intentionally about attracting followers. California's unique position in US clean air regulation allowing it to set air pollution standards higher than federal level mandates means that, de facto, California can make policy in pursuit of its own interests and still offer a model or rule for other states to adopt (Vogel, 2018; Mazmanian *et al.*, 2019).

Individual state initiatives included the adoption of renewable portfolio standards to help drive investments into renewable energy production, energy efficiency standards, building code changes, state-level purchasing incentives and a host of other measures. Some states also introduced restrictions on CO_2 emissions from the energy sector. Individual leadership efforts varied substantially by state and region of the country, and were often supported by a series of personal networks among activists, administrators and policy-makers (Selin and VanDeveer, 2009). Ideas and information about state initiatives often proliferated and diffused across states via processes that continued into the 2010s (Bang *et al.*, 2017). Increasing polycentrism, with different states leading in several different areas of policy development, became more common over time. Since 2000, a growing set of state-level policies became normed, and states inclined to lead in some areas of climate change and energy policy follow each other's leadership, as they enact similar policies related to renewable energy, energy efficiency, carbon markets and so on. In this sense, more polycentric governance is emerging among the states.

Individual states such as California, New York, Massachusetts, Oregon and Washington emerged over the first two decades of the 21st century as the most consistent subnational climate change entrepreneurial leaders. They are among the states that enacted the broadest suit of policy initiatives, including GHGE reduction goals that rival those of international leaders in Northern Europe (Karapin, 2016; Vogel, 2018). California has repeatedly strengthened its policies and made its climate change and renewable energy goals more ambitious, driving a large expansion of renewable energy investments, mandating and incentivising a more rapid shift to low emission vehicles and regulating energy efficiency of products and buildings. Together with Germany (see Chapter 9 in this volume) and China (see Chapter 2 in this volume), California policies and consumer choices helped drive the global development of solar technologies (Mulvaney, 2019). California's policies have survived state and federal court challenges, electoral challenges in referenda, private-sector opposition and occasional condemnation by the federal government. Rare in US politics, climate change and renewable energy leadership in California, Massachusetts and New York has been offered by both Democrats and Republican officials, illustrating more broad-based support among voters and other stakeholders in these parts of the country.

Some states joined collaborative endeavours that combine structural and entrepreneurial leadership and seek to increase pressure on federal policy-makers to act. Early GHGE reduction goals were formulated by the six New England states (Vermont, New Hampshire, Maine, Massachusetts, Rhode Island and Connecticut), together with several Canadian provinces (Selin and VanDeveer, 2009). Later, New York and Massachusetts played leadership roles building the Regional Greenhouse Gas Initiative (RGGI), which began in 2009 with ten members (Maine, Vermont, New Hampshire, Massachusetts, Connecticut, Rhode Island, New York, New Jersey, Delaware and Maryland) as a regional CO_2 cap-and-trade scheme covering large power generators. As such, RGGI creators displayed important cognitive and entrepreneurial leadership in promoting and developing emissions trading schemes (Raymond, 2016). New Jersey left RGGI in 2012, but rejoined later. Virginia and Pennsylvania plan to join soon, further expanding RGGI into what were once among the most coal-depended states in the country. RGGI claims credit for helping reduce GHGE from electricity generation in participating states by 47% and raising over $3 billion in auction revenues, most of which are invested in environment and energy-related public benefits (Acadia Center, 2019).

A more recent collaboration between US states occurs through the United States Climate Alliance, founded in 2017. In 2019, the United States Climate Alliance had 25 member states, which are home to over half of the US population (see Figure 7.1).[5] The 25 members include the states with some of the most explicit and stringent climate change policies, including early leaders like California and Massachusetts, as well as other states that more recently joined the list of those attempting to work with – and catch up to – the early leaders (for example, Colorado, Illinois, Nevada and Virginia). The state members of the United States Climate Alliance pledge to implement policies toward meeting

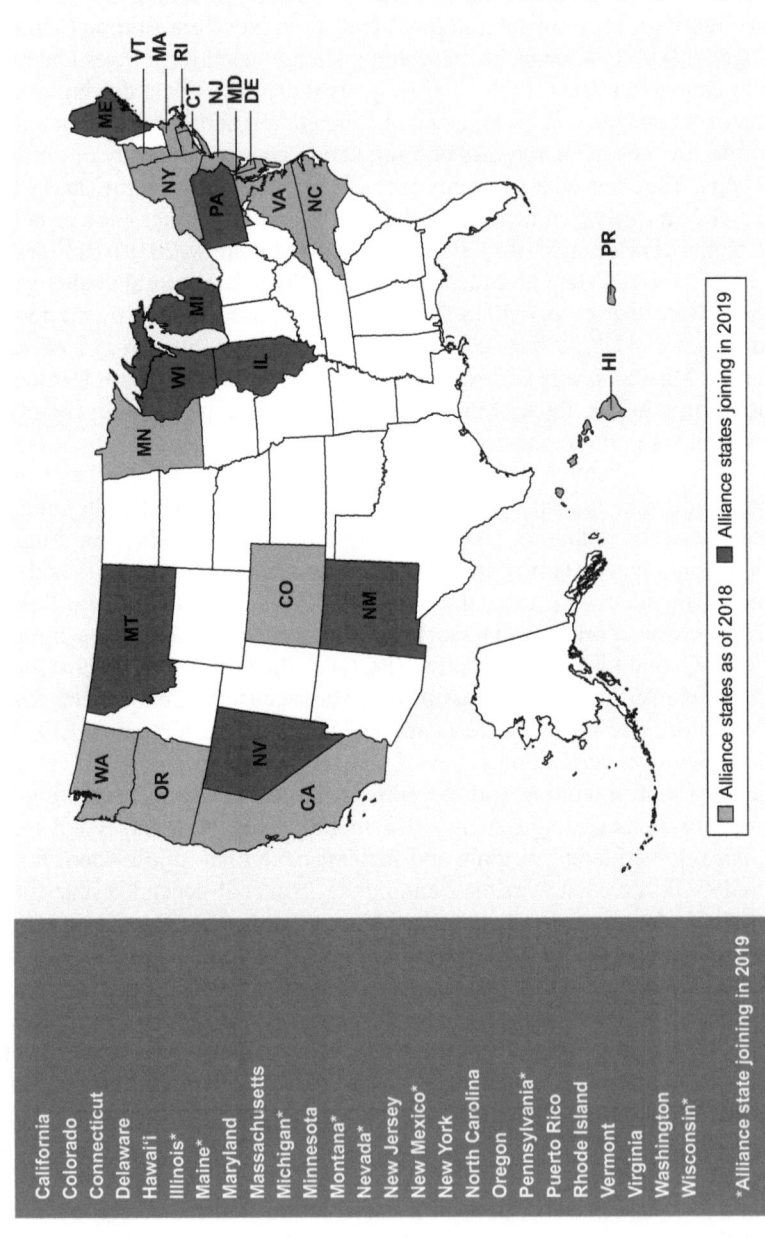

Figure 7.1 US states joining the United State Climate Alliance.

the goals of the Paris Agreement, to track and publicly report progress (including at international UNFCCC fora), to reduce GHGE and to increase clean energy deployment at state and federal levels.

Leading US states show cognitive and entrepreneurial leadership by regularly sending governors, environmental and energy department heads and members of state legislatures to high profile international climate change meetings. The United States Climate Alliance is explicit about the need for state members to do so. In this sense, the United States Climate Alliance is a manifestations of the expanding role of subnational actors in the UNFCCC process – a development very much on display at the 2015 Paris conference and in subsequent meetings and activities, and increasingly clear in its impact on policy developments and emissions outcomes around the world (Hale, 2016; Hsu *et al.*, 2019; Kuramochi *et al.*, 2020). Policy positions promoted by US states in international fora may diverge significantly from those of US presidential administrations, including the Trump administration decision to leave the Paris Agreement and reversing Obama era initiatives. The many individual and joint state actions illustrate that the Trump administration withdrawal from the Paris Agreement will not curb US subnational leadership, nor does it mean that all of the country will stop working to achieve the goals of the Paris Agreement.

Cities, firms and social movements

US states are not the only contemporary sources of climate change leadership in the country. Leaders from both large- and medium-sized urban areas – unilaterally and acting jointly – have been increasingly active since the 1990s. Many cities began by joining climate-centred networks like Local Governments for Sustainability (ICLEI) and the C40 Cities Climate Leadership Group, or pushing for more climate change action through older organisations like the US Conference of Mayors (Bulkeley and Betsill, 2003; Selin and VanDeveer, 2009; Bulkeley *et al.*, 2014; Davidson and Gleeson, 2015). The main goals of such networks have been to increase learning among cities and accelerate the diffusion of ideas and action. Examples abound from the west coast to the east coast, including Los Angeles, Seattle and Portland, to Denver, Chicago, New York, Boston and Miami. Beyond such large metropolitan areas that function as local, national and international economic centres, the trend grew to include dozens of medium-sized cities – often those home to major universities – such as Fort Collins (Colorado), Madison (Wisconsin), Austin and San Antonio (Texas), and so on.

Big US-based consumer-facing firms – as well as many American financial services and insurance firms – have also grown increasingly active around climate and energy policy and actions. Private sector leaders have tended to champion their own GHGE goals and mitigation records, and increasingly call on public officials to be more aggressive on climate change, as well. Global brands, like Apple, Microsoft and Walmart champion their renewable energy investments and their support of climate advocates, while financial service firms like CitiBank, Wells Fargo and Blackstone trumpet their sustainability records and increasingly encourage their corporate clients to become more engaged in climate change action. However, such private sector efforts have had limited impact on reducing US national GHGE and done little to change the underlying consumerist model of US society. They also had scant impact on policy actions taken by those federal policy-makers or states where fossil-fuel-based interests remain strong.

While the US has long had national and local environmental non-governmental organisations advocating for climate policy, entrepreneurial leadership has more recently been in evidence among a set of emergent social movements pushing for fossil fuel divestments and organising voter registration campaigns to help elect candidates who support more ambitious climate change policy. While contemporary social movements are not the focus of this chapter, they help to explain increasing action by some universities, state and local governments and elected officials. In short, the politically dynamic forms of leadership offered by city government, a growing set of high-profile corporations and a host of overlapping and interconnected environmental justice and climate change-themed social movements illustrate the growing polycentrism of US climate politics. Yet, these more diverse forms and sources of leadership have, to date, achieved little success in changing national climate change and energy policy.

Future leadership or catastrophic failure?

This chapter's focus on climate change leadership in the US should not obscure the continuing existence of many well-funded and politically powerful opponents of climate change action in America. The Trump administration's ongoing roll backs of environmental and energy policy standards, and its hundreds of political appointments in the executive branch and in the federal court system, are evidence that climate change policy opponents are not defeated or going away. And to date, they have paid little political price for their opposition to even minor, incremental policies to reduce GHGE. Private sector money continues to flow to candidates who oppose climate change policy and organisations that promote climate science scepticism and disinformation. The Trump administration's championing of fossil fuel firms has many local and state allies across the country. Despite high profile climate change and renewable energy leadership among some states, cities, firms and universities, many of the most powerful private sector industrial and commercial organisations remain opposed to serious federal action.

As the chapter concludes, it is useful to reflect on concepts such as federalism, multilevel governance and leadership, and their relationships to US domestic and foreign climate change policies and GHGE trajectories. On the one hand, few countries illustrate more clearly the complex multilevel nature of climate change politics than does the United States of America. Federal and subnational governmental actors have played many critical roles, both in advancing and stalling domestic action and global cooperation. Also, few other countries illustrate the ways in which global climate change governance venues, like the UNFCCC, can function as highly contested fora for divisive and high stakes domestic politics in the area of climate change. US foreign climate change and energy policies and politics are inseparable from domestic politics. It is not possible to understand one of these spheres of politics and governance without paying attention to the other. In fact, using language that suggests two separate spheres may not even be appropriate.

The US regularly seeks (or 'demands') to shape global climate change agreements to accommodate its domestic politics and institutions. The reality of US domestic and foreign climate change policies being inextricably linked and posing impediments to global cooperation is not unique. For example, it echoes years of US opposition to the 1948 Convention on the Prevention and Punishment of the Crime of Genocide after it became entangled in domestic politics around race, civil and human rights and desegregation issues. Similar domestic politisation impeded US ratification of the 1989 Convention on the Rights of the Child and the 1999 Convention on the Prohibition of the Use, Stockpiling, Production and Transfer of Anti-Personnel Mines and their Destruction, among other treaties. During the Obama administration, the US championed the voluntary NDCs approach to climate change mitigation in part because this approach avoided the need for Senate ratification of the Paris Agreement. Domestic politics shape both US global leadership and engagement, and the resulting institutions constructed in global negotiations.

Several multilevel governance tensions are in evidence in the US case. First, the hierarchical nature of federal structures shapes political and leadership opportunities in particular ways. Climate change leadership by US states, cities and firms can often take advantage of the lack of federal action, even as their progress can be constrained by national institutions. Second, what leadership means and its general direction or goals can change substantially with presidential administrations or congressional majorities. The US case clearly demonstrates that multilevel governance need not engender only competition for leader status. Opponents of climate change action compete for influence too and they use their resources to enhance division and contestation. Third, polycentricism is increasing over time, partly in reaction to the failure of climate advocates to overcome powerful, national-level opponents in federal-level policy-making. Lastly, after three decades of climate change politics, leadership- engendered innovation and competition have not yet led to serious national-level climate change action or substantial national GHGE reductions.

Almost 30 years after the UNFCCC was adopted by the US and other countries at the 1992 Rio 'Earth Summit', US domestic and foreign policies related to climate change and energy slipped through the looking glass during the Trump administration. As climate change impacts accelerate and some climate change leaders push more ambitious mitigation and adaptation initiatives– at national and subnational scales, and across public, private and civil society spheres – many national and foreign policies related to climate change mitigation and adaptation are being walked back or abandoned. Ongoing efforts include encouraging coal extraction and burning, reducing automobile fuel efficiency standards, rolling back methane emissions standards, the expansion of oil and gas extraction, and attempting to use funding authorisations and legal challenges to constrain state-level leadership in places like California. It remains unclear how far and fast global climate change governance can develop under the leadership of others until the US wakes up and returns from the illogical and fantastical world into which it has fallen.

Notes

1 See EPA https://www.epa.gov/ghgemissions/inventory-us-greenhouse-gas-emissions -and-sinks-fast-facts
2 See World Resources Institute's CAIT 2.0 data at https://cait2.wri.org
3 See https://www.thenewatlantis.com/publications/will-climate-change-the-courts
4 For US state emissions data, see https://www.eia.gov/environment/emissions/state/
5 See http://www.usclimatealliance.org/

Bibliography

Acadia Center (2019) 'The regional Greenhouse gas initiative: 10 years in review', https ://acadiacenter.org/wp-content/uploads/2019/09/Acadia-Center_RGGI_10-Years-in-R eview:2019-09-17.pdf

Bang, G., Victor, D.G. and Andresen, S. (2017) 'California's cap-and-trade system: diffusion and lessons', *Global Environmental Politics*, 17(3): 12–30.

Brulle, R.J. (2018) 'The climate lobby: a sectoral analysis of lobbying spending on climate change in the USA, 200–2016', *Climatic Change*, 149: 289–303.

Bulkeley, H. and Betsill, M. (2003) *Cities and climate change*, London: Routledge.

Bulkeley, H. et al. (2014) *Transnational climate change governance*, Cambridge: Cambridge University Press.

Davidson, K. and Gleeson, B. (2015) 'Interrogating urban climate leadership: toward a political ecology of the C40 network', *Global Environmental Politics*, 15(4): 21–38.

Derthick, M. (2010) 'Compensatory federalism', In B. Rabe, (ed.) *Greenhouse governance: addressing climate change in America*, Washington, DC: Brookings Institution Press, 58–72.

Dimitrov, R.S. (2010) 'Inside Copenhagen: the state of climate governance', *Global Environmental Politics*, 10(2): 18–24.

Dorsch, M.J. and Flachsland, C. (2017) 'A polycentric approach to global climate governance', *Global Environmental Politics*, 17(2): 45–64.

Downie, C. (2013) 'Three ways to understand state actors in international negotiations: climate change in the clinton years (1993–2000)', *Global Environmental Politics*, 13(4): 22–41.

Downie, C. (2014) *The politics of climate change negotiations*, Cheltenham: Edward Elgar.

Downie, C. (2017) 'Fighting for king coal's crown: business actors in the US coal and utility industries', *Global Environmental Politics*, 17(1): 21–39.

Duane, T.P. (2013) 'Courts, legal analysis and environmental policy', In S. Kamienieck and M.E. Kraft (eds.) *The Oxford handbook of U.S. environmental policy*, Oxford: Oxford University Press, 259–279.

Dunlap, R. and Jacques, P. (2013) 'Climate change denial books and conservative think tanks: exploring the connection', *American Behavioral Scientist*, 57(6): 699–731.

Dunlap, R. and McCright, A. (2011) 'Organized climate change denial', In J. Dryzek, R. Norgaard and D. Schosberg (eds.) *The Oxford handbook of climate change and society*, Oxford: Oxford University Press, 144–160.

Dunlap, R., McCright, A. and Yarosh, J. (2016) 'The political divide on climate change: Partisan polarization widens in the US', *Environment: Science and Policy for Sustainable Development*, 58(5): 4–23.

Farrell, A.E. and Hanemann, W.M. (2009) 'Field notes on the political economy of California climate policy', In H. Selin and S.D. VanDeveer (eds.) *Changing climates in North American politics: institutions, policy making and multilevel governance*, Cambridge, MA: MIT Press, 111–135.

Fisher, D.R. (2010) 'COP-15 in Copenhagen: how the merging of movements left civil society out in the cold', *Global Environmental Politics*, 10(2): 11–17.

Fisher, D.R. (2011) 'The limits of civil society's participation and influence at COP 15', *Global Environmental Politics*, 11(1): 8–11.

Funk, C. and Hefferon, M. (2019) *US public views on climate and energy*. Washington DC: *Pew Center Research*.

Gallagher, K. and Xuan, X. (2019) *Titan of the climate: explaining the policy process in the US and China*, Cambridge, MA: MIT Press.

Hale, T. (2016) 'All hands on deck: the Paris agreement and nonstate climate action', *Global Environmental Politics*, 16(3): 12–22.

Harrison, K. (2010) 'The United States as outlier: economic and institutional challenges to US climate policy', In K. Harrison and L.M. Sundstrom (eds), *Global commons, domestic decisions*, Cambridge, MA: MIT Press, 67–104.

Hoffmann, M. (2011) *Climate governance at the crossroads*, Oxford: Oxford University Press.

Hermville, L. and Sanderink, L. (2017) 'Make Fossil Fuels great again? The Paris agreement, Trump and the US Fossil Fuel industry', *Global Environmental Politics*, 19(4): 45–62.

Houle, D., Lachapelle, E. and Purdon, M. (2015) 'Comparative politics of sub-federal cap-and-trade: implementing the western climate initiative', *Global Environmental Politics*, 15(3): 49–73.

Houser, T. and Pitt, H. (2020) *Preliminary US emissions estimates for 2019*, Rhodium Group, https://rhg.com/research/preliminary-us-emissions-2019/

Hsu, A. *et al.* (2019) 'A research roadmap for quantifying non-state and subnational climate mitigation action', *Nature Climate Change*, 9: 11–17.

Jordan, A. *et al.* (2015) 'Emergence of polycentric climate governance and its future prospects', *Nature Climate Change*, 5(11): 977–982.

Jordan, A. *et al.* (eds.) (2018) *Governing climate change: polycentricity in action?*, Cambridge: Cambridge University Press.

Kamieniecki, S. and Kraft, M.E. (2013) *The Oxford handbook of U.S. environmental policy*, Oxford: Oxford University Press.

Karapin, R. (2016) *Political opportunities for climate policy: California, New York and the federal government*, Cambridge: Cambridge University Press.

Kindcaid, G. and Roberts, J.T. (2013) 'No talk, some walk: Obama administration first term rhetoric on climate change and US international climate budget commitments', *Global Environmental Politics*, 13(4): 41–60.

Klein, N. (2014) *This changes everything: capitalism vs. the climate*, New York: Simon & Schuster Paperbacks.

Kraft, M.E. and Kamieniecki, S. (eds.) (2007) *Business and environmental policy: corporate interests in the American political system*, Cambridge, MA: MIT Press.

Kuramochi, T. *et al.* (2020) 'Beyond national climate action: the impact of region, city, and business commitments on global Greenhouse gas emissions', *Climate Policy*, 20(3): 275–291.

Layzer, J. (2012) *Open for business: conservatives' opposition to environmental regulation*, Cambridge, MA: MIT Press.

Liefferin, D. and Wurzel, R.K.W. (2017) 'Environmental leaders and pioneers: agents of change?', *Journal of European Public Policy*, 24(7): 651–668.

MacNeil, R. and Paterson, M. (2019) 'Trump, US politics and the evolving pattern of global climate governance', *Global Change, Peace and Security*, 32(1): 1–18.

Mazmanian, D.A., Jurewitz, J.L. and Nelson, H.T. (2019) 'State leadership in US climate and energy policy: the California experience', *Journal of Environment and Development*, 29(1): 51–74.

McGregor, I.M. (2011) 'Disenfranchisement of countries and civil society at cop 15 in Copenhagen', *Global Environmental Politics*, 11(1): 1–7.

Mildenberger, M. (2020) *Carbon captured: how business & labor control climate politics*, Cambridge, MA: MIT Press.

Mufson, S. (2020) 'U.S. Greenhouse gas emissions fell slightly in 2019', *Washington Post*, https://www.washingtonpost.com/climate-environment/us-greenhouse-gas-emissions-fell-slightly-in-2019/2020/01/06/568f0a82-309e-11ea-a053-dc6d944ba776_story.html

Mulvaney, D. (2019) *Solar power: innovation, sustainability and environmental justice*, Oakland, CA: University of California Press.

O'Leary, R. (2019) 'Environmental policy and the courts', In V. Norman and M.E. Kraft (eds.) *Environmental policy: new directions for the twenty-first century*, London: SAGE/CQ Press, 144–167.

Popovich, N., Albeck-Ripka, L. and Pierre-Louis, K. (2019) '95 environmental rules rolled back under Trump', *New York Times*, https://www.nytimes.com/interactive/2019/climate/trump-environment-rollbacks.html

Rabe, B. (2004) *Statehouse and Greenhouse*, Washington, DC: Brookings Institution Press.

Rabe, B. (2009) 'Second generation climate policies in the states: proliferation, diffusion and regionalization', In H. Selin and S.D. VanDeveer (eds.) *Changing climates in North American politics*, Cambridge, MA: MIT Press, 67–86.

Rabe, B. (ed.) (2010) *Greenhouse governance: addressing climate change in America*, Washington, DC: Brookings Institution Press.

Rabe, B. (2018) *Can we tax carbon?* Cambridge, MA: MIT Press.

Raymund, L. (2016) *Reclaiming the atmospheric commons*, Cambridge, MA: MIT Press.

Scheberle, D. (2013) 'Environmental federalism and the role of state and local governments', In S. Kamienieck and M.E. Kraft (eds.) *The Oxford handbook of U.S. environmental policy*, Oxford: Oxford University Press, 394–412.

Selin, H. and VanDeveer, S.D. (eds.) (2009) *Changing climates in North American politics: institutions, policy making and multilevel governance*, Cambridge, MA: MIT Press.

Selin, H. and VanDeveer, S.D. (2010) 'Multilateral governance and transatlantic climate politics', In B. Rabe (ed.) *Greenhouse governance: addressing climate change in America*, Washington, DC: Brookings Institution Press, 336–352.

Selin, H. and VanDeveer, S.D. (2012) 'Federalism, multilevel governance and climate change action across the Atlantic', In P. Steinberg and S.D. VanDeveer (eds.) *Comparative environmental politics*, Cambridge, MA: MIT Press, 341–368.

Selin, H. and VanDeveer, S.D. (2013) 'U.S. Climate change politics: federalism & complexity', In S. Kamieniecki and M.E. Kraft (eds.) *The Oxford handbook of U.S. environmental policy*, Oxford: Oxford University Press, 164–183.

Selin, H. and VanDeveer, S.D. (2019) 'Global climate change governance: where to go after Paris?', In N. Vig and M.E. Kraft (eds.) *Environmental policy*, 10 edition, London: SAGE/CQ Press, 322–346.

Setzer, J. and Byrnes, R. (2019) *Global trends in climate change litigation: 2019 snapshot*. London: Grantham Research Institute on Climate Change and Center for Climate Change Economics and Policy, London School of Economics and Political Science.

Sovacool, B. and Van de Graaf, T. (2018) 'Building or stumbling blocks: assessing performance of polycentric energy and climate governance networks', *Energy Policy*, 118: 317–324.

Stokes, L. (2020) *Short circuiting policy: interest groups and the battle over clean energy and climate policy in the American States*, Oxford: Oxford University Press.

Tobin, P. (2017) 'Leaders and laggards: climate policy ambitions in developed states', *Global Environmental Politics*, 17(4): 28–47.

VanDeveer, S.D. and Henrik, S. (2010) 'Re-engaging international climate governance: challenges and opportunities for the United States', In B. Rabe (ed.) *Greenhouse governance: addressing climate change in America*, Washington, DC: Brookings Institution Press, 313–335.

Vig, N.J. (2013) 'The American presidency and environmental policy', In S. Kamienieck and M.E. Kraft (eds.) *The Oxford handbook of U.S. environmental policy*, Oxford: Oxford University Press, 306–328.

Vig, N.J. (2019) 'Presidential powers and environmental policy', In N. Vig and M.E. Kraft (eds.) *Environmental policy: new directions for the twenty-first century*, London: SAGE/CQ Press, 88–116.

Vig, N.J. and Kraft, M.E. (eds.) (2019) *Environmental policy: new directions for the twenty-first century*, London: SAGE/CQ Press.

Vogel, D. (2018) *California greening: how the golden state became an environmental leader*, Princeton, NJ: Princeton University Press.

Wurzel, R.K.W., Liefferink, D. and Torney, D. (2019) 'Pioneers, leaders and followers in multilateral and polycentric governance', *Environmental Politics*, 28(1): 1–21.

8 European Union leadership before, during and after the Paris Conference of the Parties

Paul Tobin[1] and Nicole M. Schmidt

Megaconferences can be lightning rods for climate action. The Paris Conference of the Parties (COP) in December 2015 was hailed as a diplomatic success (Kinley, 2017), although the predicted emissions reductions resulting from the Agreement are expected to produce global temperature increases above 2°C (Buxton, 2016). Prior to the 2015 Paris conference, which constituted the 21st Conference of the Parties (COP21) to the United Nations Framework Convention on Climate Change (UNFCCC), Parker, Karlsson and Hjerpe (2015) had found there to be increasingly fragmented leadership within the UNFCCC, representing a diminishing leadership status for the European Union (EU). The submission by each state of a target (Nationally Determined Contribution – NDC) that is then evaluated by non-state actors, amongst others (van Asselt, 2016), hints at an increasingly polycentric style within the UNFCCC[2]. Although the EU submitted a single NDC in the run-up to the 2015 Paris summit on behalf of all of its then 28 Member States, this apparently top–down policy approach masked an increasingly 'polycentric' (Ostrom, 2010) approach within the EU (for earlier analysis, see Rayner and Jordan, 2013).

This chapter employs the leadership/pioneership/followership analytical framework pursued throughout this book (see also Chapter 1 in this volume). The EU represents an important actor when examining leadership in a context of polycentricity (Torney, 2019), and of course, multilevel governance (MLG) (Marks and Hooghe, 2004; Stephenson, 2013). Whereas MLG emphasises the role of the state, and has been widely used to analyse the EU, polycentricity focuses more on societal actors (Wurzel, Liefferink and Torney, 2019: 2–3). Much has been said about MLG and the EU regarding climate governance (Jänicke and Wurzel, 2019); indeed, MLG was developed as a means of providing greater clarity for understanding the EU. Here, though, we particularly highlight more polycentric modes of governance, due to the increased emphasis placed on non-state actors in the run-up to the Paris conference. Tosun (2018) identified climate policy diffusion within the EU as helping to foster greater polycentric practices. How then can we categorise and explain the EU's polycentric leadership performance before, during and after Paris?

Through its hybrid supranational-intergovernmental nature, the EU provides multiple examples of leadership, pioneership and followership at any one time;

indeed, its multilevel structure enables the ratcheting up of standards (Schreurs and Tiberghien, 2007). As a result, analysing its leadership performance is inherently complex; the EU should not be considered as a monolithic actor, and yet assessing its performance using ideal-typical leadership models requires a degree of doing just that. According to Liefferink and Wurzel's (2017) distinction, leaders seek to attract followers, while pioneers seek greater ambition for internal reasons. As such, in this chapter we seek to highlight examples of leadership and followership within the EU, as well as assessing the EU's collective performance on the global stage. However, doing so required difficult decisions to be made regarding which key actors to analyse. Furthermore, a particular challenge when examining a multi-actor organisation such as the EU is determining the objectives underpinning behaviour, in order to assess whether followers were actively sought or not. In other words, was the EU a leader, pioneer or neither? On the whole, we align with the stance taken by Torney (2019: 176) that the EU was closer to a leader than a pioneer around the time of the 2015 Paris COP21, although its level of ambition and influence were lower than its climate leadership heyday in the mid-2000s.

At COP21, the EU most closely reflected the ideal type known as 'substantive leadership', whereby it demonstrated both high internal and external environmental ambitions (Wurzel, Liefferink and Torney, 2019: 7). 'Substantive leadership' is divided by Wurzel, Liefferink and Torney (2019) into two categories: around the 2015 Paris conference, the EU most closely resembled a 'conditional pusher' in contrast to the more 'constructive pusher' tendencies that it demonstrated during the mid-2000s. That is, the EU created goals with the expectation that these were conditional on other actors matching their ambition (conditional pusher), rather than assuming a leadership position regardless of the actions of others (constructive pusher; see also Chapter 1 in this volume). The imperative of conditionality was arguably strengthened by the 'conglomerate of crises' (Falkner, 2016) faced by the EU at the time. Regarding the style of conditional pusher leadership exhibited by the EU in Paris, we find examples of structural, entrepreneurial and intentional exemplary. This combination of characteristics is not a surprise; Wurzel, Liefferink and Torney (2019: 11) note that leaders often combine examples of different leadership types and styles.

Importantly, the EU's ability to exert structural leadership on climate policy – i.e. hard power that in this context may be shaped by 'a state's contribution to a particular environmental problem' (Liefferink and Wurzel, 2017: 957) – is diminishing in part because the EU has already reduced its greenhouse gas emissions significantly. As such, the EU is no longer seen as a policy obstacle to pursuing climate mitigation, thus shrinking its structural leadership potential at international climate conferences in contrast to, say, the India or USA (see also Chapters 3 and 7 in this volume). Therefore, climate policy represents a fascinating case for examining structural power: the more an actor reduces its emissions, the less influential in the global arena it becomes. This political reality is, however, in stark contrast to the logic behind NDCs and their five-year ambition mechanism cycle, which is designed to incentivise countries to reduce emissions over the years. Yet, it may prove to be the case in the years to come that as structural leadership diminishes in line with emissions reductions, capacity for cognitive leadership increases, as actors are able to share examples of best practice. At Paris, though, and in line with the expectation that shifts in cognitive leadership must be viewed over a long timeframe, there were few examples of the EU demonstrating cognitive leadership by going beyond the conceptual and ideological status quo in Paris, as we explore below.

This chapter proceeds as follows. First, we examine the existing literature on the EU's tradition of climate leadership prior to the 2009 COP15, the impact of crises and the facets of polycentric and multilevel governance most applicable to the EU. Second, we analyse and compare the EU's 2015 climate target against every other country's target (see Tobin *et al.*, 2018). Third, we critically assess the key actors within the EU that shaped its leadership status, touching briefly on Environmental Non-Governmental Organisations (ENGOs), before looking at the European Parliament, Commission and then Council, where we emphasise the roles played by Member States France, Germany, Sweden and Poland. Next, we touch upon the EU's leadership after Paris. Finally, we discuss our findings and then conclude with a summary that highlights the COVID-19 pandemic as a further crisis to which the EU must respond.

Existing literature

The EU's pre-Paris climate credentials

Having opposed emissions trading at the 1997 Kyoto climate conference (COP3), the EU's position pivoted starkly during the early-2000s, by introducing its own Emissions Trading System (ETS) covering around 40% of the EU's total CO_2 emissions (Damro and Méndez, 2003). Although the ETS struggled to facilitate significant emissions reductions due to low carbon permit prices, it is still one of the EU's most significant climate policy innovations, and thus an example of cognitive leadership. Indeed, during the mid-2000s, the EU emphasised its global climate credentials as a means of cultivating a green reputation that underpinned its international identity (Lenschow and Sprungk, 2010). For example, despite the EU playing a relatively minor role in shaping the Kyoto negotiations compared to the USA (Oberthür and Roche Kelly, 2008: 36), the entry of the Kyoto Protocol into force in 2005 represented a diplomatic victory for the EU (Vogler and Bretherton, 2006). Furthermore, in the run-up to the 2009 Copenhagen COP15, the EU Commission prioritised climate change through its ambitious 20-20-20 initiative: a 20% reduction on emissions and 20% renewables production by 2020 (see Parker and Karlsson, 2010). Yet, even then, the EU was noted for viewing its policies through the lens of competitiveness and economic liberalisation (see Rayner and Jordan, 2013: 79), rather than placing climate protection at the apex of its decision-making, as would be expected of an ideal-typical constructive, rather than conditional, pusher (Wurzel, Liefferink and Torney, 2019). Notwithstanding this conditionality, the EU was seen as coherent and credible leader by other states at the December 2008 COP in Poznan, Poland (Kilian and Elgström, 2010).

However, by the close of the 2009 Copenhagen COP, the picture had changed dramatically for the EU. Whilst the EU tried to act as a constructive pusher with its ambitious 20-20-20 climate and energy package, the conference had seen the EU marginalised by the structural power held by the USA and China at the negotiations, alongside the looming threat posed by the burgeoning economic crisis. This summit was a wake-up call for the EU to realign and reinterpret its international climate diplomacy role, subsequently growing more akin to a leader-mediator or 'leadiator' (Bäckstrand and Elgström, 2013). The conference also highlighted – due to the growing division between older and new Member States, and the onset of the economic crisis – the start of a challenging new policy context for the EU (Skovgaard, 2014).

Indeed, the economic crisis produced a stasis effect on the EU's environmental policy, including instances of policy development but also policy dismantling (Burns, Eckersley and Tobin, 2019; Burns and Tobin, 2020; Burns, Tobin and Sewerin, 2019; Steinebach and Knill, 2017). Gravey and Moore (2019), though, find that between 2009 and 2013, climate and energy policy intensity caught up with the level of ambition already present in environment-industry and air quality legislation. Similarly, Slominski (2016: 352) posits that 'the crisis has not fundamentally changed the trajectory of EU energy and climate policy'. Indeed, for the 2014–2020 period, 20% of the EU budget was to be spent on climate-related activities, a target the EU has not always achieved (European Commission, 2018a). As such, the primary impact of the crises prior to the 2015 Paris conference was greater fragmentation of the stances taken by the EU's Member States (Skovgaard, 2014), rather than altering the trajectory of policy output ambition straightaway. Yet, this fragmentation may have stymied the degree of ambition that the EU may otherwise have assumed internationally. Groen, Niemann and Oberthür (2012) identify a diminution of ambition between the 2009 and 2010 COPs, and Parker, Karlsson and Hjerpe (2015) note a reduction amongst actors in the perception that the EU represented a leader between the 2008 and 2011 COPs, ceding ground to the USA, China and G77 (see Chapter 7 on the USA and Chapter 2 on China). As the UNFCCC's flagship conference was once again on European soil, the 2015 Paris COP offered a means for the EU to assume a new leadership style, in a more polycentric policy-making context.

Polycentricity

It is valuable to highlight several foundations of polycentricity that are particularly relevant when analysing the EU or the UNFCCC (see also Chapter 1 in this volume). First, regarding innovation, the multiple levels of governance within the structure of the EU may be expected to provide more opportunities for policy experimentation (Abbott, 2012). However, simultaneously, the more polycentric a system of governance, the more likely that constituent sub-units act in an incoherent manner with one another (Rayner and Jordan, 2013: 80). Later, when we analyse leaders and followers within the EU, we demonstrate the wide-ranging plurality of voices competing to shape the EU's overall climate stance (e.g. France and Poland). Second, trust is a cornerstone of effective polycentric practice (see Jordan *et al.*, 2018), and yet, regarding the UNFCCC, Cole (2015: 115) notes that highly structured mega-conferences involving thousands of people are unlikely to facilitate trust-building or cooperation. While negotiations primarily take place in more close-knit settings, the desire for broad-based support from many states means that the bulk of diplomacy occurs before or after COPs, rather than at the conferences themselves. Third, orchestration – which is a 'light coordination mechanism' (Pattberg, 2010) and is particularly relevant to the EU as well as the UNFCCC – is a key feature of polycentric governance (Abbott, 2012). Here, we can see the possible utility of the UNFCCC system as a means of coordinating and steering action via similar guidelines and timetables, without necessitating that every actor takes the same stance according to their needs and political contexts. As such, the UNFCCC system is just one facet of a complex network of interaction; indeed, if the UNFCCC were the sole locus of climate governance, 'the extent of learning, and prospects for improving policies, would be quite narrow' (Cole, 2015: 115). We emphasise these features here in order to add greater analytical clarity to the multi-tiered analysis we offer, which otherwise would have focused on state actors alone, as is standard in MLG studies.

The EU's NDC in a global context

The request that every state submit an NDC prior to the 2015 Paris COP represented an important change to the landscape of international climate governance, from a more monocentric structure to a more polycentric model. This shift reflected a departure from the 'Common But Differentiated Responsibilities' principle that since 1992 had placed the obligation for mitigation on developed states, such that every country, no matter its economic status, produced an NDC for its individual climate efforts. Unsurprisingly, the NDCs varied markedly in composition, ranging from Chile's brochure with full-page photographs, to Pakistan's straight-to-the-point single-sided page. Previously, we (alongside Jale Tosun and Charlotte Burns – see Tobin *et al.*, 2018) mapped all 162 NDCs according to the formats of their mitigation targets, the inclusion of adaptation commitments, and considerations of gender issues. We identified six types of mitigation targets, such as an 'absolute reduction compared to baseline year' (e.g. Australia, Japan, Tuvalu and 35 others), an 'explicit emissions target based on Business as Usual (BAU)' (e.g. Algeria, Paraguay, Saint Lucia and 81 others) and 'no explicit emissions reduction target' (e.g. Kuwait, Nepal, Somalia and 27 others). It is impossible to label any one of these formats more ambitious than another, as a highly ambitious BAU target may produce greater emissions reductions than a weak absolute target based on a baseline year, and vice versa. However, targets structured around a baseline year that seek to reduce emissions in absolute terms should reduce emissions below existing levels, which may not be the case for, say, BAU targets that only seek to flatten the trajectory of future emissions *growth*.

The EU submitted an absolute emissions reduction target of 40% by 2030, compared to a 1990 baseline, which is broadly ambitious, in relative if not absolute terms. However, the choice of any baseline is always a political decision. In the EU's case, 1990 is a 'useful' baseline as it benefits from the emissions reductions resulting from the UK's 'dash for gas' away from coal, and from Germany's reunification (which resulted in the closure of many uncompetitive, high energy-intensive industries in the former East Germany (see Schreurs and Tiberghien, 2010: 47–50; see also Chapter 9 and 10 in this volume). As such, any claims regarding the ambitiousness of the EU's target must be viewed with that context in mind. Yet, when we compare this baseline with the baselines of other developed states with targets for 2030, such as Australia (a 26–28% reduction on 2005 levels), New Zealand (a 30% reduction on 2005 levels) (see also Chapter 5 in this volume) or Japan (a 26% reduction on 2013 levels), we can see that the EU's target can still be understood to be relatively ambitious. Australia, New Zealand and Japan each increased their annual emissions between 1990 and 2005/2013, and so, their reduction targets are easier to achieve having selected more recent baselines than 1990.

The EU's NDC, at only five pages long, is concise compared to, for example, China's 36-page document, but follows exactly the guidance on providing quantifiable information agreed under the 2014 Lima Call for Climate Action. The EU's goal is to be achieved jointly by all 28 Member States, including the UK, whose 2016 'Brexit' referendum had yet to take place but led to the UK leaving the EU on 31 January 2020). The Member States are named as the Parties to the commitment, reflecting the assumption underpinning MLG studies that the state is the primary locus for action. Yet, we should not neglect the multiple sectors that are included within the document, such as energy, manufacturing industries and so on, which from a polycentric perspective may – or may not – be encouraged to take a lead on achieving emissions reductions within and between the EU Member States. The EU's pledge contains all six greenhouse gases prioritised by the UNFCCC, including those with high warming potential, such as sulphur hexafluoride. Not all states (particularly economically developing states) included such gases in their NDCs[3]. An example of intentional exemplary leadership is shown by the EU's commitment to achieve all of its reductions domestically, without international market-based credits. In contrast, the USA merely states in its NDC that it does 'not intend to utilise international market mechanisms to implement its 2025 target' (USA, 2015); but the USA subsequently withdrew its commitment to this pledge (Rhodes, 2017; see also Chapter 7 in this volume).

However, the ambition of the EU's NDC could have been greater in two areas. First, the integration of land use into climate policy has been a complex challenge (Schmidt, 2019), and the EU did not commit to how land use would be considered as part of its goal. Although in May 2018, the EU agreed on a new Regulation on the topic, at the time of COP21, the lack of clarity in the EU's NDC sparked frustration from the ENGOs. Second, the EU target makes no reference to adaptation, even though the Paris Agreement put adaptation on an equal footing with mitigation (Fleig, Schmidt and Tosun, 2017). The EU's omission is partially explained by the multi-state nature of the EU's pledge, which would make adaptation commitments challenging, and because the EU has fewer powers for climate adaptation compared with climate mitigation. Also, as a collection of *economically developed* states, the EU's responsibilities at the global level are focused more around climate mitigation. Regardless, the EU is developing adaptation measures, and its 2013 Adaptation Strategy is evaluated every year, and every Member State must submit a document tracking its progress. Despite these omissions, the EU's target was relatively ambitious; how, then, can we characterise and explain the EU's leadership performance in Paris?

Analysing the EU's leadership performance

This section is divided into two parts: the first section analyses the types of leadership shown by the EU in Paris; the second section then seeks to determine the key actors that were responsible for shaping the EU's performance.

The types of leadership shown by the EU in Paris

The EU demonstrated multiple types of leadership in the run-up to, and during, the Paris COP. In 2014, it agreed its new 2030 Climate and Energy Framework, strengthening claims that it was taking the conference seriously and consolidating its structural leadership potential. Furthermore, the EU's structural leadership was underlined by the *followership* demonstrated by several non-EU European states, such as Iceland and Norway, which committed to achieving their NDCs via collective delivery with the EU. Additionally, the EU played a vital entrepreneurial leadership role by submitting its NDC early, second only to Switzerland (see also Chapter 13 in this volume). Specifically, the EU demonstrated entrepreneurial leadership by urging 'all other Parties, in particular major economies, to communicate their NDCs by the end of March 2015' (Latvia, 2015: 5). This timeline was, though, only met by Norway, Mexico and the USA, suggesting the EU's ability to attract followers may not be as strong as it would like. The early submission of the NDC highlighted the EU's determination to at least attempt to be a leader in Paris though, and enabled the EU to demonstrate intentional exemplary leadership by fulfilling the Lima Call for Climate Action's guidelines on the structure of NDCs, creating pressure on other states to do likewise.

A further example of the EU's entrepreneurial leadership was its role in co-creating the High Ambition Coalition. As Brandi (2018: 223) states, the EU sought to build relationships in particular with small island states and least economically developed countries, in order to create a diverse yet united collection of states seeking to drive up ambition, in marked contrast with the failure to exhibit such entrepreneurial leadership at the 2009 Copenhagen COP. This coalition was ultimately joined by major emitters, such as Brazil and the USA, highlighting the effectiveness of coalition-building as a diplomatic tool. The coalition demonstrated cognitive leadership by insisting on limiting temperature increases to

1.5°C, partially as a negotiating strategy for making the more-often cited target of 2°C more palatable.

Leaders and followers within Europe

Considering the breadth of actors comprising the networked governance model that underpins the EU-27's almost 450 million citizens, it is impossible to fully analyse all of the leaders and followers within Europe. Here, we touch briefly upon the role of ENGOs due to our focus on polycentry, before analysing the roles of the European Parliament, Commission and Council, and key individuals within them. France, Germany and Sweden are highlighted as manifesting differing forms of climate leadership, while Poland was noteworthy amongst less economically-developed EU Member States to voice concern over the costs of any Agreement.

After the disappointment of the 2009 Copenhagen COP, ENGOs changed their campaigning tactics. As Jacobs (2016) notes, in the run-up to Paris, many ENGOs shifted their focus away from campaigning on climate change to instead focus upon opposing fossil fuel consumption, resulting in new divestment campaigns, for example targeted at universities and big corporate companies. This shift was important: by reframing the issue away from the complex and relatively abstract phenomenon that is climate change and onto specific actors, a new narrative, based around the capacity for individuals and organisations to express their own agency, was developed. Many European ENGOs were 'followers' of the bigger and wealthier campaigning groups based in the USA. For instance, People's Climate Marches took place in September 2014 in Berlin, London, Paris and elsewhere in Europe, in unison with the largest ever in New York City. As the COP drew closer, several European ENGOs – as well as development NGOs, which have increasingly been involved in climate change issues since the mid-2000s (Wurzel, Connelly and Monaghan, 2017) – provided extra capacity to support the drafting of economically developing states' NDCs. Moreover, at the 2015 Paris COP, European ENGOs, such as *Friends of the Earth Europe* and *Climate Action Network Europe*, attempted to shape the negotiations in real-time by attending *en masse*. For instance, Allan and Hadden (2017) identify the important role played by ENGOs in ensuring a separate article on 'loss and damage' was included in the final Agreement; this achievement is a clear example of leadership by ENGOs, even if 'loss and damage' was mentioned in less detail than ENGOs had hoped. As such, as well as the cognitive leadership shown by these ENGOs in pushing for a 1.5°C maximum temperature increase (alongside a coalition of other actors, such as small island states), they also facilitated the development of greater structural leadership on the part of the EU's institutions by pressuring European governments to be more ambitious.

The European Parliament played only a minor role at the Paris negotiations, despite both its reputation for being the greenest EU institution (Burns, 2005; 2017) and its increased foreign policy significance since the 2009 Lisbon Treaty (Delreux and Burns, 2019). The Parliament's role should be understood as helping to strengthen the EU's climate credentials in the years running up to the 2015 Paris conference. For example, it supported amendments to the ailing ETS in February 2015 (see Upton, 2015), and called for upfront information about the EU's target, including regarding land use, which was not clarified in the final NDC, as discussed above. Thus, the Parliament strengthened the EU's foreign policy structural leadership credentials, whilst also continuing its mid-2000s hallmark by providing entrepreneurial leadership where possible (Burns and Carter, 2011).

In contrast to the Parliament, the European Commission, as might be expected from the EU's executive arm, played an important leadership role at the 2015 Paris COP. During the 2010s, the Commission's Directorate-General (DG) for Energy became more prominent compared to more explicitly pro-climate voices, such as DG Environment and DG Climate Action (Dupont, 2016). However, the Commission took a prominent stance at the summit, particularly through key individuals such as President Jean-Claude Juncker, High Representative for Foreign Affairs and Security Policy Federica Mogherini and Climate Commissioner Miguel Cañete. Cañete was the figurehead of the EU's entrepreneurial leadership in Paris, as he personally drove the coordination with other states outside the EU (Teffer, 2015), most notably the creation of a High Ambition Coalition, which Cañete (2015) labelled 'the masterplan of Europe and its allies'.

The remainder of this section explores the role played by the European Council, comprising the Heads of State of Government (i.e. the highest political representatives) of the EU's Member States. The existence of many voices that compete to shape the EU's climate negotiating system poses 'Herculean' challenges to coordination (Grubb and Yamin, 2001: 285), yet the European Council had previously played a key role in enabling the EU's international climate leader status (Oberthür and Dupont, 2017). However, it had also taken a less ambitious stance on climate change following the 2009 Copenhagen failure (Skovgaard, 2014: 13). Key Member States explored below are Germany (see also Chapter 9 in this volume), France, Sweden (see also Chapter 11 in this volume) and Poland[4]. First though, we should highlight Latvia, which hosted the Council Presidency when the EU submitted its NDC, and Luxembourg, which held the role during the negotiations. Indeed, Luxembourg's Presidency of the Council resulted in some laughter at the summit when Luxembourg's Environment Minister Carole Dieschbourg was mistakenly introduced as 'President of the EU' (Teffer, 2015). This example demonstrates the utility of the six-month rotating Council Presidency for elevating *every* Member State into positions of entrepreneurial leadership (Liefferink and Wurzel, 2017: 956), but also the complexity of analysing the performances of key actors within the EU, embodying Henry Kissinger's reflection on who to call when wishing to speak to Europe.

Despite its low-carbon electricity sector, France had struggled to become a climate leader in the decade before Paris (Szarka, 2008). However, as hosts, and under the leadership of highly experienced Foreign Minister Laurent Fabius, France played an important entrepreneurial leadership role in Paris, and was praised for its sophisticated diplomatic abilities, in slight contrast to the host in

2009, Denmark. As Christoff (2016: 769) notes, 'French diplomatic efforts prior to the Paris negotiations were intensive, persuasive, and globally extensive'.

Germany has been frequently identified as a structural leader within the EU (Jänicke, 2011; Rayner and Jordan, 2013; see also Chapter 9 in this volume). Yet, domestic climate ambition slowed in the wake of nuclear phase-out prevarication and the energy transition (*Energiewende*), resulting in a temporary pause in the phase-out of coal for electricity. However, Germany played an important structural and entrepreneurial leadership role by using its sizeable capacity to support economically developing states in creating their own NDCs, through initiatives such as the NDC Partnership, of which Germany is one of the Steering Committee members, alongside other EU countries, including Denmark, Sweden and the UK. For instance, within their NDC submission, The Gambia thanked 'the Government of Germany, the CDKN [Climate and Development Knowledge Network], GIZ [the German Corporation for International Cooperation] and Climate Analytics of Germany for the financial and technical support' (The Gambia, 2015), reflecting the Gambia's status as a 'follower' of German leadership.

Sweden also provided entrepreneurial leadership by holding bilateral meetings with economically developing states throughout the conference, which built on its extensive aid donation networks (see Teffer, 2015). Like many EU Member States, Sweden strengthened the structural leadership underpinning the EU's NDC pledge by making extensive emissions reductions, and was also a particular pioneer regarding the development of more polycentric governance practices. For example, Sweden created *Fossilfritt Sverige* ('Fossil-free Sweden') in the run-up to the 2015 Paris COP to enhance interconnections between the state and more local initiatives (see Pattberg *et al.*, 2018: 182; see also Chapter 11 in this volume).

Finally, although this book focuses on pro-climate leadership, high-profile laggards are also noteworthy. Although Jankowska (2011: 171) argues that Poland was not a blocker of climate outputs but a provider of compromises prior to Paris, during the 2010s (in the context of economic crisis), Poland repeatedly emphasised the costs of greater ambition, and led fellow East European states in opposing the tightening of the EU's ETS (Skovgaard, 2014: 347–350). In the rare instances where Polish politicians discussed climate change, parliamentarians often stated resentment towards EU climate initiatives (Marcinkiewicz and Tosun, 2015). Poland took the most explicitly antagonistic stance to greater ambition in the run-up to Paris, emphasising the need for targets to be universal and individual states' needs to be respected (see Cienski and Kureth, 2015). These red lines were agreed in the final text, highlighting Poland's status as straddling the divide between economically 'developed' and 'developing' states. As we see later in this chapter, though, Poland continued to lean more closely to the latter side of this categorisation since the Paris conference.

EU leadership after Paris

Immediately after the 2015 Paris COP, the EU continued to demonstrate relative leadership on climate policy compared to other developed states. In February

2016, the EU's Foreign Affairs Council launched its climate diplomacy action plan for 2016, highlighting climate change as a strategic diplomatic priority and emphasising the need to start implementing the submitted NDC. In October 2016, the European Council agreed to ratify the Paris Agreement, thus enabling the EU's NDC to enter into force. Poland once again exhibited reluctance towards the Paris Agreement, having emphasised in the days before ratification that the state would continue to rely upon coal for its electricity supply (see Mirowicz, 2016). Furthermore, the timing of the EU's ratification was more akin to a follower than leader, as Papua New Guinea's NDC had already entered into force over six months earlier by the time the EU ratified its own target. However, the EU's timing did enable the Paris Agreement to enter into force in time for the 2016 COP in Marrakesh, Morocco. Moreover, the EU exhibited structural leadership by agreeing to increase its international climate finance contribution on 2 November 2016.

Since the immediate aftermath of the 2015 Paris conference, the EU has continued to be a relative climate leader and pioneer. In June 2017, the EU maintained that the Paris Agreement could not be renegotiated (European Commission, 2018a), despite the USA's announced intention to withdraw. In doing so, the EU emphasised its status as an entrepreneurial and structural leader. In December 2018, COP24 took place on European soil, this time in coal-mining hub, Katowice, Poland. Despite hosting the conference, though, Poland was less keen to drive ambition than COP21 hosts France. For instance, just before the conference took place, Poland announced it had opened a new coal mine in Silisia (Evens and Timperley, 2018), and Poland's President Andrzej Duda opened the COP by stating that under no circumstances did they plan on giving up coal (Evans and Timperley, 2018). Indeed, Poland's political elite was absent from COP24, nominating the unknown and relatively young Secretary of State Michał Kurtyka as conference President (Mathiesen and Apparicio, 2018). Despite these limitations, Poland showed some ambition during the negotiations, and the Paris Rulebook for implementing the Paris Agreement was agreed at the conference. This outcome reinforced the notion that Poland wanted to present itself as a follower of international climate action, while at the same time falling short of expectations in the transition from coal, thus underlining the EU's challenge of unifying its Member States to drive forward future climate ambition in.

Discussion

The EU exhibited fluctuating leadership in the run-up to, during and after the Paris conference. First, prior to COP21 in 2015, the EU introduced a number of innovative policy instruments, most notably the ETS. Despite some problems, it is the most vital pillar of the EU's overall climate policy, covering all EU Member States in addition to Iceland, Norway and Liechtenstein. Furthermore, the ambitious 20-20-20 Climate and Energy Package introduced in 2008 prioritised climate policy. These developments fall in line with the general trend during the 2000s of prioritising climate issues, giving the EU the reputation of a coherent and credible leader with a willingness to move forward with ambitious goals.

Second, however, the 2009 Copenhagen COP saw the EU marginalised by the structural power of the USA and China. The disappointing performance of the EU heralded an uncertain time: the inability to foster a follow-up agreement with binding targets designed for the post-Kyoto period revealed the stark ambition levels between countries. This division was especially visible between old and new EU Member states, with some, such as Sweden and France, eager to adopt more ambitious plans, and others, such as Hungary and Poland, against such advances. In part, this shift is also due to the impact of the economic crisis (Burns, Tobin and Sewerin, 2019; Burns and Tobin, 2020). Yet, although the crisis generally led to greater fragmentation (Skovgaard, 2014), it did not alter the EU's policy ambition entirely (Burns, Eckersley and Tobin, 2019) and in 2010, the EU had morphed into a 'leadiator' role, albeit with less ambitious goals compared to the Copenhagen negotiations (Groen, Niemann and Oberthür, 2012; Oberthür and Dupont, 2017).

Third, the adoption of the Paris Agreement at COP21 in 2015 was a diplomatic success, driven by the French Presidency's efforts prior to the conference (Christoff, 2016), even if proposed emissions reductions are inadequate to prevent 2°C temperature increases. The EU's leadership performance was galvanised once the Paris Agreement was ratified and moved to the implementation phase (Brandi, 2018). The absolute emissions target provided by the EU's single NDC seemingly set an example. In comparison with other NDCs from around the world, the target is not only broadly ambitious, but the EU also made a case in point to submit its NDC early, incentivising other countries to follow suit with similarly ambitious targets.

Simultaneous to these events, the international climate governance landscape was rapidly becoming more polycentric. The EU's ETS – orchestrated by the Commission but designed to facilitate more cost-efficient emissions reductions wherever they may occur within Europe – started to perform more effectively following development of the Commission's Market Stability Reserve in mid-2018. However, the carbon price will need to continue to increase in the years to come if it is to reduce emissions in time to help the achievement of the EU's 2030 climate targets. In addition, the EU's Covenant of Mayors has continued to expand since its creation in 2008, bringing thousands of local governments together as a means of facilitating cooperation and the sharing of best practice. Indeed, from an MLG perspective, Jänicke and Wurzel (2019: 34) argue that climate innovation within the EU is increasingly being achieved across multiple levels.

Lastly, since 2015, the EU has continuously stated its commitment to the Paris Agreement, as well as to the United Nations Sustainable Development Goals. In line with both aims, the Commission pledged its efforts towards climate mainstreaming and commits itself to spend 25% of EU expenditure contributing to climate objectives during the EU budget 2021–2027 period (European Commission, 2018a), with entrepreneurial and cognitive leadership from eight leading states, such as France and Sweden, but notably not Germany. Regarding the second round of NDC submissions in 2020, it is likely that we will see an increase in ambition from the EU, especially following the selection of Ursula von der Leyen as Commission President. Von der Leyen began her Presidency by emphasising climate change, including the creation of a 'European Green Deal' and declaring an environment and climate emergency. In March 2020, the Commission proposed a regulation to increase the EU's greenhouse gas emission reduction target for 2030 to at least 50% and towards 55% compared with 1990 levels (European Commission, 2020), in order to facilitate the achievement of climate neutrality by 2050. This increase in ambition has the potential to both demonstrate the EU's leadership in climate policy-making, and crucially also to bolster the NDC architecture globally. However, the EU was also faced with the early challenge of eight member states failing to submit their 'National Energy and Climate Plans' before the agreed deadline of December 2019, before the much greater challenge of the COVID-19 pandemic struck. Furthermore, the ratchet mechanism for updating NDC targets is solely based on each individual country's willingness to contribute to combatting climate change with targets determined by each state individually, and so is fragile. Taken together, current NDCs are collectively not nearly enough to achieve the 2°C maximum temperature increase, let alone the 1.5°C goal. Therefore, it is important that the EU exhibits leadership to protect the medium-term viability of this more polycentric approach in a context where the time for alternative governance models is running out and urgent simultaneous challenges could reduce capacity.

In 2019, debates in Europe centred on whether to increase the emissions reduction targets for 2030 and/or 2050 (Dröge and Rattani, 2019). The need to decide on the EU's short-to-mid-term engagement towards climate change and the EU's long-term climate strategy are revealing its capacity to move forward with its vision of a climate neutral Europe by 2050 (European Commission, 2018b). However, the EU's performance continues to rely upon the ambitions of several leading Member States – which as of 2020 no longer includes the UK – and the willingness of laggards to follow their lead, or at least, not drag their feet too much.

Conclusion

Looking ahead, two notable developments have the potential to elevate the EU to its original climate leadership status. First, the Paris Rulebook adopted in December 2018 is an important cornerstone in the NDC architecture because parties agreed to structure their upcoming NDCs by agreed-upon guidelines. As the guidelines will make NDCs more coherent, they will also make them more comparable. Here, the EU has the potential to revise its targets and push forward with more ambitious goals in the years to come.

Second, the European election results of May 2019 may be crucial for redirecting enthusiasm and resources towards high-reaching EU climate policy again. The clear focus on climate- and environment-related issues during the campaign resulted in a substantial gain of 19 more seats for the Greens in comparison to the 2014–2020 period, but also a swing towards right-wing populist parties, of which some tend towards climate denial (European Parliament, 2019). The spate of states declaring 'climate emergencies' since 2019 added new urgency to policy-making, including within states that have not been traditional climate leaders, such as Ireland (see Chapter 12 in this volume). Furthermore, the expansion of Extinction Rebellion from being a social movement in the UK to a pan-European and increasingly global action reflects the growing role of non-state action in the demand for greater ambition. If the EU can build on this leadership momentum while promoting a more polycentric governance model, it could yet accomplish greater unity amongst its Member States, and consequently reduce emissions across its members, despite numerous possible crises facing it in the years to come. As we saw following the global financial crisis, post-crash carbon emissions may fall due to reduced productivity, only to bounce up as states prioritise economic growth. With time running out to act on climate change, it is imperative the same pattern does not follow the COVID-19 pandemic.

Notes

1 The support of the Economic and Social Research Council (ESRC) is gratefully acknowledged, having funded Paul Tobin via grant ES/S014500/1 during the writing of this chapter.
2 Although, the UNFCCC has never reflected an ideal-typical 'monocentric' governance model (van Asselt and Zelli, 2018).
3 On the one hand, these omissions reflect the current lack of production of these gases in those states. On the other hand, however, their absence in the states' NDS opens the door for future emissions increases, if they begin to produce these gases in the future.
4 Prior to Brexit, the UK was often also referred to as a dominant member state within the EU's climate policy. Mostly this status is due to its pioneering 2008 Climate Change Act, which laid out a framework for the UK's long-term transition to a low carbon economy.

Bibliography

Abbott, K.W. (2012) 'The transnational regime complex for climate change', *Environment and Planning C: Government and Policy*, 30: 571–590.
Allan, J.I. and Hadden, J. (2017) 'Exploring the framing of NGOs in global climate politics', *Environmental Politics*, 26(4): 600–620.
van Asselt, H. (2016) 'The role of non-state actors in reviewing ambition, implementation, and compliance under the Paris agreement', *Climate Law*, 6(1): 91–108.
van Asselt, H. and Zelli, F. (2018) 'International governance: polycentricity governing by and going beyond the UNFCCC'. In: A. Jordan et al. (eds) *Governing Climate Change: Polycentricity in Action?* Cambridge: Cambridge University Press, 29–46.
Bäckstrand, K. and Elgström, O. (2013) 'The EU's role in climate change negotiations: from leader to 'leadiator'', *Journal of European Public Policy*, 20(10): 1369–1386.
Brandi, C. (2018) 'EU climate leadership? Europe's role in global climate negotiations'. In: C. Leggiwie and F. Mauelshagen (eds) *Climate Change and Cultural Transition in Europe*. Leiden: Brill, 219–244.
Burns, C. (2005) 'The European Parliament: The European Union's environmental champion?' In: A. Jordan (ed.) *Environmental Policy in the EU: Actors, Institutions and Processes*. London: Routledge, 87–105.

Burns, C. (2017) 'The European Parliament and climate change: a constrained leader?' In: R. Wurzel, J. Connelly and D. Liefferink (eds) *The European Union in International Climate Change Politics: Still Taking a Lead?* London: Routledge, 52–65.

Burns, C. and Carter, N. (2011) 'The European Parliament and climate change: from symbolism to heroism and back again'. In: R. Wurzel and J. Connelly (eds) *The European Union as a Leader in International Climate Change Politics*. Abingdon: Routledge, 78–93.

Burns, C., Eckersley, P. and Tobin, P. (2019) 'Environmental policy in times of crisis', *Journal of European Public Policy*, 27(1): 1–19.

Burns, C. and Tobin, P. (2020) 'Crisis, climate change and comitology: policy dismantling via the backdoor?' *JCMS: Journal of Common Market Studies*, 58(3): 527–544.

Burns, C., Tobin, P. and Sewerin, S. (eds) (2019) *The Impact of the Economic Crisis on European Environmental Policy*. Oxford: Oxford University Press.

Buxton, N. (2016) 'COP 21 Charades: spin, lies and real hope in Paris', *Globalizations*, 13(6): 934–937.

Cañete, M.A. (2015) 'Historic climate deal in Paris: speech by Commissioner Miguel Arias Cañete at the press conference on the results of COP21 climate conference in Paris', *European Commission*. Available from: http://europa.eu/rapid/press-release_SPEECH -15-6320_en.htm

Christoff, P. (2016) 'The promissory note: COP 21 and the Paris climate agreement', *Environmental Politics*, 25(5): 765–787.

Cienski, J. and Kureth, A. (2015) 'Poland takes a tough line ahead of COP21'. *Politico*. Available from: https://www.politico.eu/article/poland-tough-line-cop21-paris-cli mate-summit/

Cole, D.H. (2015) 'Advantages of a polycentric approach to climate change policy', *Nature Climate Change*, 5: 114–118.

Damro, C. and Méndez, P.L. (2003) 'Emissions trading at Kyoto: from EU resistance to Union innovation', *Environmental Politics*, 12(2): 71–94.

Delreux, T. and Burns, C. (2019) 'Parliamentarizing a politicized policy: understanding the involvement of the European Parliament in UN climate negotiations', *Politics and Governance*, 7(3): 339–349.

Dröge, S. and Rattani, V. (2019) *After the Katowice Climate Summit: Building Blocks for the EU Climate Agenda*. Berlin: Stiftung Wissenschaft und Politik, Deutsches Institut für Internationale Politik und Sicherheit. Available from: https://doi.org/10.18449 /2019C09

Dupont, C. (2016) *Climate Policy Integration into EU Energy Policy. Progress and Prospects*. London: Routledge.

European Commission (2018a) 'Supporting climate action through the EU budget'. Available from: https://ec.europa.eu/clima/policies/budget/mainstreaming_en

European Commission (2018b) '2050 long-term strategy'. Available from: https://ec.euro pa.eu/clima/policies/strategies/2050_en

European Commission (2020) 'Proposal for a regulaton of the European Parliament and of the council establishing the framework for achieving climate neutrality and amending regulation (EU) 2018/1999 (European Climate Law)'. Available from: https://ec.euro pa.eu/info/sites/info/files/commission-proposal-regulation-european-climate-law-march-2020_en.pdf

European Parliament (2019) 'European Parliament, provisional results'. Available from: https://www.election-results.eu/european-results/2019-2024/

Evans, S. and Timperley, J. (2018) 'COP24: key outcomes agreed at the UN climate talks in Katowice', *Carbon Brief.* Available from: https://www.carbonbrief.org/cop24-key -outcomes-agreed-at-the-un-climate-talks-in-katowice

Falkner, G. (2016) 'The EU's current crisis and its policy effects: research design and comparative findings', *Journal of European Integration*, 38(3): 219–235.

Fleig, A., Schmidt, N.M. and Tosun, J. (2017) 'Legislative dynamics of mitigation and adaptation framework policies in the EU', *European Policy Analysis*, 3: 101–124. doi: 10.1002/epa2.1002

Galaz, V., Crona, B., Osterblom, H., Olsson, P. and Folke, C. (2012) 'Polycentric systems and interacting planetary boundaries: emerging governance of climate change–ocean acidification–marine biodiversity', *Ecological Economics*, 81: 21–32.

Gravey, V. and Moore, B. (2019) 'Full steam ahead or dead in the water? European Union Environmental Policy after the economic crisis'. In: C. Burns, P. Tobin and S. Sewerin (eds) *The Impact of the Economic Crisis on European Environmental Policy*. Oxford: Oxford University Press, 19–42.

Groen, L., Niemann, A. and Oberthür, S. (2012) 'The EU as a global leader? The Copenhagen and Cancun UN climate change negotiations', *Journal of Contemporary European Research*, 8(2): 173–191.

Grubb, M. and Yamin, F. (2001) 'Climate collapse at the Hague: what happened and where do we go from here?', *International Affairs*, 77(2): 261–276.

Jacobs, M. (2016) 'High pressure for low emissions: how civil society created the Paris climate agreement', *IPPR*. Available from: https://www.ippr.org/juncture/high-press ure-for-low-emissions-how-civil-society-created-the-paris-climate-agreement

Jänicke, M. (2011) 'German climate change policy: political and economic leadership'. In: R. Wurzel and J. Connelly (eds) *The European Union as a Leader in International Climate Change Politics*. London: Routledge, 129–146.

Jänicke, M. and Wurzel, R.K.W. (2019) 'Leadership and lesson-drawing in the European Union's multilevel climate governance system', *Environmental Politics*, 28(1): 22–42.

Jankowska, K. (2011) 'Poland's climate change policy struggle: greening the east?' In: R. Wurzel and J. Connelly (eds) *The European Union as a Leader in International Climate Change Politics*. London: Routledge, 163–178.

Jordan, A., Huitema, D., Van Asselt, H. and Forster, J. (eds) (2018) *Governing Climate Change: Polycentricity in Action?* Cambridge: Cambridge University Press.

Kilian, B. and Elgström, O. (2010) 'Still a green leader? The European Union's role in international climate negotiations', *Cooperation and Conflict*, 45(3): 255–273.

Kinley, R. (2017) 'Climate change after Paris: from turning point to transformation', *Climate Policy*, 17(1): 9–15.

Latvia (2015) 'Submission by Latvia and the European Commission on behalf of the European Union and its Member States', *Latvian Presidency of the Council of the European Union*. Riga, 6th March 2015. Available from: https://www4.unfccc.int/sites/ ndcstaging/PublishedDocuments/Sweden%20First/EU%20First%20NDC.pdf

Lenschow, A. and Sprungk, C. (2010) 'The myth of a green Europe', *Journal of Common Market Studies*, 48(1): 133–154.

Liefferink, D. and Wurzel, R. (2017) 'Environmental leaders and pioneers: agents of change?' *Journal of European Public Policy*, 24(7): 951–968.

Liefferink, D. and Wurzel, R. (2018) 'Leadership and pioneership: exploring their role in polycentric governance'. In: A. Jordan et al. (eds) *Governing Climate Change: Polycentricity in Action?* Cambridge: Cambridge University Press, 135–151.

Marcinkiewicz, K. and Tosun, J. (2015) 'Contesting climate change: mapping the political debate in Poland', *East European Politics*, 31(2): 187–207.

Marks, G. and Hooghe, L. (2004) 'Contrasting visions of multi-level governance'. In: I. Bache and M. Flinders (eds) *Multi-Level Governance*. Oxford: Oxford University Press, 15–30.

Mathiesen, K. and Apparicio, S. (2018) 'Poland appoints deputy minister to run biggest climate talks since Paris'. Available from: https://www.climatechangenews.com/2018 /04/30/poland-appoints-deputy-minister-run-biggest-climate-talks-since-paris/

Mirowicz, A. (2016) 'Will Poland change the course of the Paris agreement?' *Sandbag*. Available from: https://sandbag.org.uk/2016/09/29/will-poland-change-the-course-of -the-paris-agreement/

Oberthür, S. and Roche Kelly, C. (2008) 'EU leadership in international climate policy: achievements and challenges', *The International Spectator: Italian Journal of International Affairs*, 43(3): 35–50.

Oberthür, S. and Dupont, C. (2017) 'The council and the European Council: stuck on the road to transformational leadership'. In: R. Wurzel, J. Connelly and D. Liefferink (eds) *The European Union as a Leader in International Climate Change Politics*. Abingdon: Routledge, 66–80.

Ostrom, E. (2010) 'Polycentric systems for coping with collective action and global environmental change', *Global Environmental Change*, 20: 550–557.

Parker, C.F. and Karlsson, C. (2010) 'Climate change and the European Union's leadership moment: an inconvenient truth?', *Journal of Common Market Studies*, 48(4): 923–943.

Parker, C.F., Karlsson, C. and Hjerpe, M. (2015) 'Climate change leaders and followers: leadership recognition and selection in the UNFCCC negotiations', *International Relations*, 29(4): 434–454.

Pattberg, P. (2010) 'Public–private partnerships in global climate governance', *WIREs Climate Change*, 1: 279–287.

Pattberg, P., Chan, S., Sanderink, L. and Widerberg, O. (2018) 'Linkages understanding their role in polycentric governance'. In: A. Jordan et al.(eds) *Governing Climate Change: Polycentricity in Action?* Cambridge: Cambridge University Press, 169–187.

Rayner, T. and Jordan, A. (2013) 'The European Union: the polycentric climate policy leader?' *WIREs Climate Change*, 4: 75–90.

Rhodes, C.J. (2017) 'US withdrawal from the COP21 Paris climate change agreement, and its possible implications'. *Science*, 100(4): 411–419.

Schmidt, N.M. (2019) 'Late bloomer? Agricultural policy integration and coordination patterns in climate policies', *Journal of European Public Policy*. doi: 10.1080/13501763.2019.1617334

Schreurs, M.A. and Tiberghien, Y. (2007) 'Multi-level reinforcement: explaining European Union leadership in climate change mitigation', *Global Environmental Politics*, 7(4): 19–46.

Schreurs, M.A. and Tiberghien, Y. (2010) 'European Union leadership in climate change: mitigation through multilevel reinforcement'. In: K. Harrison and L. Sundstrom (eds) *Global Commons, Domestic Decisions: The Comparative Politics of Climate Change*. Cambridge, MA: Massachusetts Institute of Technology Press, 23–66.

Skovgaard, J. (2014) 'EU climate policy after the crisis', *Environmental Politics*, 23(1): 1–17.

Slominski, P. (2016) 'Energy and climate policy: does the competitiveness narrative prevail in times of crisis?', *Journal of European Integration*, 38(3): 343–357.

Steinebach, Y. and Knill, C. (2017) 'Still an entrepreneur? The changing role of the European Commission in EU environmental policy-making', *Journal of European Public Policy*, 24(3): 429–446.

Stephenson, P. (2013) 'Twenty years of multi-level governance: where does it come from? What is it? Where is it going?' *Journal of European Public Policy*, 20(6): 817–837.

Szarka, J. (2008) 'France's troubled bids to climate leadership'. In: H. Compston and J. Bailey (eds) *Turning Down the Heat: The Politics of Climate Policy in Affluent Democracies*. Basingstoke: Palgrave Macmillan, 132–148.

Teffer, P. (2015) 'Who speaks for the EU at Paris climate summit?' *EU Observer*. Available from: https://euobserver.com/environment/131492

The Gambia (2015) 'Intended nationally determined contribution of The Gambia'. Available from: https://www4.unfccc.int/sites/submissions/INDC/Submission%20Pages/submissions.aspx

Tobin, P., Schmidt, N., Tosun, J. and Burns, C. (2018) 'Climate policy innovation: mapping states' Paris climate pledges', *Global Environmental Change*, 48: 11–21.

Torney, D. (2019) 'Follow the leader? Conceptualising the relationship between leaders and followers in polycentric climate governance', *Environmental Politics*, 28(1): 167–86.

Tosun, T. (2018) 'Diffusion: an outcome of and an opportunity for polycentric activity?' In: A. Jordan et al. (eds) *Governing Climate Change: Polycentricity in Action?* Cambridge: Cambridge University Press, 152–168.

Upton, J. (2015) 'Europe lays out its vision for climate change', *Scientific American*. Available from: https://www.scientificamerican.com/article/europe-lays-out-vision -for-climate-change1/

USA (2015) 'Intended nationally determined contribution', Available from: https://www4 .unfccc.int/sites/submissions/NDC/Submission%20Pages/submissions.aspx

Vogler, J. and Bretherton, C. (2006) 'The European Union as a protagonist to the United States on climate change', *International Studies Perspectives*, 7: 1–22.

Wurzel, R., Connelly, J. and Monaghan, E. (2017), 'Environmental NGOs: Pushing for leadership', in: R. Wurzel, J. Connelly and D. Liefferink (eds), *The European Union in International Climate Change Politics. Still Taking A Lead?* London: Routledge, 221–236.

Wurzel, R.K.W. and Connelly, J. (eds) (2011) *The European Union as a Leader in International Climate Change Politics*. Abingdon: Routledge.

Wurzel, R.K.W., Liefferink, D. and Torney, D. (2019) 'Pioneers, leaders and followers in multilevel and polycentric climate governance', *Environmental Politics*, 28(1): 1–21.

9 Climate policy in Germany

Pioneering a complex transformation process

Sibyl Steuwer and Julia Hertin[1]

Introduction

Much of the literature on pioneers and leaders is focused on technological innovations and ways of promoting them through government policy, for example, through traditional environmental regulation, market-based instruments or 'soft' policy approaches. As a result, we know a lot about success conditions for pioneer and leader states with regard to single innovations. However, there is now a widespread view that the threat of climate and environmental change requires deep socio-technical transformations, which go far beyond the development of isolated technological solutions and which affect a wide range of sectors such as energy, resources and agriculture (IPCC, 2018). Such ecological transformations are complex processes. Little research has been undertaken on what either pioneering and/or leadership looks like in complex transformation processes.

There is no consistent use of terms such as pioneer and leader in the academic literature, with some authors making use of these terms interchangeably (e.g. Weidner, 2008). In line with the general framework of the book, we distinguish between pioneers and leaders (see Chapter 1 in this volume; Liefferink and Wurzel, 2017; Wurzel *et al.*, 2019; Steinbacher and Pahle, 2015). Pioneers are characterised by a high level of ambition in relation to domestic policy change. They are not concerned about other jurisdictions following their ambitious pathway. Leaders, in contrast, seek to actively promote their model of change to potential followers. Leaders with high levels of both internal and external ambition will be referred to as constructive pushers, while those with high external leadership ambitions low domestic ambitions and/or poor domestic record will be called symbolic leaders (Liefferink and Wurzel, 2017). Actors without either internal or external ambition are referred to as laggards (ibid.). We also distinguish between four leadership types: structural leadership (which requires structural power), entrepreneurial leadership (which relies on bargaining and negotiating skills), cognitive leadership (which necessitates knowledge resources) and exemplary leadership (see also Chapter 1 in this volume; Liefferink and Wurzel, 2017). In addition, we aim at contributing with our case study additional insights into the nature of transformational leadership while following Liefferink and Wurzel (2017) who differentiate between transactional and transformational *styles* of leadership (in addition to the above mentioned four *types* of leadership). Wurzel, Liefferink and Torney (2019) mainly differentiate transformational from transactional leadership, with the former emphasising the long-term perspective and the quality of change ('revolutionary'). We differentiate also between different phases of transformation and claim that transformational leadership requires different kinds of actions and faces different types of challenges to overcome according to the respective transformational phase.

The German energy transition (*Energiewende)* provides an excellent opportunity to study a deep socio-technical transformation of a (self-proclaimed) leader in climate policy: Germany was one of the first countries to submit its long-term strategy for climate protection to the United Nations as required by the 2015 Paris Agreement. It aims to become largely greenhouse gas-neutral by 2050 while generating at least 80% of its electricity and 60% of total energy consumption from renewables. The German government has also decided to phase-out nuclear power by 2022.

The chapter will draw on the literature on large-scale socio-technical transformation processes – including the more political accounts of this literature – to examine conceptually what policy interventions, instruments and processes are required to pioneer an environmental transformation at the national level. The second part of the chapter will apply these theoretical insights to the German *Energiewende* with a focus on electricity generation and the built environment. Here, we analyse: 1) to what extent and in which respects Germany can rightfully claim to be a domestic pioneer in climate policy; 2) whether Germany has made efforts to promote its energy transition abroad; and 3) challenges of pioneership and leadership in relation to socio-technical transformations (as opposed to simple technical innovations).

Pioneering environmental transformations

There is now wide agreement that radical and pervasive change is needed to stabilise the global climate at conditions that are safe for human civilisation (e.g. IPCC, 2018). Decarbonising the economy will affect virtually all sectors – not only the generation of electricity, but also housing, transport and industry. The need to drastically reduce greenhouse gas emissions (GHGE) equally poses a big challenge for agriculture, forestry and other land use systems.

In response, there is a growing literature not only on the technical and economic feasibility of transformations, but also on their political and social governance. The varied literature has emerged from several disciplines, in particular economics, political science, sociology and engineering. Fields of research include transition management, innovation and diffusion research, post-growth and eco-sufficiency research, and change management (e.g. Kemp, Schot and Hoogma, 1998; Geels and Schot, 2007; Scoones, Leach and Newell, 2015).

Innovations are key in transformation processes as they are both drivers for and constituents of a transformed system. For innovations to take off, they need to be promoted by agents of change. An ideal-typical model of transformation dynamics shows that transformation processes reach a decisive stage when niche innovations take on broader social, political and economic relevance, and thus also influence the relevant ground rules, standards, institutions and constellations of power (see Geels, 2004; Geels and Schot, 2007). Previous technological and institutional paths have, however, become entrenched over time. Therefore, promoting a new socio-technical paradigm is likely to meet with strong resistance (Howlett, 2014). Established industrial and other actors are likely to adopt strategies aimed at thwarting innovative practices (Szarka, 2012). Socio-technical transformations thus differ from the diffusion of relatively simple technical innovations in that they are the subject of intensive social conflict (Newell, 2015).

Early on, social science research on transitions to sustainability mainly took on a governance and institutional perspective (e.g. Partzsch, 2015; Weiland and Partzsch, 2015). Recently, researchers have looked at the political dimension of transitions with a focus on interests and power (e.g. Geels, 2014; Scoones, Leach and Newell, 2015; Newell, 2015; Jordan and Matt, 2014). This research defines transformation as a fundamental restructuring of the balance of power within a strategic action field (Fligstein and McAdam, 2011; Newell, 2015; Geels, 2014;). According to this analytical perspective, most drivers of innovation are increasingly at odds with the logic of existing action fields since they question current practices and thrive towards replacing them. As innovation develops from niche to mainstream technologies, it alters the balance of power (albeit often only incrementally) between coalitions with a vested interest in the status quo and alliances of change-orientated actors. The resulting conflicts of interest determine what and how much change can occur (Fligstein and McAdam, 2011). They may emerge, for example, through trade-offs between transformations in different action fields, competing transformational paths within one particular action field or a time lag between technical and social innovations (see Figure 9.1). Technology and social innovations (such as user behaviour, social institutions and cultural practices) co-evolve mutually adaptively within manifold interwoven processes (Geels, 2004). Technical niche innovations face incumbents' strong and established interests when expanding into mainstream technical solutions. Similarly, social innovations also face strong path-dependent habits that need to be overcome to achieve cultural and other changes. In addition, conflict and uncertainty, in combination with increasing lack in transparency, are exacerbated in a second stage of innovation by the non-congruity of technological and social innovation, at least temporarily. In those situations, incumbents typically employ various strategies to counteract the diffusion of innovation (for the strategies see Schneider and Veugelers, 2010; Smink, Hekkert and Negro, 2015). In such situations, government action can support innovative action and its diffusion.

The role of state actors in transformation processes

The role played by state actors in such transformations is contested. Some authors have taken a sceptical view of the ability of governments to manage these various processes of change (Rotmans, Kemp and Asselt, 2001; Colander and Kupers,

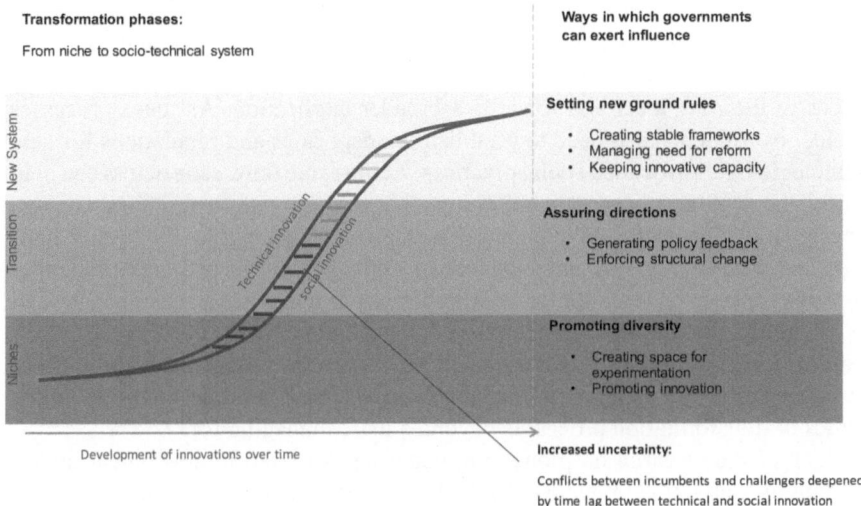

Figure 9.1 The role of state actors in socio-technical transformations.

2014). Some have argued that transformations evolve predominantly as a series of sometimes conflicting polycentric processes of change (e.g. Rotmans, Kemp and Asselt, 2001). However, other scholars have put more emphasis on the role of state actors. For them, government action is of major importance for successful transformation, because no other type of actor has comparable resources that are necessary to coordinate processes, scale individual innovations, and reduce uncertainty (Mazzucato, 2015; Grießhammer and Brohmann, 2015).

We adopt the latter position and emphasise the role that state actors play in all phases of transformation processes in multilevel governance systems. This does not, however, imply that other actors or polycentric governance are irrelevant in environmental transformations or that state actors will be able to initiate and guide transformation processes on their own. Altering stable political routines and overcoming path dependencies and instrumental lock-in effects is particularly difficult (e.g. Jordan *et al.*, 2012). Policy makers are faced with a steering paradox. They need to respond flexibly to change and have to have the capacity to adjust their actions as necessary, while at the same time they need to reduce uncertainty in order to provide potential investors with a suitable investment framework (cf. SRU, 2016). Will the multilevel governance system provide opportunities for the mutual reinforcement of ambitious policy-making (Schreurs and Tibergien, 2010) also in the case of complex transformation processes?

The key tasks of government steering vary depending on the phase of the transition (Figure 9.1; SRU, 2016). The role of governments in the initial phases of transformation processes is to promote the development of innovations. They can also support innovative niches created by societal actors by affording niche players space to experiment, granting them funding or fostering the establishment of networks (Geels and Schot, 2007; Kemp, Schot and Hoogma, 1998). Governments can steer both on the supply side (development of innovations) and on the demand side (adoption of innovations). Moreover, they can promote the diffusion of innovations to the next level and towards a broader application. As these processes unfold, governments also need to establish or adapt rules and regulations for new technologies and associated social practices. At the same time, state actors can help address the problems of actors and groups that lose out due to structural change. Especially in this transformational phase with increasing struggles between incumbents and challengers, governments need to offer structural and entrepreneurial leadership to keep on track the transformation. Once a new socio-technical system has emerged, governments can stabilise it by maintaining an appropriate policy framework while ensuring that the system stays open for further innovation. Thus, governmental actors can play a crucial role in all phases and get involved in the politics of transformation processes beyond a mere managing role.

Much of the research on pioneering countries is rooted in a technical understanding of innovation (Weidner, 2008; Jänicke and Jacob, 2004) although there are some notable exceptions (Andersen and Liefferink, 1997; Underdal, 1994). Technical innovations and individual environmental policy measures lend themselves more readily to globalization than may be the case with complex socio-technical transformations. Whether and in what way the different types of transformations (e.g. incremental or 'revolutionary') influence opportunities for both pioneership and leadership is a key question to be addressed by the subsequent analysis. In addition, we will examine to what extend actors in Germany make use of the multilevel governance system to mutually reinforce transformative approaches.

German climate policy: a pusher risking becoming a symbolic leader

Introduction

This section will not provide an encompassing historical account of the *Energiewende*. Instead, we provide a critical analysis of pioneership and leadership in German climate policy which draws attention to the particular characteristics and challenges of socio-technical transformations. Covering the period from 1990 to 2019, the analysis will highlight key events and driving forces of the political process around the German energy transition. We will start with a general background on German climate policy. The sections that follow will examine two major areas of German climate policy: the electricity sector and the building sector.

Germany as a self-proclaimed leader in climate policy

Traditionally, Germany has been both a pioneer and leader in environmental policy (SRU, 2002; Jänicke, 2017). It is therefore not surprising that successive governments also aimed to adopt a similar role for climate policy (Weidner, 2008). German Chancellors from Helmut Kohl (1982–1998) to Angela Merkel (since 2005) have stressed Germany's climate leadership role, for example, at international climate summits. Germany also hosted the first United Nations Framework Convention on Climate Change (UNFCCC) Conference of the Parties (COP1) in Berlin in 1995. The claim to be a domestic pioneer was substantiated by early national institutions including a Parliamentary Commission of Inquiry (1987) and an inter-ministerial working group on CO_2 mitigation (1990). The work of these institutions led to the ambitious target to reduce CO_2 emission by 25% between 1990 and 2005 as well as to a wide range of related policy measures. The Red–Green coalition, which was made up of the Social Democratic Party of Germany (*Sozialdemokratische Partei Deutschlands* – SPD) and the Green Party (*Bündnis 90/Die Grünen*), that came into power in 1998 gave a new impetus to climate policy which was made a political priority. Most prominently, the Red–Green coalition government adopted an eco-tax and drastically improved the existing funding mechanism for the deployment of renewable electricity. A gradual phase-out of nuclear power was agreed with industry in 2000. Other climate policy measures, for example, in the area of transport and buildings, remained mostly incremental. Between 1990 and 2009, GHGE in Germany fell by around 27.4% (Federal Ministry of Economic Affairs and Energy, 2019, 2019). Even though this was largely due to closure of inefficient industrial installations in former East Germany following the reunification of Germany in 1990 (so-called 'wall-fall profits'), the reunified Germany's climate policy was, at the time, considered to be 'amongst the most ambitious and effective' in the world (Weidner, 2008: 1).

Germany has committed itself to reducing GHGE by 80–95% until 2050 (compared to 1990) and to generating at least 80% of its electricity and 60% of total energy consumption from renewables (see Table 9.1). These core climate policy objectives are accompanied by a wide range of additional targets relating to different sectors, technologies and timescales. Overall, German national climate targets were, at the time, fairly ambitious in an international comparison.

As time went on, a gap between Germany's international climate leadership ambitions and its domestic performance emerged (Kemfert, 2017). Most importantly, German climate protection has not been successfully integrated as a cross-sectoral task into all areas of policy-making. Progress in both the transport and building sectors has been slow. There has, for example, not been any serious attempt to reduce environmentally harmful subsidies. While the funding mechanism for renewable electricity is still in place, the deployment of installations has been considerably reduced (see below). As a result of a lack of ambitious climate policy, GHGE have not been significantly reduced between 2009 and 2016 (see Figure 9.2). Germany will not only miss its national 2020 targets, but also the legally binding European Union (EU) obligations for sectors not covered by the EU Emissions Trading System (ETS).

Despite the relatively modest track record at home, the German government maintains the ambition to be an international climate leader. For example,

Table 9.1 German Climate Targets.

	2017	2020	2030	2040	2050
Greenhouse gas emissions					
Greenhouse gas emissions (compared with 1990)	-27.5%	At least -40%	At least -55%	At least -70%	Largely greenhouse-gas-neutral -80% to -95%
Renewable energy					
Share of gross final energy consumption	15.9%	18%	30%	45%	60%
Share of gross electricity consumption	36%	At least 35%	At least 50%* Renewable Energy Sources Act 2017: 40–45% by 2025	At least 65% Renewable Energy Sources Act 2017: 55–60% by 2035	At least 80%
Share of heat consumption	13.4%	14%			
Efficiency and consumption					
Primary energy consumption (compared with 2008)	-5.5%		20%	→	-50%
Final energy productivity (2008–2050)	1.0% per year (2008–2017)		2.1% per year (2008–2050)		
Gross electricity consumption (compared with 2008)	-3.3%		-10%	→	-25%
Primary energy consumption in buildings (compared with 2008)	-18.8%			→	-80%

| Heat consumption in buildings (compared with 2008) | -6.9% | -20% |
| Final energy consumption in the transport sector (compared with 2008) | 6.5% | -10% ——————— ➤ -40% |

Source: Federal Ministry for Economic Affairs and Energy (2019: 16)

Notes: * *Targeted, efficient, grid-synchronised and an increasingly market-driven expansion of renewable energy sources is a prerequisite for successful energy transition and climate protection policies. The Federal Government aims to increase the share in renewables in the energy sector – especially in the context of the challenges of better synchronisation of renewable energy sources and grid capacity – in order to reach the goal set by the Coalition Agreement of approximately 65% by 2030. The capacity of the power grid to transport energy is key. Increasing expansion of renewable energy is ultimately necessary for covering the additional demand for electricity, so that climate protection goals in transport, in buildings and in industry can be achieved.*

Germany has joined the High Ambition Coalition, which was set up in 2015, to push for an ambitious and strong implementation of the Paris Agreement. In 2019 and despite not yet having adopted a coal phase-out, Germany joined the Powering Past Coal Alliance (PPCA), a group of countries and regions committed to accelerating the phase out of coal power. Thus, some entrepreneurial leadership ambition of the German government can still be documented.

In 2018, a new dynamic was created by the climate strikes by pupils (Fridays for Future), a protest which has quickly expanded to other groups such as parents, scientists and students. Large demonstrations across Germany pushed the issue of climate change high up on the political agenda. In 2019, the German government adopted a Climate Protection Plan and a Climate Protection Law in an attempt to regain the initiative – both domestically and internationally. Part of this package was a governance mechanism which aims to ensure that climate targets are met, a national price mechanism for fossil fuels and a wide range of individual measures such as a ban on oil-fired heating and increased funding for energetic refurbishment. The climate package was, however, strongly criticised by domestic climate experts as being unable to get Germany on track to reach its GHGE targets and as failing to provide sufficient social compensation (Edenhofer *et al.*, 2019). Thus, polycentric ambitions and resulting bottom–up activities were fostering political action although they were not fully successful in setting an adequate ambition level.

The electricity sector

As mentioned above, under the first Red–Green coalition government, which took office in 1998, the *Energiewende* became a very important government project. The decision to phase-out nuclear power and the adoption of the Renewable Energy Act (*Erneuerbare-Energien-Gesetz* – EEG) marked the beginning of the *Energiewende*. The EEG has been the key driver of the German energy transition. Building on the previous Feed-in Law, the main innovation of the EEG was that it guaranteed investors fixed preferential tariffs for all power generated by a renewable installation for 20 years. It thereby sharply reduced investment uncertainty, thereby providing long-term or 'patient capital' (Mazzucato, 2015). The EEG has enabled renewable energy to emerge from its technological niche.

In many respects the EEG is an embodiment of a successful 'pusher strategy'. As a result, the costs of solar and wind power have declined steeply. Moreover, the EEG has also altered the balance of power between energy policy actors. With the increasing economic importance of the renewable energy industries, new players have emerged (Ragwitz and Huber, 2005). They have increasingly challenged the incumbents, subsequently becoming part of a more diverse status quo regime (e.g. SRU, 2013). The EEG has been emulated by many other countries (Solorio *et al.*, 2014). In summary, the EEG has: 1) massively increased the installation of renewable capacities in Germany, 2) strongly contributed to reducing the costs of renewable technologies (especially wind and photovoltaics (PV)) thus enabling other countries to begin using these technologies, 3) contributed

in million tonnes CO_2 equivalent

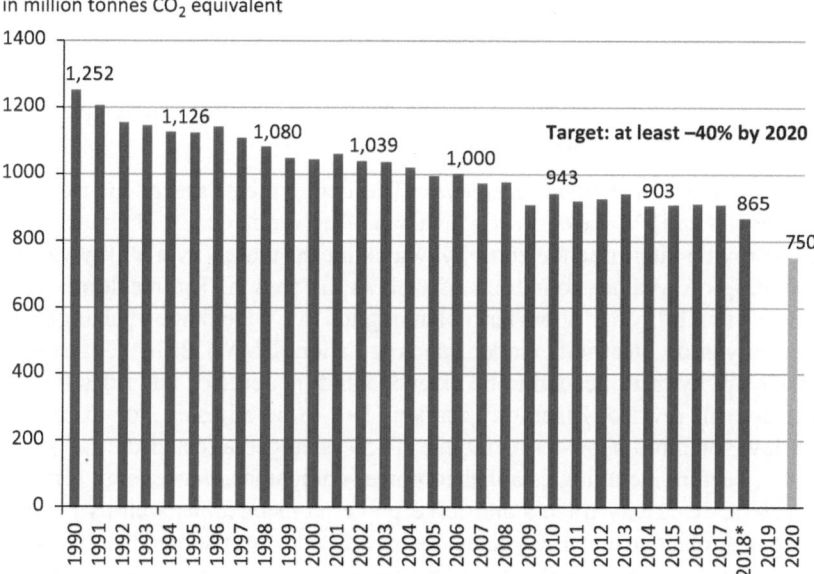

Figure 9.2 Greenhouse gas emissions in Germany.

to the diffusion of the guaranteed feed-in tariff as a popular instrument for the deployment of renewables, and 4) changed the political economy around energy policy in Germany.

The term *Energiewende* became known to a wide international audience after the 2011 Fukushima nuclear disaster. Within days after the event, Chancellor Angela Merkel (Christian Democratic Union – CDU) announced a review of nuclear policy. A few months later the centre-right CDU/CSU-FDP[2] coalition government returned to the previous pathway of nuclear phase-out by reversing its earlier decision to extend the time period for the phase-out as adopted by the Red–Green coalition government. It subsequently decided that the transition towards a largely renewables-based, efficient energy system needed to be accelerated in order to compensate for the closure of nuclear power stations. During this time Germany intensified its efforts to promulgate the *Energiewende* also as an industrial policy agenda and has staked its reputation for innovative policy on the energy transition's success (Steinbacher and Pahle, 2015; Altmaier, 2012). Due to its economic strength and power, Germany was capable of taking strong decisions internally vis-à-vis incumbent actors, both in relation to the support of feed-in tariffs but also with regard to the nuclear phase-out regime after Fukushima. Since these decisions have been promoted widely internationally, it can be claimed that Germany acted to some extent as a structural leader.

Challenges of the energy transformation

Mainly as a result of the guaranteed feed-in tariff, renewable electricity moved from a niche innovation to a mainstream technology (see Figure 9.1). This led, however, to a wide range of structural challenges and profound changes which, in turn, triggered complex political, social and economic conflicts. These can only be sketched out briefly here as follows:

- *Integration of intermittent electricity:* The deployment largely focused on wind power and PV, the costs of which had fallen dramatically. Both technologies provide intermittent electricity, which means that its availability varies depending on weather conditions as well as the season. The resulting need for more flexibility and short-term planning led to a wide range of changes in electricity grids, electricity markets and regulation (BMWi, 2014). For example, electricity grids needed to be extended and modernised and rules for its management revised. Electricity markets had to be adjusted, for example, to allow for more short-term trading. These changes affected the interests of many actors including established power companies, new energy service providers, electricity traders and large power consumers as well as people living near planned electricity lines. Status quo interests claimed that because of intermittency, renewables were unable to serve as a basis for a reliable electricity supply (INSM, 2017).
- *Changing cost structures*: The specific cost structure of wind and PV – high investment costs and virtually no operational costs – fundamentally changed the conditions of the electricity markets (Sensfuß *et al.*, 2008; BMWi, 2014). Most importantly, it dampened the prices on the wholesale power market, which is largely based on operational costs. As a result, emission-intensive coal power plants continued to operate at high capacity while efficient gas plants – which are also important from a security of supply perspective – stood idle. This meant that GHGE did not fall as much as could be expected given the increase in renewables and that there were concerns about the medium-term security of supply. This led to an intense political debate about the need for an additional scheme for reliable generation capacity.
- *Costs to consumers and economy*: The lower wholesale prices also increased the level of the EEG surcharge. Even though the surcharge is not a reliable indicator of the cost of renewable expansion (Weber and Hey, 2012), the annual increase of the surcharge visible on household bills contributed to an intense political discussion of the cost of the *Energiewende*. This conflict obtained a social and regional dimension because the origin and destination of payments under the EEG surcharge are unevenly distributed within Germany.
- *Effects on landscape, biodiversity and noise levels:* The widespread deployment of renewables and the supporting infrastructure, such as transmission grids, had a range of negative environmental impacts (Schuler *et al.*, 2017). Examples include the biodiversity impacts of the large-scale cultivation of energy plants, the effect of wind turbines on local landscapes and the impact

of offshore wind parks on marine animals. New renewable projects are often met with a considerable degree of local opposition.

As a response to concerns of various actors, the government slowed the pace of the energy transition in the power sector. The deployment of wind energy has reduced dramatically due to changes in the funding mechanism at federal level but also due to planning law decisions at regional level as well as local resistance against onshore wind power and grid extensions. It also decided to phase-out funding for small- and medium-sized PV once an overall capacity of 52 GW is reached (expected by around 2021). The government delegated the contentious issue of coal-power – traditionally a major source of electricity in Germany – to an independent 'coal commission' (KWSB, 2019) without being clearly committed to implementing its recommendations.

The EU multilevel governance dimension of pioneering a transformation

The EU has played an important role in shaping the German *Energiewende*. It has provided both barriers and opportunities. Two barriers should be highlighted. First, the EU's Single European Market policies have restricted the ability of Member States to develop national policies for the promotion of renewable energy (Tews, 2015). The European Commission has enforced Single European Market rules against national support schemes for renewable energy (e.g. Kahl, 2015). State aid rules are mainly prompted by the fear that statutory feed-in tariffs will distort competition. Second, the EU ETS has not only failed to provide investment incentives for renewable energy because of an excess of emission allowances, but it has also proved to be a disincentive for the adoption of pioneering policies at national level. When national measures brought about a reduction in CO_2 emissions from industries covered by the ETS, emission allowances became available at a relatively low price that could be used by other emitters. Hence national reduction activities in sectors covered by the EU ETS did not always result in a reduction of the EU's overall CO_2 emissions – a phenomenon known as the 'waterbed effect' (e.g. SRU, 2015). Thus, critics of the energy transition argued that the EEG does not contribute to climate protection because emissions from the European electricity sector are capped under the EU ETS (INSM, 2017). There are ways to avoid this effect, especially the cancellation of emission allowances made available through additional mitigation. This was, for example, proposed by the Economics Ministry in connection with a planned (but never adopted) climate levy, for which the Ministry offered cognitive leadership (BMWi, 2015). The waterbed effect has later been neutralised by the Market Stability Reserve (Agora Energiewende and Öko-Institut, 2018).

On the other hand, the EU multilevel governance structures have offered Germany numerous opportunities to advance climate policy at home and abroad. In collaboration with similarly-minded stakeholders and/or countries, Germany has championed ambitious emissions reduction goals and the expansion of renewable energy and thus showed entrepreneurial leadership. In the past, progressive forces at the EU and Member State levels often strengthened each other (Schreurs and Tiberghien, 2010). Europe's transnational electricity grids and electricity markets not only help to reduce costs, but also enhance Europe's security of supply and ease the task of balancing the fluctuating amounts of solar and wind power electricity that are fed into the grid. Countries and regions such as the Scandinavian countries and the Alpine region, with their sizeable upside potential for storage capacities, can also help handle such fluctuations: in off-peaks, water is pumped uphill with low-cost electricity and can be stored until used as hydroelectric power. European institutions, such as the Pentalateral Energy Forum, constitute additional platforms for close cooperation in areas such as cross-border calculation of guaranteed output and efficient use of cross-border interconnectors. The EU is an influential actor also on the global level and can thus bring about progress in international climate governance (Jänicke, 2017).

Climate policy in the building sector

The transformation of the German building stock is another cornerstone of the *Energiewende*. It is mainly concerned with reducing heat demand in the building sector and the shift towards a fully decarbonised heating system by 2050.

The climate protection target for the building sector is deemed quite ambitious. According to the Climate Action Plan 2050 and the proposal for a Climate Protection Law, the sector will be allowed to emit only 70–72 million tonnes of CO_2 equivalent by 2030. These targets are in line with the sector targets fixed at the EU level. To reach the target and thus to be truly transformative, Germany's buildings policy will need to:

- Accelerate emission reductions
- Reach radical technological change (e.g. sustainable building material and circular economy in the building sector)
- Realise social innovation (e.g. by establishing new business models, incentivising behavioural change)
- Treat the building sector as part of the wider energy system (e.g. by integrating flexibility in the electricity system)

National strategies targeting the building sector such as the National Action Plan on Energy Efficiency NAPE, 2014 or the Energy Efficiency Strategy on Buildings (*Effizienzstrategie Gebäude*, 2015) are comprehensive and ambitious. The Energy Efficiency Strategy for Buildings is especially transformative in its core ideas. It develops a corridor for different combinations of energy efficiency and renewable energy measures able to reach the long-term climate protection target. It also stresses the need for sector coupling (e.g. PV electricity needed to power electric vehicles). Since it is solely focusing on Germany, the strategy displays transformative cognitive pioneership.

Besides minimum performance standards for new buildings, the most successful policy instrument – both in terms of impact and visibility – has been the state-owned development bank's (*Kreditanstalt für Wiederaufbau* – KfW) support programmes for building renovation. Since 1996, its intentions has been to save CO_2 in the building sector. Internationally, the KfW support programmes for new energy-efficient buildings and energy efficiency renovation are regarded as examples for cognitive leadership. The German government actively used this support programme to show internationally its high climate protection ambitions in the building sector (Rosenow *et al.*, 2013).

Despite this rather successful policy instrument and ambitious targets, the current policies for the building sector are insufficient for reaching the climate protection target. To achieve the target, a stepping up of 'deep energy renovation', rather than merely 'shallow renovation' which would lead to lock-in effects at a low ambition level, is necessary. Moreover, at least a doubling of the renovation rate would be required. In addition, the building's function for the energy transition (e.g. its energy storing and production capacity) will need a policy framework that systematically takes those synergies into account (BPIE, 2017). However, defining and implementing measures and policy instruments as well as transposing EU directives into national law lack ambition, innovation and commitment (Weyland and Steuwer, 2018). Here, Germany acted in some cases even as a laggard.

Germany's building policy from a multilevel governance perspective

The EU Energy Performance of Buildings Directive (EPBD) has been an important instrument to guide Member States in setting up ambitious policies. It obliges Member States to define a nearly zero energy building (nZEB) standard, to implement Energy Performance Certificates (EPC) and to set up ambitious long-term renovation strategies. However, until now, the German government has failed to define the nZEB standard, while other Member States (e.g. Denmark) are pioneering the definition of ambitious standards (D'Agostino *et al.*, 2017). Although Germany is complying with the required EPC schemes, analysis has shown that the implementation and the lack of comprehensive data collected mean that Germany is a laggard (Li *et al.*, 2019). Germany is among only three Member States that have not complied with the requirements of implementing the long-term renovation strategies (Castellazzi *et al.*, 2019).

With regard to the negotiation of the amendments of the EPBD in 2017, the question of whether Germany acted as a laggard, follower, pioneer or even a leader cannot to be answered easily. While German policy-makers defended the status quo of the EPC scheme, they were keen on introducing new policy approaches that had already been discussed at the national level. For example, German policy-makers advocated the introduction of the Building Renovation Passport, a policy instrument that would allow to plan individual renovations in view of long-term climate protection targets. It has been developed at the subnational level by the German state (*Land*) Baden-Württemberg and is now also available at the national level. Baden-Württemberg therefore acted as a cognitive leader. There are several examples for its cognitive leadership at the subnational level including its 2007 Renewable Heat Act, which was the first of its kind. The example of a Building Renovation Passport has not only inspired policy-makers at the German national level but also the development of similar schemes in France and Flanders, with the latter establishing a digital logbook and thus advancing the idea a step further (BPIE, 2018). Although France and Flanders initially acted as followers, they are now regarded as cognitive leaders with several research projects evaluating success factors and transferability to other EU Member States. Its uptake in the EPBD, although in a non-binding way, provides evidence for Germany's active leadership in promoting this new approach across the EU. In other words, Germany was leading by example. The multilevel governance system helped not only to generate followers but also to spur regulatory competition and promote the follower's leadership.

A second example, where German policy-makers took a leading role during the negotiations of the EPBD, relates to the attempt to allow for so-called district approaches. It would allow the balancing of energy performance requirements across a district. The German government was heavily criticised for the district approach by climate protection stakeholders who argued that it would be an excuse for many existing buildings to not undergo renovations because a new building might compensate for it (Gebäude-Allianz, 2019). This example shows how an innovative idea became the subject of a debate between challengers and incumbents. Incumbents were embracing the idea of a district approach to possibly dilute the level of renovation ambition. Challengers were then in a position to fight against an innovative idea with transformative potential.

At the national level, the government failed repeatedly to adopt the Energy in Buildings Law (GEG). The first draft of the law was developed during the 2013–2017 legislature when the building sector was part of the Environmental Ministry's responsibility. It included a rather ambitious nZEB definition which, however, was stopped by the CDU/CSU before it could be adopted at the end of the legislature. Following the 2017 elections, responsibility for buildings became part of the Interior Ministry. The change in ministerial responsibilities for the building sector resulted in a watering down of the draft law. Accordingly, the new draft proposal failed to gain the required majority among the different ministries involved and was still not adopted by early 2020.

A transformative climate policy would have required innovative approaches for renovation, ambitious minimum standards and tax breaks as promised in the 2017 coalition treaty. When the Finance Minister dropped the idea to introduce tax breaks for energy renovation, the ministries in charge – Interior Ministry and Economics Ministry – proposed the introduction of a *Baukindergeld* to support families with children who want to build new houses. It was heavily criticised by economic experts for supporting additional vacancy in rural areas and making construction works more expensive in highly populated urban areas (Deutscher Bundestag, 2018). Some reports have claimed that the *Baukindergeld* can be regarded as windfall profits for the construction industry (ZIA, 2019).

The negotiation and policy-making process of the GEG since 2017 has shown that transformative challengers may face considerable opposition from incumbent interests whose policies are also labelled as innovative although they do not support a sustainable transformation of the heat sector. Some of the *Länder* (the 16 states within Germany) became leaders by contributing innovations for heat system transformation. Baden-Württemberg not only initiated and implemented the Building Renovation Passport, but also implemented a new and innovative support scheme to finance 'deep energy' retrofits in combination with prefabricated facades and roof modules which allowed for innovative business models (such as the Dutch *Energiesprong* model) to be tested also in Germany. This scheme supports the transformation of the whole logic of building renovation and lifting solutions from its current niche into the mass market. Baden-Württemberg was following international good practice and showed exemplary leadership by establishing the first financial support programme and to help scale up this innovative instrument from niche to mass market. While the German government followed some ideas of Baden-Württemberg and even promoted them at the EU level, the level of ambition when negotiating EU directives and subsequently transposing them into national legislation clearly lacked structural or entrepreneurial leadership, both of which would have been required in the transformation phases during which the struggles between incumbents and challengers become more intense.

Conclusion: German climate policy – from leader to symbolic leader

Germany has been a pioneer in climate policy and energy technology. In some areas it has also acted as a leader in climate policy. In particular, Germany pushed for the diffusion of technology innovations – predominantly wind energy and PV – and its feed-in tariff model is a policy instrument innovation. Germany's strategy was clearly one of creating a lead market at home while, at the same time, diffusing a political innovation that in turn encouraged the diffusion of the technology (see Jänicke and Jacob, 2004).

The overarching strategies for the building sector in Germany – the Energy Efficiency Strategy for Buildings – and other political initiatives and instruments are also innovative and may unfold transformative potential. District approaches are an attempt to realise integrated energy transitions on a smaller scale with implications not only for technological, but also for social change. Germany actively promoted at European level not only the well-known KfW support programmes but also other policy instruments including the Building Renovation Passport and the new support programme for prefabricated renovation solutions.

During the transition phase (see Figure 9.1) it became clear that the *Energiewende* was much more than the deployment of a discrete set of technologies – it is also about a fundamental redesign of a pervasive socio-technical system. With a broadening of the *Energiewende* within and beyond the electricity sector, the transition came under attack by a range of actors for different reasons. In the electricity sector, the government made adjustments concerning the direction of change and slowed the pace of transition.

Germany has also changed its leadership strategy when experiencing that a complex sociotechnical transition cannot follow a masterplan but must be flexible and adaptive. Countries also have different starting positions when it comes to energy technologies and potentials. Therefore, policy transfer cannot mean that one country provides a comprehensive blueprint for adoption by others. The leadership strategy has therefore shifted from 'selling' renewable technologies and the feed-in tariff to more modest and more polycentric ideas of exchange and joint learning. The German government has developed a large number of international energy dialogues and partnerships (Foreign Office, 2017) and thus provided entrepreneurial and cognitive leadership. While admitting that other countries may be able to learn not only from Germany's successes but also from its 'course corrections' (Foreign Office, 2020), the German government has become notably more restrained with regard to its ability to offer ready-made solutions.

The heat system transformation in Germany is facing a political lock-in situation that lacks an overall strategy for climate protection in the building sector. With responsibility for the building sector having been transferred from the Environment Ministry to the Interior Ministry, incumbents have been strengthened which tried to water down the policy ambition level. To counteract this, the Environment Ministry tried to challenge the other resorts by transferring to them the responsibility for reaching specific sectoral climate targets. The Climate Protection Programme 2030 provides new impulses such as a carbon price. However, overall the measures are insufficient to fully reach the long-term climate protection target of the building sector. Especially new minimum standards for new and existing buildings are still missing. This leads to an overall situation where not only the status quo is preserved but Germany can even be regarded as a laggard (e.g. for the pending GEG and its non-compliance with EU regulation) or at best as a follower. Important transformative topics, such as sustainable buildings and the circular economy, are largely niche activities carried out by private actors. They have not yet been pushed to a great extent by the German government.

Transformational leadership requires different interventions and support from governments and state actors depending on the particular transformational phase. While the support of innovations and the protection of niches for development and experimentation is comparatively easy, the subsequent phase of transformation requires structural and entrepreneurial leadership to overcome increasing incumbent-challenger struggles. Germany has not exerted this leadership at equal intensity domestically and externally. While it aimed at maintaining its international climate leadership position, the German government was often not able to solve increasing struggles about climate protection between different actors at the national level.

The EU has provided barriers to and opportunities for transformational leadership for Germany's *Energiewende*. Member States can only support innovations in accordance with the Single European Market. Also, the EU ETS had partly adverse effects on national initiatives to promote energy transition policies. On the other hand, alliances and partnerships at the EU level and the wider European transnational grids and electricity policies brought about cost reductions and platforms to promote ambitious climate protection.

In addition, transformational dynamics can arise in a bottom–up fashion also from the subnational level, inspiring both national and EU policy-making. In the building sector, leadership ambitions by the German government originated from innovative policies at the *Länder* level, in particular with regard to the Building Renovation Passport and support programmes for industrial renovation, which have been taken up first at the national level and were then promoted at the EU level by the German government.

With Chancellor Angela Merkel (CDU) promoting the Climate Protection Programme 2030 internationally, Germany is still trying to maintain leadership role. While there are still examples for innovative policies at the *Länder* and national levels which may qualify for 'leadership by example' and to some extent also as cognitive leadership, in early 2020 Germany's climate policy was not on track to meeting the 2030 national target. National targets are also outdated because Germany has so far failed to raise its overall climate policy goals in line with the ambition of the Paris Agreement, resulting in a lack of ambition and an implementation gap (SRU, 2019). Instead of becoming a truly transformational leader, Germany risks becoming a symbolic leader with decreasing credibility.

Notes

1 Julia Hertin is Managing Director of the German Advisory Council on the Environment (SRU). Sibyl Steuwer is Head of Buildings Performance Institute Europe (BPIE) Berlin Office. The article expresses personal views of the authors and does not reflect an official position of the SRU or BPIE.
2 The Christian Social Union (CSU) stands for election only in the state of Bavaria while the Christian Democratic Union (CDU) competes in all German states apart from Bavaria. CDU and CSU form one party faction in parliament. The Free Democratic Party (FDP) is a centre-right liberal party.

Bibliography

Agora Energiewende and Öko-Institut (2018) *Vom Wasserbett zur Badewanne. Die Auswirkungen der EU-Emissionshandelsreform 2018 auf CO₂-Preis, Kohleausstieg und den Ausbau der Erneuerbaren.* Berlin: Agora Energiewende, Öko-Institut e.V.

Altmaier, P. (2012) *Mit neuer Energie – 10 Punkte für eine Energie- und Umweltpolitik mit Ambition und Augenmaß.* Berlin: BMU.

Andersen, M. and Liefferink, D. (1997) *European Environmental Policy. The Pioneers.* Manchester: Manchester University Press.

BMWi (2014) *Ein Strommarkt für die Energiewende (Grünbuch).* Berlin: Bundesministerium für Wirtschaft und Energie.

BMWi (2015) *Der nationale Klimaschutzbeitrag der deutschen Stromerzeugung. Ergebnisse der Task Force "CO₂-Minderung".* Berlin: Bundesministerium für Wirtschaft und Energie.

BPIE (2017) *Das smarte Gebäude in der Energiewende.* Brussels: Buildings Performance Institute Europe.

BPIE (2018) *The Concept of the Individual Building Renovation Roadmap. An In-Depth Case Study of Four Frontrunner Projects*. Brussels: Buildings Performance Institute Europe.

Castellazzi, L., Zangheri, P., Paci, D., Economidou, M., Labanca, N., Ribeiro Serrenho, T., Panev, S., Zancanella, P. and Broc, J.-S. (2019) *Assessment of Second Long-Term Renovation Strategies under the Energy Efficiency Directive*. Luxembourg: Publications Office of the European Union.

Colander, D.C. and Kupers, K. (2014) *Complexity and the Art of Public Policy: Solving Society's Problems from the Bottom Up*. Princeton: Princeton University Press.

D'Agostino, D., Zangheri, P. and Castellazzi, L. (2017) Towards Nearly Zero Energy Buildings in Europe: A Focus on Retrofit in Non-Residential Buildings. *Energies*, 10. Available at: https://www.mdpi.com/1996-1073/10/1/117/pdf [accessed last 26 February 2020].

Deutscher Bundestag - Wissenschaftliche Dienste (2018) *Sachstand. Baukindergeld. WD 4 - 3000 - 071/18*. Berlin: Deutscher Bundestag.

Edenhofer, O., Flachsland, C., Kalkuhl, M., Knopf, B. and Pahle, M., (2019) *Bewertung des Klimapakets und nächste Schritte. CO$_2$-Preis, sozialer Ausgleich, Europa, Monitoring*. Berlin: Mercator Research Institute on Global Commons and Climate Change.

Federal Ministry for Economic Affairs and Energy (2019) Second Progress Report on the Energy Transition: The Energy of the Future. Reporting Year 2017. Berlin: Federal Ministry for Economic Affairs and Energy.

Fligstein, N. and McAdam, D. (2011) 'Toward a general theory of strategic action fields', *Sociological Theory*, 29: 1–26.

Foreign Office (2017) *The Energiewende: Secure, sustainable and affordable energy for the 21st century*. Berlin: Federal Foreign Office.

Foreign Office (2020) *Energiewende International*. Available at: https://www.auswaert iges-amt.de/de/aussenpolitik/themen/energie/energiewende/238782 [Last accessed 26 February 2020].

Gebäude-Allianz (2019) *Energiewende im Gebäudesektor voranbringen. Gemeinsame Position der Gebäude-Allianz zum Entwurf des Gebäudeenergiegesetzes GEG*. Berlin: Gebäude-Allianz.

Geels, F.W. (2004) 'From sectoral systems of innovation to socio-technical systems. Insights about dynamics and change from sociology and institutional theory', *Research Policy*, 33: 897–920.

Geels, F.W. (2014) 'Reconceptualising the co-evolution of firms-in-industries and theirenvironments: developing an inter-disciplinary Triple Embeddedness Framework', *Research Policy*, 43: 261–277.

Geels, F.W. and Schot, J. (2007) 'Typology of sociotechnical transition pathways', *Research Policy*, 36: 399–417.

Grießhammer, R. and Brohmann, B. (2015) *Wie Transformationen und gesellschaftliche Innovationen gelingen können. UFOPLAN-Vorhaben - FKZ 37121113*. Freiburg: Öko-Institut.

Howlett, M. (2014) 'Why are policy innovations rare and so often negative? Blame avoidance and problem denial in climate change policy-making', *Global Environmental Change*, 29: 395–403.

INSM (2017) *Die Fehler der Energiewende*. Berlin: INSM Initiative Neue Soziale Marktwirtschaft.

IPCC (2018) *Global Warming of 1.5°C. An IPCC Special Report on the Impacts of Global Warming of 1.5°*. Geneva: Intergovernmental Panel on Climate Change.

Jänicke, M. (2017) 'Germany: innovation and climate leadership'. In: R. Wurzel, J. Connelly and D. Liefferink (eds) *The European Union in International Climate Change Politics. Still Taking a Lead?* London: Routledge, 138–154.

Jänicke, M. and Jacob, K. (2004) 'Lead markets for environmental innovations. A new role for the Nation State', *Global Environmental Politics*, 4: 29–46.

Jordan, A., Asselt, H.V., Berkhout, F., Huitema, D. and Rayner, T. (2012) 'Understanding the paradoxes of multilevel governing: climate change policy in the European Union', *Global Environmental Politics*, 12: 43–66.

Jordan, A. and Lenschow, A. (2008) *Innovation in Environmental Policy? Integrating the Environment for Sustainability*. Cheltenham: Edward Elgar.

Jordan, A. and Matt, E. (2014) 'Designing policies that intentionally stick: policy feedback in a changing climate', *Policy Sciences*, 47: 227–247.

Kahl, H. (2015) 'Viele Wege führen nach Rom: Die Preisfindung bei der Förderung erneuerbarer Energien im Beihilferecht der EU und Subventionsrecht der WTO', *Zeitschrift für Umweltrecht*, 26: 67–72.

Kemfert, C. (2017) 'Germany must go back to its low-carbon future', *Nature*, 549: 26–27.

Kemp, R., Schot, J. and Hoogma, R. (1998) Regime shifts to sustainability through processes of niche formation', *Technology Analysis & Strategic Management*, 10: 175–195.

KWSB (2019) *Kommission "Wachstum, Strukturwandel und Beschäftigung". Abschlussbericht*. Berlin: Bundesministerium für Wirtschaft und Energie, Kommission Wachstum Strukturwandel und Beschäftigung.

Li, Y., Kubicki, S., Guerriero, A. and Rezgui, Y. (2019) 'Review of building energy performance certification schemes towards future improvement', *Renewable and Sustainable Energy Reviews*, 113: 109244.

Liefferink, D. and Wurzel, R.K.W. (2017) 'Environmental leaders and pioneers: agents of change?' *Journal of European Public Policy*, 24(7): 951–968.

Mazzucato, M. (2015) *The entrepreneurial state: Debunking public vs. private sector myths (Vol. 1)*, London Anthem Press.

Newell, P. (2015) 'The politics of green transformations in capitalism'. In: I. Scoones, M. Leach and P. Newell (eds) *The Politics of Green Transformations*. Abingdon: Routledge, 86–103.

Partzsch, L. (2015) 'Umweltpolitik: Welche Macht führt zum Wandel? In: L. Partzsch and S. Weiland (eds) *Macht und Wandel in der Umweltpolitik*. Baden-Baden: Nomos, 7–26.

Ragwitz, M. and Huber, C. (2005) *Feed-In Systems in Germany and Spain and a Comparison*. Karlsruhe: Fraunhofer Institut für Systemtechnik und Innovationsforschung.

Rosenow, J., Eyre, N., Rohde, C. and Bürger, V. (2013) 'Overcoming the upfront investment barrier - comparison of the German CO_2 Building Rehabilitation Programme and the British Green Deal', *Energy & Environment*, 24: 83–103.

Rotmans, J., Kemp, R. and Asselt, M.V. (2001) 'More evolution than revolution: transition management in public policy', *Foresight*, 3: 15–31.

Schneider, C. and Veugelers, R. (2010) 'On young highly innovative companies: why they matter and how (not) to policy support them', *Industrial and Corporate Change*, 19: 969–1007.

Schreurs, M.A. and Tiberghien, Y. (2010) 'European Union leadership in climate change: mitigation through multilevel reinforcement'. In: K. Harrison and L.M. Sundstrom (eds) *Global Commons, Domestic Decisions: The Comparative Politics of Climate Change*. Cambridge, MA: MIT Press.

Schuler, J., Krämer, C., Hildebrandt, S., Steinhäußer, R., Starick, A. and Reutter, M. (2017) *Kumulative Wirkungen des Ausbaus erneuerbarer Energien auf Natur und Landschaft*. Bonn: Bundesamt für Naturschutz.

Scoones, I., Leach, M. and Newell, P. (eds) (2015) *The Politics of Green Transformations*. Abingdon: Routledge.

Sensfuß, F., Ragwitz, M. and Genoese, M. (2008) 'The merit-order effect: a detailed analysis of the price effect of renewable electricity generation on spot market prices in Germany', *Energy Policy*, 36: 3086–3094.

Smink, M.M., Hekkert, M.P. and Negro, S.O. (2015) 'Keeping sustainable innovation on a leash? Exploring incumbents' institutional strategies', *Business Strategy and the Environment*, 24: 86–101.

Solorio, I., Öller, E. and Jörgens, H. (2014) 'The German energy transition in the context to the EU renewable energy policy. A reality check!' In: A. Brunnengräber and M.R. Di Nucci (eds) *Im Hürdenlauf zur Energiewende. Von Transformationen, Reformen und Innovationen*. Wiesbaden: Springer VS.

SRU (2002) *Umweltgutachten 2002. Für eine neue Vorreiterrolle*. Stuttgart: Metzler-Poeschel.

SRU (2013) *Shaping the Electricity Market of the Future. Special Report*. Berlin: German Advisory Council on the Environment.

SRU (2015) *The Future of Coal trough 2040*. Berlin: German Advisory Council on the Environment.

SRU (2016) *Umweltgutachten 2016. Impulse für eine integrative Umweltpolitik*. Berlin: Erich Schmidt.

SRU (2017) *Kohleausstieg jetzt einleiten*. Berlin: Sachverständigenrat für Umweltfragen.

SRU (2019) *Offener Brief zum Klimakabinett: Umsetzungs- und Ambitionslücke schließen*. Berlin: Sachverständigenrat für Umweltfragen.

Steinbacher, K. and Pahle, M. (2015) *Leadership by Diffusion and the German Energiewende*. Potsdam: Potsdam Institut für Klimafolgenforschung. Available: https ://www.pik-potsdam.de/members/pahle/dp-ew-leadership-2015.pdf [Accessed 24 June 2015].

Szarka, J. (2012) 'Climate challenges, ecological modernization, and technological forcing: policy lessons from a comparative US-EU analysis', *Global Environmental Politics*, 12: 87–109.

Tews, K. (2015) 'Europeanization of energy and climate policy: the struggle between competing ideas of coordinating energy transitions', *Journal of Environment and Development*, 24: 267–291.

Underdal, A. (1994) 'Leadership theory: rediscovering the arts of management'. In: IIASA (ed.) *International Multilateral Negotiating: Approaches to the Management of Complexity*. San Francisco: International Institute for Applied Systems Analysis, 178–197.

Weber, M. and Hey, C. (2012) 'Energiewende: Kosten wirklich zu hoch?', *Wirtschaftsdienst*, 92: 360.

Weidner, H. (2008) *Klimaschutzpolitik: Warum ist Deutschland ein Vorreiter im internationalen Vergleich? Zur Rolle von Handlungskapazitäten und Pfadabhängigkeit*. Berlin: Wissenschaftszentrum Berlin für Sozialforschung.

Weiland, S. and Partzsch, L. (2015) 'Zum Nexus von Macht und Wandel'. In: L. Partzsch and S. Weiland (eds) *Macht und Wandel in der Umweltpolitik*. Baden-Baden: Nomos, 225–238.

Weyland, M. and Steuwer, S.D. (2018) 'Germany's struggle for energy efficiency', *GAIA*, 27: 216–221.

Wurzel, R.K.W., Liefferink, D. and Torney, D. (2019) 'Pioneers, leaders and followers in multilevel and polycentric climate governance', *Environmental Politics*, 28(2): 1–21.

ZIA (2019) *Frühjahrsgutachten Immobilienwirtschaft 2019 des Rates der Immobilienweisen.* Berlin: Zentraler Immobilien Ausschuss.

10 Lessons from climate action in the UK

The limitations of state leadership

Jeremy F.G. Moulton

Introduction

The UK's reputation of having a paradoxical record on climate action, wherein ambitious commitments have not always been matched with comparable action (Rayner and Jordan, 2017: 173), is one that came into particularly stark contrast in the years surrounding the 2015 Paris Agreement, finalised at the Conference of Parties (COP) 21 of the United Nations Framework Convention on Climate Change (UNFCCC). This tension between moments of global leadership and the subsequent undermining of UK climate action became a hallmark of the country's approach to climate change in the 2010s. This marks a continuation of a pattern that has marked the UK's approach to climate leadership since the issue first rose to international prominence. This chapter will critically explore that record through the lens of the different types of leadership that the UK has utilised. Such an understanding of the UK's leadership types is important for the message that it sends not only to potential follower states but also to other governance actors that the UK might lead towards more ambitious climate actorness, e.g. the business sector.

The UK's leadership record is one that deserves particular attention for a number of reasons. Firstly, the UK has attempted to establish itself as a climate leader. Yet, within these attempts at leadership there are tensions between claimed climate leadership and governance reality, as the concerns that the UK risks not meeting future carbon budgets demonstrate. Secondly, the UK's role as a leader on climate change has been brought into particularly sharp focus in the lead-up to its hosting of COP26, which had been due to take place in Glasgow in November 2020. In April 2020, the Conference was postponed to 2021 due to the COVID-19 pandemic. COP26 has the capacity to be a key moment for the UK to demonstrate its entrepreneurial leadership, as it has in the past at previous COPs (Rayner and Jordan, 2017: 173). Thirdly, the UK's actorness on climate change is especially necessary due to the country's withdrawal (herein: Brexit) from the European Union (EU) (Moulton and Silverwood, 2018). As discussed below, Brexit led to tensions around the future of UK climate leadership during the late 2010s. These factors provide a case for why the UK's leadership record deserves assessment.

The following analysis finds that the UK has developed a credibility gap between its symbolic leadership on climate action and the extent and results of policy outputs and outcomes to reduce greenhouse gas emissions (GHGE). In particular, the UK fell short of its declaratory commitments to combat climate action after the 2008 *Climate Change Act* (CCA), prioritising economic preferences over environmental necessities. Whilst the CCA legislated for an ever-expanding extent of climate actorness in order to reduce GHGE, the 2010s experienced periods of laggardly commitment to action that include evidence of rollback of key domestic climate policy commitments. Yet, within this context, it is argued that the UK has employed a range of different leadership types in order to provide leadership. Therefore, whilst there are clear shortcomings in policy commitments and tensions between the country's leadership ambitions and the realities of governance, the UK's record as a climate leader remains strong overall.

A limited record of state leadership

The story of the UK's record on climate action is one that is riddled with tensions between demonstrations of leadership, in a variety of forms, and markedly unambitious additional actions, or even the withdrawing of leadership actions. The picture that is painted of the UK through an examination of this timeline is one wherein the UK wants to be a leader much more than it wishes to be a pioneer. This is a distinction that is most usefully made by Liefferink and Wurzel (2017) and Wurzel *et al.* (2019): while a leader intends to attract followers, this is not normally the case for a pioneer. The evidence suggests that within UK governance circles there is no desire to see the UK 'go it alone' on climate action (Liefferink and Wurzel, 2017: 954). In the following section, the tension in the UK's bid to climate leadership will be analysed through the lens of the four types of leadership (exemplary, entrepreneurial, cognitive and structural) as detailed by Liefferink and Wurzel (2017) and Wurzel *et al.* (2019). This analysis will detail both the leadership types that the UK has demonstrated in climate action from the late-1980s onwards and the inherent tension that exists between moments of claimed leadership and realities of the country's track record.

Exemplary leadership

The exemplary leadership type is one that features highly in the UK's record. However, it is also one within which one can most easily see the emergence of a credibility gap between different climate governance actions. It therefore makes an important starting point for the analysis of UK leadership. Wurzel *et al.* define exemplary leadership as the 'intentional setting of examples for others' (Wurzel *et al.*, 2019: 11). Exemplary leadership necessitates a constructive pusher role where '[c]onstructive pushers intentionally put forward domestic policies as models for others' (*ibid.*). As the examples below detail, the UK has certainly adopted this role over the last three decades of its climate action.

In the early stages of UK climate actorness, this tension between claimed leadership and governance reality is well detailed by Rayner and Jordan (2017) (see more below on symbolic leadership). This initial moment of actorness is key to note as the UK was an early state actor on climate change. Prime Minister Margaret Thatcher was the first UK leader to put climate change onto the UK agenda and committed the country to take on a leadership role in a speech made at the UN General Assembly in 1989:

> Britain has some of the leading experts in this field and I am pleased to tell you that the United Kingdom will be establishing a new centre for the prediction of climate change… Every nation will need to make its contribution to the world effort, so I want to tell you how Britain intends to contribute, either by improving our own national performance in protecting the environment or through the help that we give to others.
>
> (Thatcher, 1989)

The subsequent creation of the Hadley Centre for Climate Prediction and Research, along with calls for stabilising emissions levels and the institution of market-based instruments, were early moves towards climate leadership that were exemplary in nature. However, whilst they were moves towards leadership, these actions did not transform into a serious attempt at leadership. In the following decade the largest GHGE reduction achieved by the UK was the result not of conscious climate action but of the shift from coal to gas-fired power generation on economic grounds – the so-called 'dash for gas'. Certainly, the UK was offering a more transactional (i.e. piecemeal, slower paced and without sizeable impact) approach to climate action than it was transformational (i.e. holistic, fast-paced and with significant impact) (Liefferink and Wurzel, 2017). The UK opposed the EU Commission's 1992 carbon dioxide tax proposal largely because it was opposed to any taxes on the supranational level (e.g. Rayner and Jordan, 2017: 175). Rather than being an early signal of the UK's unwillingness to act on climate change, the opposition to the EU carbon tax has been concluded as being more firmly understood as the result of UK ideology and political preferences in its relationship with Europe (Weale, 1999: 43), including its refusal to accept any taxes on the supranational level.

Whilst these early actions were not marked by the setting of example through the adoption of ambitious policy, the UK has made attempts at being an exemplary leader through the creation and implementation of climate legislation (Wurzel *et al.*, 2019: 11). The 2008 CCA made the UK the first country in the world to pass legally binding GHGE reduction targets (DEFRA, 2010). That the UK was taking an international lead is something that David Miliband, Secretary of State for Environment, Food and Rural Affairs, emphasised in a YouTube video publicising the draft *Climate Change Bill* (the video's creation itself is a sign of the enthusiasm in governance circles to communicate leadership on this issue). As Miliband (2007) stresses in the video: 'This Bill is a world first, Britain has become the first country to set itself on a road towards a legislative requirement to reduce its carbon emissions'. As well as this legislation marking the UK as an exemplary leader because of its novelty, it also contained an innovative governance arrangement that could be repeated in other states and that has been labelled a 'radical institutional change in environmental governance' (Lorenzoni and Benson, 2014). As well as instituting an unconditional 2050 GHGE reduction target of 80%, the CCA included the introduction of the Climate Change Committee (CCC) and the Adaptation Sub-Committee. These two independent committees' purpose is to advise on the UK's climate action in both mitigation and adaptation and to create 'carbon budgets' (five-year interim targets that progress the UK to the ultimate 2050 target). As the following section on entrepreneurial leadership details, the UK was also keen to use its sizeable diplomatic power in order to maximise the potential of becoming a constructive pusher with the example of the CCA being promoted to other nations (especially those within Europe).

It should also be noted that environmental advocacy groups have developed a strong track record of positively influencing the UK's climate action, especially contributing to the policy that has made the UK able to lead as an example. Groups such as Friends of the Earth (FoE) were central to raising public awareness of climate change as a policy problem and to pushing government to adopt specific policy solutions, such as the CCA and the Renewable Energy Obligation (Hale, 2010: 264; Carter and Childs, 2018). The environmental law charity ClientEarth built on the success of the CCA's creation by promoting the notion of legally binding climate legislation, drawing on the UK case study, to other EU countries (ClientEarth, 2009). This reveals the nature of leadership within the UK's climate governance field, wherein the leadership of the UK vis-à-vis other states has often been the result of leadership pressure from environmental advocacy groups.

A more contemporary example of this leadership relationship came in May 2019 as the UK government, under pressure from Extinction Rebellion protests across the country, became the first country to declare a 'climate emergency' (Turney, 2019). This resulted in the furthering of UK climate policy commitments. In June 2019, the UK continued its leadership track record by becoming the first major economy to adopt a 2050 net zero carbon emissions target for 2050 (Hook and Sheppard, 2019). This was a clear increase in the ambition laid out in the CCA over a decade before, despite other leading economies, such as the USA (see Chapter 7 in this volume), reducing their level of climate action in the same period.

Despite these 'flagship' and attention-grabbing examples of leadership through ambitious target setting, the UK has come under criticism for these measures' results. Reviews of the CCA indicated that a more prescriptive, command-and-control approach to themes such as transition to renewables would have provided a more consistent and stronger message to the energy sector and investors (Lockwood, 2013; Fankhauser *et al.*, 2018: 25). Lockwood (2013: 1341) argues that the CCA, despite including targets and budgets that are in principle legally binding, is better recognised as an attempt to ensure future climate action through political mechanisms, e.g. ensuring transparency, accountability and long-term pressure. Similarly, there are concerns that the original legislation was not strong enough to avoid the reversal of climate action and policy commitments, leading to the worry that 'the gap is widening between the emissions targets set in law and the policies put in place to deliver them' (Fankhauser *et al.*, 2018: 25). This concern is one that the UK government has admitted to, stating that there are 'projected shortfalls against the fourth and fifth carbon budgets of 139 and 245 MtCO$_{2e}$ respectively' (Department for Business, Energy & Industrial Strategy, 2019). These numbers demonstrate that while the UK has been furthering its symbolic commitment to climate action, there is a widening credibility gap when it comes to the reality of GHGE reductions in the country. One example of this is the long-term fight between climate activists and the government on the issue of airport expansion (Cooke, 2020). This reality is one that undermines the UK's role as an exemplary leader.

The third theme that is useful to explore in reference to the UK's record as exemplary leader on climate action is the country's support for and use of the renewables sector. The UK has, at times, been keen to position itself as a world leader in the renewables sector, especially in relation to offshore wind, despite having previously been seen as a laggard in this field (Toke, 2011; Kern *et al.*, 2014). That the UK wanted to be a world leader on offshore wind became particularly apparent during the 2010–2015 Coalition Government between the Conservative and Liberal Democrat Parties. In 2013, at the opening of a 270 MW offshore wind farm, the Liberal Democrat Deputy Prime Minister, Nick Clegg, made clear the intentionality behind this strategy towards leadership, stating:

> The race is now on to lead the world in clean, green energy. As an island nation, and with our weather, the UK is ideally placed to make the most of offshore wind energy – you could say it was a technology designed for us... This strategy will keep Britain as the world leader in one of the most important industries of the 21st century.
>
> (Clegg quoted in *Climate Home News*, 2013)

Looking at the results of the offshore wind sector in the UK, it is apparent that the hopes laid out by the Coalition Government have been borne out. In the 2020 *Energy Trends*, the UK government's annual publication of energy sector statistics, it was stated that in 2019 renewables had reached a record high of 36.9% of electricity generation. In total, the UK ended 2019 with 47.4 GW of installed renewables capacity, a 6.9% increase on the previous year with half of the increase coming from offshore wind (Department of Business, Environment and Industrial Strategy, 2020: 3). This expansion of offshore wind has made the UK the world leader in that sector of the renewables market in terms of total installed capacity (RenewableUK, 2020). As of 2019, the UK had 9,945 MW of installed capacity, compared to Germany with 7,445 MW and Denmark with 1,703 MW (WindEurope, 2020: 15; see also Chapters 9 and 11 in this volume). This leadership record has been promoted by successive UK governments (e.g. UK Trade & Investment, 2015). The country has also hosted international delegations to demonstrate the feasibility and benefits of the adoption of offshore wind as a renewable energy source (e.g. renews, 2018). This stands as evidence that the UK has developed credibility on offshore wind and wishes to stand as an exemplar to others. One can also see, in the UK's embrace of offshore wind, the move the country has made from a follower into a leader on certain aspects of climate action.

The UK's strong record on offshore wind and certain other renewables (e.g. solar photovoltaics) is, however, not one that has always easily developed nor been marked by a wholesale commitment to renewables and state-level financial support. In 2013, concern about the government's support for renewables was raised when the UK opposed an EU level 2030 renewable energy target (Harvey, 2013). That the UK was so committed to offshore wind in fact originated in a 2016 moratorium from Prime Minister David Cameron that excluded onshore wind turbines from applying for state subsidies for low-carbon energy (Pickard, 2020). This moratorium, primarily driven by strong opposition to onshore wind turbines (largely for aesthetic reasons) among many supporters of the Conservative Party, especially in rural areas, had severely limited the expansion of onshore wind in the UK, which fell to its lowest level in 2019 (*ibid.*). This decision was then reversed by the Boris Johnson-led Conservative government in March 2020 as part of the drive to demonstrate some forms of leadership ahead of COP26. The financial support for offshore wind itself has also suffered setbacks. The post-Coalition, Conservative-majority government of 2015–2017 lost some of its climate credentials as part of a 'policy reset' including withdrawing financial support for renewables whilst simultaneously offering tax breaks for oil and gas extraction (Rayner and Jordan, 2017: 177) – a move that was clearly distinct from ambitious climate actorness. Similarly, the 2017 sale of the Green Investment Bank – an innovative state-created, low-carbon lender that was launched by the Coalition government in 2012 – by the Conservative government has been criticised as unnecessary and detrimental to the expansion of the renewables sector in the UK (Environmental Audit Committee, 2018; Cumbo, 2019). These moves and others have led to questions around the credibility of the apparent enthusiasm in the UK governance circles for renewables – though the extent of offshore wind installed capacity does provide a strong defence of the UK's record of leadership here.

Overall, it is apparent that the UK has attempted to develop an international role for itself that does, at times, reflect the exemplary leadership type defined by Liefferink and Wurzel (2017). In the above examples, it is made clear that UK governments have 'actively use[d] experiential knowledge gained at the domestic level in their efforts to convince others of the feasibility of their preferred external policy solution' (Liefferink and Wurzel, 2017: 960). However, this leadership is most notable in the offshore wind sector. Whilst the UK has become the global leader (a role that it has been keen to promote) in this sector, it is just one sector of both renewables and of potential climate action. As also noted in this section, the UK's exemplary leadership record has been mixed due to occasions of policy reversal, inconsistent behaviour and an emerging credibility gap centred on ambitious climate pledges not being matched with reduced GHGE reductions. These factors contribute to the critique of the UK's climate action record and have resulted in its reputation as a paradoxical leader. Whilst the CCA, its global leadership in offshore wind and its declaration of a climate emergency all cast the country as a leader, these have yet to materialise holistic and transformative approaches to climate action. In fact, the CCA and the commitment to reduce the UK's emissions to net zero by 2050 can be noted as measures that ultimately set targets for others to meet in the future, rather than necessitating radical action by contemporary governments. Nevertheless, this leadership type, demonstrated by those key 'flagship' projects and moments of initiative, is undoubtedly central to the perception of the UK and the UK's track record of being a climate leader.

Entrepreneurial leadership

In order to maximise the UK's role as an exemplar, the country deploys strategies that can be recognised as reflecting entrepreneurial leadership. Liefferink and Wurzel (2017: 957) define the entrepreneurial leadership type as one which 'involves diplomatic, negotiating and bargaining skills in facilitating compromise solutions and agreements'. The UK as a state with a high degree of structural power (see the section on structural leadership below) is one that is well placed to exploit this power in a bid to entrepreneurial leadership and has a track record of attempting to play a central role in international climate politics (Sindico, 2007). This type of leadership is particularly important to take note of as the UK heads towards its chairing of the delayed COP26 in 2021.

Entrepreneurial leadership is best defined by Young (1991: 300) as 'an agenda setter and populariser who uses negotiating skills to devise attractive formulas and to broker interests'. The UK has long demonstrated its ability to use these entrepreneurial strategies, particularly in the international climate regime. The UK took an early and prominent role in the UNFCCC, a key sign of this bid towards outwards-facing leadership. In an important example of entrepreneurial leadership in the lead-up to the 1992 agreement to create the UNFCCC, it was the UK that worked hard on a deal which was eventually acceptable for the USA (Rayner and Jordan, 2017: 175).

The key international arena where the UK's diplomatic, negotiating and bargaining skills have often been put to work in an effort to increase climate action ambition is in its participation in the EU. The UK can largely be classed as having been one of the more ambitious EU Member States from the mid-1990s onwards, pressing other Member States to participate in ambitious climate action (Rayner and Jordan, 2017: 182). In a bid to facilitate compromise it has even demonstrated a willingness to exceed the ambition of most other Member States and make more of a relative effort in climate mitigation efforts. It increased its 10% GHGE reduction pledge to 12.5% after the Kyoto Agreement, making it the only state which increased its committed level of GHGE reduction (Jordan *et al.*, 2010). This example also reveals that, at times, the UK does display the hallmarks of a climate policy pioneer, being willing to act without the expectation of followers (Wurzel *et al.*, 2019). The UK attempted to use its entrepreneurial leadership to 'upload' its policy preferences to the European level in order to gain a first-mover advantage – with the most notable example of attempted upload being on the development of an EU emissions trading scheme (ETS). UK efforts were, however, largely frustrated in the establishment of the EU ETS despite the secondment of civil servants to the EU in an effort to help shape the legislation (Rayner and Jordan, 2017: 182). The UK had more success in using its 2005 EU Presidency, which involved chairing Council meetings, for giving the Commission a steer to include aviation within the EU ETS for the first time (*ibid.*). However, the country also came under some criticism for failing to join up its EU Presidency with its parallel chairing of the G8 in order to promote a common EU response on climate action (Whitman and Thomas, 2005: 5), which represented a faltering in the UK's entrepreneurial leadership.

The UK's entrepreneurial leadership has also been demonstrated in the wider field of international climate politics. As well as the work of the UK to include the USA in the UNFCCC on its inception, the country went on to play a keystone role in the ensuing Kyoto Protocol – the world's first international climate regime. In particular, the diplomatic work of John Prescott, the UK Deputy Minister, who took two round-the-world trips to meet with key actors ahead of the negotiations, was noted at the time as giving the UK a surprisingly central role in the success of the agreement (*Independent*, 1997). The example of Kyoto is important in understanding the continuity in the UK's enthusiasm for being an entrepreneurial leader in the international climate regime – an enthusiasm that continued in the lead-up to the 2015 COP21 in Paris. Ahead of COP21, the UK, along with France and Germany, engaged in coordinated outreach activities (notably through their foreign ministries, i.e. ministries with established international networks that can draw on entrepreneurial experience) (Wurzel, Liefferink and Di Lullo, 2019: 262). In 2015, the same three countries drove an increase in the EU's pledged public finance contributions to 2020 and were active in assisting nearly 100 countries develop climate action plans to submit for COP21 (Oberthür and Groen, 2018, 721). The UK, drawing on a network of partners and a willingness to coordinate with other countries, has demonstrated entrepreneurial leadership in its central participation in the UNFCCC international climate regime.

The UK has also demonstrated an enthusiasm to integrate climate action into a range of other key international policy areas, such as international security. For example, in 2007, the UK was the first state to raise the issue of climate change at the UN Security Council, an act that substantially added to the perception of the country as a diplomatic leader on the issue (Sindico, 2007). While still an EU Member State, the UK was also part of the High Ambition Coalition, a bloc within the UNFCCC negotiations that wished to push for increased climate action. However, cuts in funding that had supported climate diplomacy efforts in the years surrounding COP21 in Paris have given rise to concern that the UK has been inconsistent with its entrepreneurial leadership in climate diplomacy (Darby, 2014; King, 2015; Kolster and Smith, 2017).

Concerns about the UK's entrepreneurial leadership capacity in contemporary international climate politics have become focused around two key issues: the country's hosting of COP26 and Brexit.

As stated above, the COP26 summit was due to take place in November 2020, hosted in Glasgow by the UK and with Italy hosting a number of preparatory events. Whilst the summit was delayed until 2021 due to the COVID-19 pandemic, this event remains a vital case study of the UK's entrepreneurial leadership. That the UK, along with Italy, would make a bid to become the host of the most important conference since COP21 in Paris demonstrates that the country is still keen to play a central role in climate negotiations. However, it is important to note that a host's desire for a central role does not always correspond with promoting ambitious climate action. Poland hosted COP24 even though the country is committed to coal-fired power and has taken a laggardly approach to EU climate action (see also Chapter 8 in this volume). Despite Brexit, the UK has continued to publicise its membership of the High Ambition Coalition as part of its Presidency of COP26 (UKCOP26, 2020). Nevertheless, the UK failed to make early headway on utilising entrepreneurial leadership to develop an ambitious approach to COP26 in the months after it was awarded the position as host. Again, the tension in the UK's record as a paradoxical leader was revealed during this time as in mid-2019 the CCC warned that the country was falling short in terms of its GHGE reduction commitments, and that it may well not be able to command a leadership position at COP26. As the Chief Executive of the CCC, Chris Stark, stated:

> The government must show it is serious about its legal obligations... [its] credibility really is at stake here... There is a window of 12–18 months to do something about this. If we don't do that, I fear the government will be embarrassed at COP26.
>
> (Stark quoted in Evans, 2019)

However, in a dramatic turn, embarrassment came for the UK government some time before the Glasgow summit. On 31 January 2020, the UK government, under Prime Minister Boris Johnson, fired the UK's President of COP26 Claire O'Neill. This was soon followed by a widely publicised and highly critical public letter from O'Neill to the Prime Minister on 3 February 2020 – the day before Johnson was to deliver a speech to launch COP26. The letter detailed severe problems in the preparation for COP26 and the seriousness with which the Johnson government was treating the issue, with the warning that it would take significant work to improve the situation:

> To do that will require a whole government reset and for your team to move the vast and immediate challenge of climate recovery to the top of the Premier League of their priorities from where it is now – stuck currently somewhere around the middle of League One.
>
> (O'Neill, 2020)

This puts into contrast the reality of UK climate governance and leadership. This case shows the country's enthusiasm for flagship projects and commitments, such as hosting COP26, and wanting to be a central player in other UNFCCC negotiations. However, these high-profile pieces of action are often then undermined by a lack of commitment and follow through in practical terms.

As noted above, one of the ways in which the UK has exerted its influence as an entrepreneurial climate leader has been through its membership of the EU. The withdrawal from the Union has therefore caused many to question the future viability of UK climate leadership (Rayner and Jordan, 2017; Moulton and Silverwood, 2018). An early pledge that the UK would commit itself to a 'Green Brexit' has largely disappeared from the governance debate around Brexit, giving rise to concerns about the salience of climate action in the post-transition period. These concerns have been heightened during the post-withdrawal negotiations, with the UK resisting EU-led efforts to include Paris Agreement commitments and the EU's level playing field demands as part of a future UK-EU trade deal (Brunsden, 2020) – the EU has made a pledge to include climate action in all future trade agreements. The move would oblige the UK to meet its Paris Agreement commitments and to match the level of EU environmental standards, or else the EU would have the legal justification to withdraw a preferential trading arrangement with the former Member State. Whilst in theory the resistance to the level playing field would allow for the UK to more easily act as a climate leader, the steadfast resistance to such measures does not reflect the diplomatic and negotiating tactics of an entrepreneurial leader, nor the 'positive' behaviour that Underdal (1998: 101) asserts is central to leadership: 'a leader is supposed to exercise what might be called "positive" influence, guiding rather than vetoing or obstructing collective action'. Therefore, there is a disconnect between UK claims to leadership and its actions in relation to Brexit and climate action.

In summary, whilst the UK has a historic record of using its position as a climate actor with structural power to display the hallmarks of an exemplary leader, putting climate action onto the international agenda and facilitating compromise, this leadership role is under threat in the lead-up to COP26 and due to Brexit. This example shows again the tensions that lie at the heart of the UK's claims to climate leadership.

Cognitive leadership

Cognitive leadership is a type of leadership that 'involves defining or redefining ideas and concepts… [It] may also relate to cause-effect relations and policy solutions through the provision of scientific knowledge regarding innovative climate measures' (Wurzel *et al.*, 2019: 10). At its core, this is a type of leadership that relates to the presentation of climate change as a policy problem and ensuring presentation of a variety of preferable policy solutions.

In the climate leadership/pioneership literature, one of the foremost examples of cognitive leadership is the promotion of ecological modernisation (*ibid.*). Ecological modernisation is a concept that proposes that a synergy between environmental protection and economic growth is possible and should be a policy priority over demodernisation/anti-growth alternatives. Successive UK governments have certainly been influential in driving ecological modernisation into the main-stream. Politicians and policy-makers in the UK have a track record of making strong rhetorical commitments reflecting the ecological modernisation framework (see Revell, 2005). Likewise, claims have been made that the UK has successfully instituted a policy approach that reflects the ecological modernisation concep-tion, e.g. a statement from Margaret Beckett, UK Environment Secretary (2001–2006) and Patricia Hewett, UK Trade and Industry Secretary (2001–2006) clearly reflects this well: 'The UK has proved that economic growth does not have to lead to increased greenhouse-gas emissions' (quoted in Hayden, 2014: 1). Such is the strength of this potential 'win–win' between environmental protection and eco-nomic growth within the UK's approach to environmental action, that one envi-ronmental actor in the UK has been quoted as saying, 'ecological modernisation is the only game in town within the policy context that we work' (Hayden, 2014: 277). However, the UK's commitment to ecological modernisation has been cri-tiqued on the basis that it reveals a preference for the modernisation aspect of the concept (i.e. economic growth) over the ecological (i.e. environmental protection) (Revell, 2005; Hayden, 2014). This weak form of ecological modernisation likely limits the UK's role as a potential cognitive leader on climate action, a role that it would be more successful in undertaking if it were supporting a stronger and more consistent message.

Part of the support for and imperative behind the ecological modernisation concept came with the publication of the hugely influential *The Economics of Climate Change: The Stern Review* in 2006. *The Stern Review* was commissioned by Chancellor Gordon Brown in order to study the challenge that climate change posed to the economy both in terms of action and inaction. The report emphasised that there was still time to stop climate change, provided that strong action was taken immediately. If that action was not taken, then the economic cost would be devastating – between 5 and 20% GDP loss per annum (Stern, 2006). The report was hugely influential in reframing climate change as not only an environmental problem but as an economic problem as well (Grubb, 2015). The approach of weighing economic costs of action versus inaction is one that has continued with the advent of 'mini-Sterns' to assess local-level climate action (Wesselink and Gouldson, 2014). That the report was UK-commissioned put the country at the centre of this developed debate around conceptions of climate change as an eco-nomic policy problem.

Leading off from a period wherein climate action had been a valence issue in the UK, the Coalition government in 2010, led by Prime Minister David Cameron, made a pledge to be the 'greenest government ever' (Cameron quoted in Randerson, 2010). This could have been a serious moment of cognitive leader-ship with a major world economy making a clear and ambitious pledge to climate action. However, this is a commitment that the UK and Cameron failed to turn into a serious reimagining of government-level climate leadership. As climate action became an increasingly politicised issue with his Conservative Party, Cameron surrendered his earlier claims to green credentials in a bid to continue to appeal to the party at large but especially the right-wing of the Conservatives that does not favour climate leadership. This surrendering of ground, along with the policy rollbacks that this included, led to the conclusion by Carter (2015: 1056) that, '[g]enerally, Cameron provided weak leadership on climate change'. This in turn impacted the UK's role as a cognitive leader as the 'greenest government ever' moment pledge, which could have represented a significant shift in redefining the norms of governance in a major economy, was not followed through.

The decision by the UK government to declare a climate emergency in May 2019 is a more contemporary example of cognitive leadership. Whilst the UK was the first country to declare such an emergency, it was followed by all other EU countries declaring such an emergency and other countries including Canada, Argentina, Bangladesh and the Maldives, as well as a number of states and local authorities. Whilst this was undoubtedly the result of pressure from the Extinction Rebellion and high-profile climate activist Greta Thunberg, that the UK was again the first country to commit to the initiative again shows its willingness to redefine how such problems should be perceived and presented by governments. However, it is again necessary to note that this is a declaratory commitment that did not lead to immediate, ambitious climate action.

The UK has evidently attempted to play the role of cognitive leader and has succeeded in pushing the conjoining of economic and environmental logics, both in terms of *The Stern Review* and the promotion of ecological modernisation. However, as well as with inconsistent messaging around climate action, such as the failure of the Coalition government to fulfil its pledge to be the 'greenest government ever', the UK has also undermined its messaging and the redefining of climate action as an economically positive choice due to its prioritisation of the economy about the ecological. Therefore, given this track record, the UK's declaration of the climate emergency might also be considered through the lens of the credibility gap between stated leadership and the practical realities of UK climate governance.

Structural leadership

Structural power is defined by Liefferink and Wurzel (2017: 957) as one that relates to 'an actor's hard power and depends on material resources such as military power and economic strength'. Of course, as Liefferink and Wurzel (*ibid.*) go on to state: 'Apart from ecological conflicts about scarce resources (e.g. water), the relevance of military power tends to be low for environmental problem solving... For most environmental issues, structural power relies usually primarily on economic power'. Therefore, despite the relative unimportance of military power in this case, this still remains a relevant leadership type to examine with reference to the UK example. As one of the world's leading economic powers and a permanent member of the UN Security Council, the UK holds a large degree of influence and has sought to exploit its economic power and reach to promote its climate leadership, for example through its wide diplomatic network. Importantly, it is because of the UK's structural power that it is able to be such an influential climate leader in the types already explored in this chapter. However, it must be noted that the UK has not used its structural power in the manner that Germany (a higher emitting economy) has demonstrated (Rayner and Jordan, 2017: 184). Of the four types of leadership analysed in this chapter, structural leadership is the least committed to by the UK, arguably because of the ideological propensity for austerity during the 2010s, meaning that successive governments have been less willing to put their economic strength behind climate action.

With a sizable government budget, the UK has the capital to assist others in their transition towards climate actorness. The UK has utilised its international development work in an effort to influence other states and actors to alter behaviour in a more climate-conscious manner. In 2019, the Department for International Development (which has since been merged to join the Foreign Office) announced that the UK would double its international aid investments focused on the effort to mitigate climate change. This was an increase from £5.8 billion between 2015 and 2021 (announced before COP21) to £11.6 billion to be spent between 2021 and 2026 (Department for International Development, 2019). By funding climate mitigation and adaptation projects, the UK is using its economic strength to ensure that it has other countries that follow it on climate action.

In another example of the reach a world power can have, the UK used its structural position of power to promote the 2008 CCA as a model for other states to follow (ClientEarth, 2009). The UK was so keen to be a leader rather than a follower-less pioneer that in 2009 it hosted a series of seminars in British Embassies across Europe in a bid to share lessons from the institution of the CCA and potentially encourage uptake in other states (*ibid.*) – a move that distinguished the UK as a *pusher* for climate action (Wurzel *et al.*, 2019: 8). In fact, in their ten-year review of the CCA Fankhauser *et al.* (2018: 4) listed '[i]nternational leadership, inspiring others to act' as one of the key differences that has been made by the legislation. The UK government continues to use its network of embassies to promote both the country's track record on climate action and its preferred policy solutions to climate change (British Embassy Berlin, 2019). This demonstrates that the UK does want followers, whether it be out of climate consciousness or to avoid the risk of acting alone. It should be noted here that one might attempt to fit this leadership action within the entrepreneurial type due to Liefferink and Wurzel's (2017) assertion that this latter type is marked by diplomatic efforts. However, these examples do not, importantly, reflect the use of the UK's diplomatic power to reach agreements or text-based/legislation-based compromise. The CCA is not a policy innovation that could be uploaded to the EU level, for example (Rayner and Jordan, 2017: 182). Rather, the use of a diplomatic network, to promote the UK's climate record and preferred policy solutions, must instead be classed as one that exploits the structural power of the UK as a leading economy and a country that has used its military and diplomatic power to command an international reach.

Structural leadership has, to date, remained an underutilised leadership type in contrast to the other three types explored here on which the UK has relied more heavily. However, the UK's structural power remains vital to the country's relevance as a climate actor. Structural leadership is a type that is more strongly recognised in the EU's climate action (Scott, 2011: 28). Therefore, as the UK continues to develop its post-Brexit role in terms of climate action, it could be that this is a type it will choose to draw on more heavily now that it will be removed from the EU's structural climate leadership. Ironically, the UK may draw more heavily on structural climate leadership when its structural leadership potential has arguably been reduced due to Brexit.

Conclusion

In conclusion, although there have been tensions, sometimes severe, within claims of the UK to climate leadership, the variety of leadership types that have been employed have allowed for the UK to remain a prominent and influential actor on the climate crisis. The UK has especially employed its role as an exemplary and entrepreneurial leader in order to be a relevant leader with an extensive reach. Whilst the UK has remained a prominent and innovative leader (e.g. being the first country to institute a legally binding CCA as well as the first to declare a climate emergency) this does not always correspond with the level of practical climate action that the UK has committed to and followed through on. This chapter has detailed the UK's enthusiasm of flagship projects and commitments (in a manner that clearly goes beyond simply being a symbolic leader) whilst nearly simultaneously undermining climate action. This was a contradiction that was especially pronounced in the 2010s. Therefore, one can see that the reputation of the UK as a paradoxical climate leader is well deserved.

Because of this record, there are indications that its leadership record could be under threat. In particular, the lacklustre engagement with hosting COP26 that the Conservative government under Prime Minister Johnson demonstrated in the months after being awarded the position of host, is a warning sign that the UK might have shifting policy priorities that will put climate action further down the policy agenda. Similarly, the withdrawal from the EU has set off alarm bells for environmental activists as a promised 'Green Brexit' has ceased to be mentioned in governance circles and the issue of climate action has become a source of strain in the post-Brexit UK-EU trade deal negotiations. The emergence of the COVID-19 pandemic is only likely to further dominate the political agenda and therefore take political attention away from climate change at a particularly pressing time. COP26 was due to be the largest and most influential climate summit since the 2015 Paris Climate Change Conference (COP21). If the pledged increase in nationally determined contributions that was agreed to be undertaken every five years after the signing of the Paris Agreement does not succeed at the first hurdle, then the bottom–up approach championed at Paris (see also Chapters 1 and 5 in this volume) could itself be in danger of failure. Therefore, the success of the UK as a climate leader is especially vital for the future of international climate governance.

The lessons that one can draw from climate action in the UK are clear. Firstly, high-profile declaratory commitments and policy work well in establishing a country as a climate leader. Secondly, prominence in a key sector of climate action (in the UK's case, global leadership in offshore wind renewable energy) is important to the maintenance of that role as a climate leader. Finally, the UK case reveals that while these first two factors can give a country prominence and leadership capacity, that capacity can be severely undermined without a clear, practical commitment to climate action.

Bibliography

British Embassy Berlin (2019) *The United Kingdom - A Leader in Climate Protection.* 31 October 2019. Available at: https://www.gov.uk/government/news/the-united-king dom-a-leader-in-climate-protection [Accessed on 4 May 2020].

Brunsden, J. (2020) 'Brussels and Britain clash over climate conditions in trade deal', *Financial Times*, 6.5.2020. Available at: https://www.ft.com/content/0f09f819-77b3 -45d8-9ba3-76a3042c240c [Accessed on 6 May 2020].

Carter, N. (2015) 'The greens in the UK general election of 7 May 2015', *Environmental Politics*, 24(6): 1055–1060.

Carter, N. and Childs, M. (2018) 'Friends of the earth as a policy entrepreneur: 'The big ask' campaign for a UK climate change act', *Environmental Politics*, 27(6): 994–1013.

ClientEarth (2009) 'The UK Climate Change Act 2008 – Lessons for national climate laws', *An Independent Review*. Available at: https://www.documents.clientearth.org/ wp-content/uploads/library/2009-11-10-the-uk-climate-change-act-2008-xxx-lessons-f or-national-climate-laws-ce-en.pdf [Accessed on 28 June 2020].

Climate Home News (2013) 'UK outlines strategy to be 'world leader' in offshore wind', *Climate Home News*, 1.8.2013. Available at: https://www.climatechangenews.com/2 013/08/01/uk-outlines-strategy-to-be-world-leader-in-offshore-wind/ [Accessed on 4 May 2020].

Cooke, P. (2020) 'UK's Heathrow airport expansion ruled unlawful over climate change', *Climate Home News*, 27.2.2020. Available at: https://www.climatechangenews.com/2 020/02/27/uks-heathrow-airport-expansion-ruled-unlawful-climate-change/ [Accessed on 28 June 2020].

Cumbo, J. (2019) 'Green Investment Bank under fire for loss of UK focus', *Financial Times*, 6.1.2019. Available at: https://www.ft.com/content/9b1aa5e4-11a5-11e9-a581- 4ff78404524e [Accessed on 4 May 2020].

Darby, M. (2014) 'UK slashes climate diplomacy budget', *Climate Home News*, 31 July 2014. Available at: https://www.climatechangenews.com/2014/07/31/uk-slashes-clim ate-diplomacy-budget/ [Accessed on 4 May 2020].

DEFRA (2010) *Implementing the Climate Change Act.* Available at: https://webarchive.n ationalarchives.gov.uk/20130402165848/http://archive.http://defra.gov.uk/environmen t/climate/legislation/index.htm [Accessed on 4 May 2020].

Department for Business, Energy & Industrial Strategy (2019) *Updated Energy and Emissions Projections 2018.* Available at: https://assets.publishing.service.gov.uk/ government/uploads/system/uploads/attachment_data/file/794590/updated-energy-and -emissions-projections-2018.pdf [Accessed on 4 May 2020].

Department for Business, Energy & Industrial Strategy (2020) *Energy Trends: March 2020.* Available at: https://assets.publishing.service.gov.uk/government/uploads/syste m/uploads/attachment_data/file/875381/Energy_Trends_March_2020.pdf [Accessed on 4 May 2020].

Department for International Development (2019) *UK Aid to Double Efforts to Tackle Climate Change*. Available at: https://www.gov.uk/government/news/uk-aid-to-doubl e-efforts-to-tackle-climate-change [Accessed on 4 May 2020].

Environmental Audit Committee (2018) *Green Finance: Mobilising Investment in Clean Energy and Sustainable Development*. 16.5.2018. Available at: https://publications .parliament.uk/pa/cm201719/cmselect/cmenvaud/617/61702.htm [Accessed on 4 May 2020].

Evans, S. (2019) 'CCC: UK has just 18 months to avoid "embarrassment" over climate inaction', *Carbon Brief*. 10.7.2019. Available at: https://www.carbonbrief.org/ccc -uk-has-just-18-months-to-avoid-embarrassment-over-climate-inaction [Accessed on 4.5.2020].

Fankhauser, S., Averchenkova, A. and Finnegan, J. (2018) *10 years of the UK Climate Change Act*, Grantham Research Institute on Climate Change and the Environment. London: London School of Economics.

Grubb, M. (2015) 'Climate economics: the high road', *Nature*, 52: 614–615.

Hale, S. (2010) 'The new politics of climate change: why we are failing and how we will succeed', *Environmental Politics*, 19(2): 255–275.

Harvey, F. (2013) 'Britain resists EU bid to set new target on renewable energy', *The Guardian*. 25.5.2013. Available at: https://www.theguardian.com/environment/2013/ may/25/uk-blocks-eu-target-renewable-energy [Accessed on 28 June 2020].

Hayden, A. (2014) *When Green Growth is Not Enough*. Montreal: McGill-Queen's University Press.

Hook, L. and Sheppard, D. (2019) 'The UK's net zero target: what are the greatest challenges?', *Financial Times*. 12 June 2019. Available at: https://www.ft.com/content /2c212fa8-8d17-11e9-a1c1-51bf8f989972 [Accessed on 4 May 2020].

Independent (1997) 'Kyoto Summit: Prescott takes a leading role', *Independent*. 11.12.1997. Available at: https://www.independent.co.uk/news/kyoto-summit-prescot t-takes-a-leading-role-1288078.html [Accessed on 4 May 2020].

Jordan, A.J., Huitema, D., van Asselt, H., Rayner, T. and Berkhout, F. (eds) (2010) *Climate Change Policy in the European Union*. Cambridge: Cambridge University Press.

Kern, F., Smith, A., Shaw, C., Raven, R. and Verhees, B. (2014) 'From laggard to leader: explaining offshore wind developments in the UK', *Energy Policy*, 69: 635–646.

King, E. (2015) 'UK climate diplomats face axe after COP21 Paris summit', *Climate Home News*. 24.11.2015. Available at: https://www.climatechangenews.com/2015/11/24/uk -climate-diplomats-face-axe-after-cop21-paris-summit/ [Accessed on 4 May 2020].

Kolster, C. and Smith, S. (2017) 'The UK post-Brexit: a leader in climate change diplomacy?', *Imperial College London*. Available at: https://www.imperial.ac.uk/med ia/imperial-college/grantham-institute/public/publications/discussion-papers/The-UK -post-Brexit,-a-leader-in-climate-change-diplomacy.pdf [Accessed on 4 May 2020].

Liefferink, D. and Wurzel, R.K.W. (2017) 'Environmental leaders and pioneers: agents of change?', *Journal of European Public Policy*, 24(7): 951–968.

Lockwood, M. (2013) 'The political sustainability of climate policy: the case of the UK Climate Change Act', *Global Environmental Policy*, 23(5): 1339–1348.

Lorenzoni, I. and Benson, D. (2014) 'Radical institutional change in environmental governance: explaining the origins of the UK Climate Change Act 2008 through discursive and streams perspectives', *Global Environmental Change*, 29(1): 10–21.

Miliband, D. (2007) *Climate Change Bill Launch*. YouTube. Available at: https://youtu.be /IY3F9TT2jDs [Accessed on 4 May 2020].

Moulton, J.F.G. and Silverwood, J. (2018) 'Still on the agenda? The multiple streams of Brexit-era UK climate policy', *Marmara Journal of European Studies*, 26(1): 75–100.

O'Neill, C. (2020) *Letter to the Prime Minister*, 3 February 2020. Available at: http://prod-upp-image-read.ft.com/9267af30-46c1-11ea-aeb3-955839e06441 [Accessed on 4 May 2020].

Oberthür, S. and Groen, L. (2018) 'Explaining goal achievement in international negotiations: the EU and the Paris agreement on climate change', *Journal of European Public Policy*, 25(5): 708–727.

Pickard, J. (2020) 'Johnson revives onshore wind farm after 4-year ban', *Financial Times*, 2 March 2020. Available at: https://www.ft.com/content/b8ddb2f4-5c83-11ea-8033-fa40a0d65a98 [Accessed on 4 May 2020].

Randerson, J. (2010) 'Cameron: I want coalition to be 'greenest government ever', *The Guardian*. Available at: https://www.theguardian.com/environment/2010/may/14/cameron-wants-greenest-government-ever [Accessed on 4 May 2020].

Rayner, J. and Jordan, A. (2017) 'The United Kingdom: a record of leadership under threat'. In: R.K.W. Wurzel, J. Connelly, and D. Liefferink (eds) *The European Union in International Climate Change Politics: Still Taking a Lead?* London: Routledge, 173–188.

renews (2018) 'Petrobas sets out on UK offshore mission', *renews*, 22.10.2018. Available at: https://renews.biz/48652/petrobras-sets-out-on-uk-offshore-mission/ [Accessed on 28 June 2020].

RenewableUK (2020) 'Wind energy', *RenewableUK*. Available at: https://www.renewableuk.com/page/WindEnergy#:~:text=The%20UK%20is%20the%20world,of%20UK%20electricity%20by%202020. [Accessed on 21 September 2020].

Revell, A. (2005) 'Ecological modernization in the UK: Rhetoric or reality?', *European Environment*, 15(1): 344–361.

Scott, J. (2011) 'The multi-level governance of climate change', *Carbon & Climate Law Review*, 25(1): 25–33.

Sindico, F. (2007) 'Climate change: security (council) issue', *Carbon & Climate Law Review*, 29(1): 29–34.

Stern, N. (2006) *The Economics of Climate Change: The Stern Review*. Cambridge: Cambridge University Press.

Thatcher, M. (1989) *Speech to the United Nations General Assembly (Global Environment)*. 8 November. New York: United Nations Building.

Toke, D. (2011) 'The UK offshore wind power programme: A sea-change in UK energy policy?', *Energy Policy*, 39(2): 526–534.

Turney, C. (2019) 'UK becomes first country to declare a 'climate emergency', *The Conversation*, 2.5.2019. Available at: https://theconversation.com/uk-becomes-first-country-to-declare-a-climate-emergency-116428 [Accessed on 28 June 2020].

UK Trade & Investment (2015) *Building Offshore Wind in England*. Available at: https://assets.publishing.service.gov.uk/government/uploads/system/uploads/attachment_data/file/405959/CoreBrochure_2015.pdf [Accessed on 28 June 2020].

UKCOP26 (2020) *The UK-Italy Partnership*. Available at: https://www.ukcop26.org/pre-cop/ [Accessed on 4.5.2020].

Underdal, A. (1998) 'Leadership in international environmental negotiations: designing feasible solutions'. In: A. Underdal (eds) *The Politics of International Environmental Management*. Dordrecht: Springer, 101–127.

Weale, A. (1999) 'European environmental policy by stealth: the dysfunctionality of functionalism?', *Environment and Planning C: Government and Policy*, 17(1): 37–51.

Wesselink, A. and Gouldson, A. (2014) 'Pathways to impact in local government: the mini-Stern review as evidence in policy making in the Leeds City Region', *Policy Sciences*, 47: 403–424.

Whitman, R. and Thomas, G. (2005) 'Two cheers for the UK's EU presidency', *Chatham House*. Available at: https://www.chathamhouse.org/sites/default/files/public/Rese arch/Europe/bpukeupresidency.pdf [Accessed on 4 May 2020].

WindEurope (2020) *Offshore Wind in Europe: Key Trends and Statistics 2019*, February 2020. Available at: https://windeurope.org/wp-content/uploads/files/about-wind/statis tics/WindEurope-Annual-Offshore-Statistics-2019.pdf [Accessed on 28 June 2020].

Wurzel, R.K.W, Liefferink, D. and Di Lullo, M. (2019) 'The European Council, the Council and the Member States: changing environmental leadership dynamics in the European Union', *Environmental Politics*, 28(2): 248–270.

Wurzel, R.K.W., Liefferink, D. and Torney, D. (2019) 'Pioneers, leaders and followers in multilevel and polycentric climate governance', *Environmental Politics*, 28(1): 1–21.

Young, O.R. (1991) 'Political leadership and regime formation: on the development of institutions in international society', *International Organization*, 45(3): 281–308.

11 Governance, green finance and global climate advocacy of the Nordic countries

Small state syndrome or novel middle power?

Mikael Skou Andersen

Introduction

At their August 2019 joint meeting in Reykjavik, the five Nordic prime ministers passed what might qualify as an epochal joint statement on sustainable development. They decided to commit their countries (Denmark, Finland, Iceland, Norway and Sweden) and the Nordic Council to becoming 'global leaders and advocates for climate action' as well as to 'pursue climate diplomacy in international forums to deliver solutions with impact on emissions and to meet the goals of the Paris Agreement' (Nordic Cooperation, 2019).

While Nordic countries have been pioneering ambitious policies to promote energy efficiency, renewables and climate mitigation for many years, the explicit proclamation of climate leadership is something new. The conventional approach has been to provide 'a good example', nudging other countries to follow suit (Andersen and Liefferink, 1997), but Nordic countries are increasingly aiming at providing more profound leadership, not only in Europe but worldwide (Ollila, 2017; Laine *et al.*, 2019).

Considering the relatively small size of their populations, to some this may resemble the mouse quipping to the elephant 'gee, we are trampling', but as pointed out by Wetterberg (2010), favourable economic performances place the Nordics combined among the largest economies in the world (27 million citizens and a GDP of $1,450 billion). In 2016, the Nordics ranked 11th globally in their combined gross domestic product, right after Canada and before South Korea and Russia.[1]

Over the years, the existence of the Nordic Council has helped to maintain close collaborations among Nordic decision makers and civil servants, at national as well as sectoral levels, despite divided participation in other regional fora (Iceland, Norway and Denmark in the NATO; Finland, Sweden and Denmark in the EU). Established in 1953, and extended in 1971 with high-level ministerial meetings, the Nordic Council has promoted a spirit of close cooperation fuelled by shared cultures and relatively strong welfare states.[2] Thus, the multilevel governance framework of the unitary Nordic countries is exceptionally layered, featuring regional, European and Transatlantic fora, as well as relatively strong and independent local and regional authorities, with numerous opportunities for culti-vating intersecting relationships.

Sweden, which is the largest of the Nordic countries with a population base of 10 million (compared to about 5 million each in Finland, Denmark and Norway), often stands out at the global level as the most well-connected and ardent propo-nent of traditional Nordic values and perspectives. Still, Denmark has been part of the EU for almost 20 years before Finland and Sweden also joined. Denmark has self-ruling territories in Greenland and the Faroe Islands and benefits from Transatlantic ties through its NATO membership. Finland was for many years in the shadow of the Soviet Union, but since joining the EU has risen in prominence and influence, not least because it belongs to the inner circle of Member States that have adopted the Euro as their common currency. Norway has developed a high diplomatic profile by mediating in international conflicts and launching global climate policy initiatives such as the Reducing Emissions from Deforestation and Forest Degradation (REDD) programme for reducing emissions from deforesta-tion and forest degradation. Nordic countries thus display a high level of interna-tional involvement in international fora and organisations.

The question is whether Nordic countries thus are elevating their status from individual 'small states' to a coordinated 'middle power'. Among the economi-cally advanced countries, Rothstein (1968) defines small states as having an upper limit of 10–15 million inhabitants, a category that all five Nordic countries easily fit. Keohane (1969) observes that a small power is a state whose leaders consider that it can never, acting alone or in a small group, have a significant impact on the international system. In contrast, a 'middle power' is defined as a state whose leaders consider that while it cannot on its own act effectively it may be able to have a 'systemic impact' in a small group or through an international organi-sation. Middle powers differ from great powers and secondary powers, which often succeed in having an impact on the global system to a large or some extent, respectively. While small states must adjust to realities, having a systemic impact in the international system refers to the ability to exert some kind of influence affecting it. Middle powers can obtain 'significant impact' on the system by work-ing through small groups or alliances or through universal or regional interna-tional organisations' (Keohane, 1969: 295).

Due to their great power status, the EU, USA and China (see Chapters 8, 7 and 2 in this volume) have widely been perceived as the 'Big Three' when it comes to leadership in international climate negotiations, having provided entrepre-neurial and cognitive leadership in defining approaches and mechanisms to the United Nations Framework Convention on Climate Change (UNFCCC) and the 2015 Paris Agreement. However, their recognition has been waxing and waning over time, notably with the withdrawal of the USA from the Paris Agreement. Although international relations theory emphasises the significance of structural leadership, surveys of negotiators for and participants of Conference of the Parties (COP) to the UNFCCC indicate that it is 'imperative for any actor seeking rec-ognition (as leader) to be perceived as being devoted to promoting the common good' (Parker, Karlson and Hjerpe, 2015: 16). To convince others, meaningful domestic action – exemplary leadership – is required by the alleged leaders, demonstrating that they are fully committed to tackling the climate change problem.

Han (2015) explored the efforts of Korea to become a middle power in low-carbon development by promoting international green growth cooperation, finding that the lack of a credible domestic base in terms of public support and relevant technologies for decarbonisation served to undermine these ambitions. This chapter analyses the efforts of Nordic countries to exercise leadership in international climate governance both individually as well as collectively as a potential 'middle power'.

Modes and examples of leadership

An examination of the performance reviews of Nordic countries, conducted by the International Energy Agency (IEA), provides ample evidence for the practices and potentials for exerting such exemplary leadership, due to the many instances of pioneering measures, technologies and policies in place. Several of these initiatives have emerged entirely in response to domestic needs, such as district heating networks based on combined power and heat generation, the energy efficiency which is desirable in the cold climate of the Nordics. Originating in the 1920s and gradually extended, district heating networks today supply more than 50% of households in Finland, Sweden and Denmark – while in Iceland supplied by geothermal energy (IEA, 2017a, 2018, 2019; OECD, 2014). Wind turbine technology has an even longer history which can be traced back to government support for Poul La Cour, the Danish Edison, who used wind tunnels to test and develop novel blade designs while educating rural wind engineers from the 1880s. He planted the seeds of the wind power industry which emerged a century later in a quest to overcome the oil crisis without nuclear power (Nissen, 2009). It has earned Denmark a 'world leadership' role in a host of wind energy and system integration technologies (IEA, 2017a: 13).

Since the 1988 Toronto declaration on the changing atmosphere, the first to address both the science and policy of global warming, Nordic countries have been aiming for proactive policies to mitigate climate change and reduce greenhouse gas emissions (GHGE). Finland displayed both cognitive and exemplary leadership by introducing, in 1990, the very first carbon tax in the world, a measure that Sweden soon exalted into a more encompassing tax reform, lowering taxes on income in exchange for higher energy taxes with a specific carbon tax component. By the early 2000s, all Nordic countries had carbon taxes in place, achieved through tax-shifting, covering most of their non-emissions trading system (ETS) emissions and at rates that are significant in international comparison. In terms of carbon-neutral energy supply, Sweden has become a world leader (IEA, 2019: 3), as the carbon tax has supplanted fossil fuels from district heating, while underpinning the competitiveness of the vast hydro and nuclear power installations.

Finland is pioneering deployment of smart grids (IEA, 2018: 117), extending the knowledge base in the digital cell phone industry to obtain further efficiencies in the heat and power sectors. In Denmark, solar power and biogas installations are today the beneficiaries of feed-in tariffs, thus making Denmark an exemplary pioneer in large solar heating systems (IEA, 2017a: 173), as wind power has become competitive to other energy carriers, even when deployed offshore.

Entrepreneurial leadership has been possible due to the ample forestry resources which provide the Nordic region with comparative advantages in bio energy. Besides the use of biofuels for heating, blending requirements for motor fuels have resulted in Finland 'leading globally' in biodiesel (IEA, 2018: 14), while Sweden has displayed leadership in promoting bioethanol from advanced second-generation biomass (IEA, 2019: 133). With its domestic automobile industry and long road distances, Sweden has further targeted heavy-duty trucks, becoming a pioneer in the use of biogas for transport. Norway, on the other hand, has gained entrepreneurial leadership in the electrification of passenger vehicles, with electric vehicles (EVs) making up 40% of new sales. With 99% of Norway's electricity based on hydropower and generous public support to electric vehicle drivers, the country is a world leader in transforming the vehicle fleet to become carbon neutral (IEA, 2017b: 52). As a paradox, this policy owes its success partly to the revenues flowing from Norway's oil and gas exploration in the North Sea, but Norway is accumulating most of this wealth in a national oil fund that adheres to principles of divestment from pure fossil fuel companies, providing another instance of leadership[3]. Moreover, with the Sleipner project, Norway is a world exemplary leader in carbon capture and storage (CCS) technologies (IEA, 2017b: 12). Iceland, which has powerful comparative advantages with its access to geothermal energy and hydropower from which its industry and domestic heating sector benefit, has announced a ban on sales of fossil fuel cars from 2030.

Looking to the future, there is a strong emphasis on innovation of novel green technology, with considerable resources in all five countries devoted to government-funded research and development projects related to climate mitigation. Finland adheres to circular economy principles with the development of integrated bio-refineries, providing a framework for cascading uses of biomass. Based on lignin from wood, many present uses of fossil-fuel-based plastics could be substituted, e.g. for packaging. Norway has a stronger focus on the blue value chain, as the harvest of seaweed and algae ties in with its offshore activities in fish farming, with opportunities for deriving new sources of proteins to provide relief to feedstuff imports from countries with negative land use practices (e.g. clearing of rain forest). Public support and close cooperation with industry allow for high-risk-high-gain projects, such as the harvest of novel food additives from seaweed which could help significantly reduce methane emissions from livestock.

Taking stock of achievements

Between 1990 and 2017, GHGE[4] have been reduced in Denmark, Sweden and Finland by 29%, 24% and 20%, respectively. Norway has just about stabilised its emissions at 1990 levels, though with a 5% increase, while Iceland has seen an increase of 55%. This adds up to an 18% reduction (weighted) for the Nordic region as a whole. It should not be neglected that the Nordic countries have relatively high per capita GHGE (EEA, 2019). Denmark at 8.3 tonnes is close to the EU28 average of 8.5 tonnes, whereas both Finland and Norway are at 10 tonnes. Iceland has a whopping 13.9 tonnes per capita (17 tonnes with international aviation included). Sweden is leading among the Nordics with 5.2 tonnes and performs far better than the average of EU28.[5] It is not only heating needs that explain the emissions profiles, but also high levels of consumption, the prevalence of energy-intensive industries and the long distances of transport.

The transformation of the heating sector in Sweden to rely entirely on biofuels is the key to understanding its lead on the other Nordic countries. Iceland, on the other hand, has attracted energy-intensive global producers of aluminium which benefit from its low-cost energy carriers, while in Norway offshore oil and gas extraction has doubled emissions since 1990. On the other hand, the Nordic power sector, which is fully integrated via the joint Nord Pool market, has been largely decarbonised, with the share of fossil fuels down to 15%. Despite their differences, a major common challenge for all Nordic countries is to decarbonise transportation.

Denmark, Finland, Norway and Sweden have institutionalised frameworks for the long-term planning and monitoring of policies to reduce GHGE, for which the UK's 2008 Climate Change Act seems to have provided a shared model (see also Chapter 10 in this volume). In all four countries, the government is required to draw up a climate policy action plan every fourth or fifth year to demonstrate how reductions and targets will be achieved. In Denmark, Sweden and Finland an independent advisory body is tasked with overseeing climate policy-making, while informing and contributing to discussions in society. Denmark has moreover decided to establish a Climate Citizen Assembly based on the Irish model (see also Chapter 12 in this volume).

Denmark's Climate Act aims for emissions reductions of 70% by 2030, while climate neutrality is the target for 2050 (Timperley, 2019). Finland's Climate Act establishes a binding objective of 80% GHGE reductions by 2050, while a net zero goal for 2035 was announced by the government in June 2019. Sweden's climate policy also features targets that go beyond international obligations, with a net zero emission target by 2045, involving a requirement for 85% domestic reductions (excluding land use, land-use change and forestry). The milestones for 2030 and 2040 stipulate 63% and 75% reductions, respectively. Iceland has drawn up a national action plan of mitigation measures and has committed itself to a European joint 40% reduction target for 2030, aiming for carbon neutrality by 2040 (MENR, 2018).

Despite declarations by its parliament calling for more ambition, Norway remains the only country without a formal long-term neutrality target. Norway has committed itself to reduce emissions by 40% in 2030 and by 80–95% in 2050. Both Iceland and Norway intend to rely on a significant role for flexible mechanisms such as emissions trading and project-based or international cooperation on emissions reductions (IEA, 2017a, 2017b), as well as for sinks related to land use and land-use changes and forestry (LULUCF). Norway may also be opting to host sinks for carbon capture and storage to European industries, in an effort to offset its own emissions. Flexible mechanisms are available according to an agreement with the EU within the European Economic Area (EEA) framework, but there is no detailed plan yet and there are many uncertainties (Hermansen *et al.*, 2019). Denmark and Sweden had been aiming for a 40% reduction by 2020, which both countries seemed likely to miss in early 2020. All of the above-mentioned reduction targets are relative to 1990 levels.

To sum up, it is clear that the various Nordic countries are performing somewhat differently in terms of GHGE reductions, and the joint Nordic reduction of 18% is slightly below the average 20% reduction by the EU28. However, the Central and Eastern European states (CEES), which had planned economies prior to the collapse of their Communist regimes in the early 1990s, have had large reductions of 'hot air' available due to their economic transition towards market economies and the Southern EU Member States have in recent years suffered major economic declines. It therefore seems more appropriate to compare the achievements of the Nordics with a subset of high-GDP-per-capita Member States. For example, Germany has achieved a reduction of 26% which, however, shrinks to 22% in the old states (*Länder*), i.e. the former West Germany. 'Hot air' helped reach a 43% reduction in GHGE in the new *Länder*, as a result of the significant deindustrialisation in the former East Germany (Jänicke, 2017). The UK (England, Scotland, Wales and Northern Ireland) has obtained an impressive GHGE reduction of 38%, to a large part achieved by the phase-out of coal and a 'dash to gas' (see Chapter 10 in this volume). The BENELUX (Belgium, the Netherlands and Luxembourg) countries and France, on the other hand, have achieved somewhat lower reductions of 13%. Again, the reduction figures stated above are all relative to 1990. It is clear from these figures that the emissions reductions achieved by the Nordics rank in the top of the EU15; they are good but not exceptional. In a global context, it is mainly Sweden that attracts attention due to its low per capita emissions and to some extent Denmark with its 29% reduction.

As many countries around the world are considering their Nationally Determined Contributions (NDCs) to the Paris Agreement, it is nevertheless of interest to understand how Nordic countries have managed to embark on a trajectory of emissions reductions. While specific policy instruments such as carbon taxes and feed-in-tariffs for renewables have been important drivers for this process, there is also a deeper layer of formal and informal institutions at play, facilitating favourably the transition towards carbon neutrality. During a joint visit to one of the first offshore wind power sites in the mid-1990s, a professor from Japan repeatedly asked the author of this chapter: 'But how was this possible?'. Not paying much attention to the technological aspects per se, he was inquiring about explanations for the wider socio-economic and institutional conditions that had enabled the Nordic countries to take the lead on carbon neutral renewables.

Governance

Recent performance reviews by the IEA (2017a, 2017b, 2018, 2019) all stress the significance of domestic carbon pricing schemes in Nordic countries for climate mitigation policies. These schemes, which have been analysed in considerable detail elsewhere, have played a key role in stimulating the market to take account of the costs of carbon emissions (Andersen and Ekins, 2009). While carbon taxes initially applied to all sectors, the introduction of the EU's ETS in 2005 allowed for the exemption of the power sector and large industrial installations from domestic carbon pricing schemes. Consequently, the carbon tax schemes have since then targeted mainly the domestic sector, including households and smaller businesses as well as the transport sector – the so-called non-ETS sectors – and in Norway also the offshore oil and gas industry. The rates of carbon taxes, which started at low and seemingly insignificant levels, have gradually been ramped up. In 2020, Sweden was leading with a carbon tax rate of €130 per tonne CO_2.

The Nordic countries have a long tradition for taxing energy use, which dates back to the first oil crises in the 1970s. Energy taxation was able to spur energy efficiency, which is desirable in the cold climate, and thus relax the dependence on costly energy imports. Thus, when taxes on carbon were implemented, there was nothing unusual in the taxation of energy. In fact, motor fuel tax rates were partly lowered to offset the introduction of carbon tax components in a modification of the tax base penalising carbon-intensive fuels. Still, the remarkable aspect of the Nordic carbon pricing schemes is underlined by the absence of such schemes in most other countries that had signed the 1997 Kyoto Protocol which came into force in 2005. For more than 20 years only the Nordic countries and three to four other smaller European countries had implemented a domestic tax on carbon (Andersen, 2019).

In the Nordic region, the term 'tax' does not trigger the negative connotations so common in other parts of the world. These circumstances are often explained with the legitimacy of taxation in a welfare state system where health services, childcare and education are provided free of charge or below actual costs. The average tax burden on salary earners is about 40% and a progressive scale implies that more than 50% is due on incomes at the higher end, while the aggregate tax burden is between 48 and 49% of GDP (EC, 2019). While Nordic decision makers routinely are haggling over tax burdens and Nordic citizens certainly are weary of new and increasing taxes, they nevertheless feel assured that over a lifetime there will be a decent return on their tax payments in terms of provision of a range of welfare services (Partanen and Corson, 2019). The fact that carbon taxes partly substituted income taxes helped to provide support for their rate increases.

Adding to the legitimacy of the welfare state is the important role of local authorities in imposing and collecting taxes. Over the course of history, Nordic countries have institutionalised relatively strong municipalities, which are the backbone in the provision of welfare services, accounting for a large share of public expenditures and two-thirds of public investment (Andersson, 2014). Within a national framework, they have powers to decide on local tax rates for personal income and real estate, adjusted to the specific needs and welfare provision ambitions of the local community. With this distribution of responsibilities, a substantial share of total taxes (20–30%) is imposed directly by the local town hall (EC, 2019: 201). The implications of tax revenues being managed locally are that decision makers are never far away and there is a sense of providing financial support to the local community with tax payments. Welfare services in terms of kindergartens, schools and retirements homes are provided by local authorities. There is a subtle but significant difference to countries where most taxes are imposed, collected and managed by the national government, with limited revenues reaching the local community. While carbon and energy taxes belong to the excises collected and retained by the national government, these taxes nevertheless benefit from the overall legitimacy and reduced controversy over taxation in Nordic countries.

The relatively strong municipalities are also key to understanding the early extension of district heating networks in Nordic countries. The suppliers of power and heat have historically been local utilities owned by the local municipality and operated on a non-profit basis to the benefit of local citizens. District heating emerged in an effort to utilise the excess energy and heat from power generation, facilitated by the city planning competencies devoted to the local authorities. As the 20th century elapsed, mergers of the numerous local utilities into larger, regional operators became the norm, but with the municipalities often retaining their role and control via joint governing boards. The planning of heat provision is based on detailed legislation requiring the local municipalities to designate the suppliers, making provision of energy for heat and power a highly regulated and strictly planned sector of the economy. Moreover, national legislation requires local authorities to plan and designate sites for wind turbines, enabling them to sort out NIMBY (not in my backyard) controversies, well before operators move in.

Green finance

Revenues from carbon taxes have not been used to provide direct public support for renewables, energy efficiency or decarbonisation measures. An exception was Denmark's carbon tax scheme which, during its first five years, devoted 20% of revenues to support for industry to co-fund investment and advisory services. This scheme was highly successful in enhancing incentives from the carbon tax and contributed significantly to the rapid reduction of industrial carbon emissions in the first years of the Kyoto Protocol; a reduction by 24% according to Enevoldsen (2005).

The presence of strong municipalities has some interesting implications for the opportunities to secure access to finance for local low-carbon projects and thus for enabling a polycentric approach. Municipal credit facilities have historically been key to the green transition in Nordic countries, as they have been financing projects in cooperation with public utilities within energy, transport and water on generous terms, providing low-interest loans guaranteed by local municipalities, underpinned by their tax raising powers, and with lending limited strictly to the public sector.

The Nordic countries feature a rather unique national institution, based on the creditworthiness of municipalities' own tax revenues. A local government funding agency (LGFA) procures low-interest loans to municipalities for infrastructure projects with a public purpose. Loans are provided on the basis of bonds issued by the agency, and due to the solidity of municipal finance such bonds are very attractive in the domestic and international capital markets, with high credit ratings awarded by international agencies. Denmark's agency, Kommunekredit, was set up more than 100 years ago in 1898, while Norway's agency, Kommunalbanken (KBN), followed in 1926. Sweden and Finland founded their agencies more recently, in 1986 and 1989, named respectively Kommuninvest (Municipal Finance) and Kuntarahoitus Oyj (Municipality Finance). They are all rated AAA or AA+ due to creditworthy borrowers and high-quality assets.

Legally, an LGFA is a non-profit association controlled and owned by municipalities and regions in cooperation with the national government. The organisational and legal models differ slightly, but the local authorities are generally acting collectively as guarantors for the bonds issued. Capital can thus be obtained well below the commercial interest rates requested by banks and other credit providers. For the lenders there is a significant discount, as interest rates were down to 2%, less than half the market rate (Dansk Fjernvarme, 2017). Infrastructure investments eligible for financing include everything from schools, retirement homes and sport facilities to light rails, district-heating and energy-saving street lights, as long as they have a public purpose and are in the mandate of a local authority. Still, to control overall public spending the national government imposes ceilings on municipal and regional investment.

By virtue of the municipal ownership of public utilities with a natural monopoly, such loans can also be granted directly to suppliers of district heating, provided that the operators can obtain a guarantee from the local municipality. This will involve a guarantee premium payment to the local authority, currently at about 1% annually. In Denmark, district heating utilities account for 28% of the total credits negotiated by its LGFA. Municipalities per se account for 49% while the remaining loans are held by water supply utilities and municipal harbours (Dansk Fjernvarme, 2017).

Obtaining a municipal guarantee requires not only the presence of a convincing project economy but also the demonstration of socio-economic advantages in accordance with the national government's manual for cost-benefit analysis. There are certain legal requirements that must be respected too, including conformity with EU state aid rules, which prescribe the exclusion of public support that can distort competition. Still, agency loans have routinely been granted to waste incinerators and wind farms, despite their supply of power to electricity markets. The volume of lending by agencies is huge, as they essentially are the backbone in financing investment of the local governments in the Nordic countries. Their market share in subnational government lending is more than 90% in Denmark, about 50% in Finland and Sweden, and 40% in Norway (OECD, 2017: 200).

Over more than 100 years of existence, they have never defaulted. While the total loan portfolio may look modest on an international scale, it nevertheless corresponds to discounted public investment credits of about 4,000–5,000 Euro per capita.

The Nordic LGFAs have started issuing dedicated green bonds, providing proof to investors of the related environmentally and climate-friendly investment profile. Green bonds are in high demand in the market by investors that are keen to demonstrate the sustainability of their investments, such as the pension funds of labour unions. The approaches differ somewhat, but, for instance, Finland's Municipality Finance promotes green bond loans for renewables, energy efficiency, sustainable transportation and buildings as well as for waste, water and waste water management, which gives priority to low-carbon development projects. A green loan committee oversees lending decisions, while relying on project appraisals with life cycle analyses of projects where necessary (CICERO, 2019). The Norwegian State Agency for Local Government Funding offers a direct discount on green project loans in an effort to support and scale up low-carbon and sustainable development (KBN, 2019).

In a novel development reflecting a degree of cognitive leadership, green bonds can obtain the label of the Nordic Swan as proof of their high standards. The Nordic Swan is a widely recognised consumer- or eco-label – comparable to the EU Flower – managed by the Nordic Council. In 2017, criteria were defined and agreed to award the label to financial products at the bond market. The criteria for obtaining the Nordic Swan aim to exclude fossil fuels (i.e. oil and coal), tobacco, genetically modified crops, weapons and violations of certain international conventions (which may serve to rule out investment in US federal bonds, when US ratification of the Paris Agreement expires (see also Chapter 7 in this volume). Nordic Swan labelled investor funds are required to produce an annual audit and submit documentation to verify their overall compliance with the criteria (Lunde, 2017). At the inception of the Swan label, 12 investment funds had been licensed to apply it, including major financial actors such as Skandia, Swedbank, SEB, Alfred Berg and Handelsbanken (Duus, 2017).

Outside the Nordic countries, LGFAs have a track record only in the Netherlands[6]. Belgium and Austria used to have a similar agency, but following mergers, privatisations and financial turbulence they now serve local authorities on less advantageous terms. However, in response to the financial crisis, New Zealand has founded a novel LGFA in 2011 (see also Chapter 5 in this volume), while a comparable initiative is reported from a French region (*Agence France Locale*) (Andersson, 2014; OECD, 2017). Standard and Poors (2011) indicates that one can be found also in Japan, and summarises the overall strategic aims as follows:

> LGFA is intended to provide local governments with better access to funding at cheaper rates than local and regional governments can source individually, while maintaining a low risk profile. By centralising funding, LGFA benefits from asset diversification and reduces aggregate credit risk. Its higher rating

than several individual local governments should provide it with better funding rates, while larger debt issuance sizes are expected to attract a greater range of investors. Additionally, it will have the ability to raise foreign-currency funding (which it will swap back to domestic currency), further deepening the investor pool.

It is clear that the presence of such financial agencies help explain how Nordic countries have been able to leverage public investment in the transformation of district heating, away from fossil fuels, even if other sources of financing will normally be involved in a project too. The main burden for low-carbon development is the high upfront investment required for embarking on new technologies and for upscaling these in society. By enabling a significant discount on the interests to be paid over the lifetime of projects, these agencies are key to whether a project will materialise, and their decision on whether or not to grant a loan is often crucial.

The notion of 'green financing' emerged about ten years ago in the run-up to the 2009 Copenhagen Climate Change Conference (COP15) and in response to the evolving financial crisis, as a way to reconcile the needs for a new stimulus to the economy with the imperatives of a low carbon transition (WEF, 2009). It extends the legacy of the movement for socially responsible investment (SRI), while engaging with the actual mechanisms for governance of low-carbon financing (Richardson, 2009). Yet, in Nordic countries the foundations of green financing are vested to some institutions that have been in place well before the recent interest in climate finance. Fuel use for heating accounts for more than 50% of total energy use in the EU and €641 billion will be required from 2020–2040 for investments in district heating (Mathiesen *et al.*, 2019). It is difficult to see how other countries will be able to leverage them without such agencies in place.

Global engagement

As early as 2015, a Nordic prime ministers' initiative was agreed, focusing on 'Nordic Solutions to Global Challenges'.[7] The programme was described as a novel departure and a step onto the global stage, with the mission being the coordination of efforts among Nordic countries in making progress towards the UN's sustainability goals and in meeting the Paris Agreement. The explicit purpose was to showcase and promote Nordic solutions and innovations through a series of flagship projects. Three of these have been clustered under the topic 'Nordic Green' and relate to sustainable cities, energy research and climate solutions. A report from the Nordic Council on making the most of Nordic leadership for decarbonisation claims, that Nordic countries are well placed 'to create the smartest energy system in the world and to find the most cost-efficient solution in moving towards the low-carbon green economy' (Ollila, 2017: 10).

Nordic countries have long-established programmes for development aid and are present in many countries around the world. Sweden, Norway and Denmark are among the very few countries that fulfil the aim of contributing 0.7% of gross national income (GNI) in aid to developing countries; only two other countries do so. The five Nordic countries jointly spend about USD14 billion annually, which ranks them among the largest donors globally even in absolute terms, exceeded only by the USA, UK, Germany, Japan and the EU.[8] In contrast to some other donors, Nordic countries tend to provide grants rather than loans, and the share of multilateral aid is relatively high, although many bilateral projects are implemented in partner countries too (see also Chapter 3 in this volume). An increasing share of development aid is provided as climate finance, with Sweden and Norway honouring the UNFCCC guidelines to make it additional to the 0.7% of GNI in conventional development aid.

As development aid may have climate-relevant aspects, a systematic methodology is required for declaring the share of climate finance in different assistance programmes. OECD statistics suggest that the Nordic countries jointly contribute about USD2 billion in climate-related development aid, while in an independent assessment of the national reporting on climate finance to UNFCCC, Oxfam (2018) finds that the Nordic countries jointly contribute about USD1 billion in dedicated climate finance annually. Nordic climate finance is 90–98% grant-based and according to Oxfam (2018) its total is close to the grant equivalent of climate finance from the UK, and approximates about one-third of the EU's European Bank for Reconstruction and Development (EBRD) and European Investment Bank (EIB). Individually, Nordic countries contribute more than Australia, Canada and Switzerland in grant equivalent of climate finance, despite lower population numbers (Oxfam, 2018: 11). They also provide a higher share to the least developed countries. There is still some way to go before the COP15 pledge of the Annex-1 countries to the Kyoto Protocol for USD100 billion in annual climate finance is reached. While the OECD estimates that USD75 billion was reached, Oxfam puts this figure at USD16–21 billion in grant-equivalent transfers. Three of the Nordic countries (Iceland, Finland and Denmark) have not honoured the commitment to make climate finance additional to the development aid target of 0.7%. It is when considered jointly that the Nordic countries stand out in international comparison, which owes much to the contributions from Sweden and Norway.

One example of the involvement at global level by Nordic countries is the Friends of Fossil Fuel Reform, counting also Costa Rica, New Zealand, Ethiopia, Uruguay and Switzerland (see also Chapters 4, 5 and 13 in this volume). The Friends were established in 2010 to support commitments of Asian-Pacific countries to phase-out inefficient subsidies for fossil fuels. In continuation of this alliance, Nordic countries have provided technical assistance to developing countries, e.g. Indonesia, Morocco, Zambia and Bangladesh, to phase-out fossil fuel subsidies (Merrill *et al.*, 2017). Another example, directly related to the innovative climate change mitigation technologies that Nordic countries are proficient in, is the 'green-to-scale' approach. By considering what a specific country could achieve in terms of GHGE reductions and socio-economic advantages if it were to upscale well-known technologies from the Nordic region by making use of them as a blueprint for decarbonisation. For example, it has been shown that Nordic climate solutions would have a potential for emissions reductions of 23 million tonnes in Kenya, 64 million tonnes in Ukraine and of 71 million tonnes in Poland by 2030. Fifteen specific Nordic climate solution technologies could reduce overall global emissions by 4 gigatonnes (Gt) per year (made up of heat cogeneration 1.2 Gt; onshore wind 0.7 Gt; substitution of mineral fertilisers with manure 0.5 Gt; energy efficiency in buildings 0.5 Gt). Analyses of more than 300 international climate initiatives show that one or more Nordic countries are deeply involved, participating in or contributing to about 64% of these (Laine *et al.*, 2019).

Three Nordic countries (Denmark, Finland and Sweden) have a say within the EU. It is customary that the Nordic Member States meet or exchange opinions prior to meetings in the EU's Council of Ministers. There are numerous examples that Nordic countries have pushed for higher climate ambitions, especially when holding the six-monthly rotating Council Presidency. Finland's Prime Minister Antti Rinne announced that 'a key priority of Finland's [2019] Presidency is the EU's global leadership in climate action', adding that the EU should become 'the world's most competitive and socially inclusive low-carbon economy' (*ENDS Europe*, 2019). Individual members of the European Commission from Nordic countries have promoted climate action too. Initiated by the Danish EU Climate Commissioner Connie Hedegaard, the 'Greenland dialogue' helped forge consensus on the 2 degrees target with the USA and China in 2009. All three Nordic EU Member States support the Commission's proposal for climate neutrality by 2050. Still, opinions frequently differ on specific aspects of climate action, depending on the national interests, say in biofuels, and their perceived properties.

Conclusions

The commitment of Nordic countries to aspire to becoming leaders and advocates for global climate action has its roots in the Kyoto Protocol, while the Paris Agreement provides a strong impetus, with its bottom–up approach based on NDCs to emissions reductions. While it is not difficult to identify a high level of involvement of Nordic countries in global climate action, it is less evident to which extent they manage to exercise actual leadership. The litmus test of leadership is whether there are also countries that acknowledge that role, and who will respond as 'early adopters' or 'fast followers' of approaches, measures and technologies introduced.

The review of governance and green finance mechanisms in place in Nordic countries in this chapter indicates that many of their climate mitigation accomplishments are based on deeper societal structures dating back to the formation of the modern Nordic welfare states. The presence of independent local authorities, with their own revenue sources and established financial institutions, could not easily be transplanted to countries with a different past state tradition. In contrast, to maintain their secondary power status most of the larger European countries have centralised revenue flows with local authorities frequently deprived of money and mandates. To exercise leadership Nordic countries will have to reflect more carefully on how green technologies can be transferred to different societal settings, whether in developing countries, emerging economies or Annex-1 countries, that may not have similar opportunities for exploiting the dynamics of fluid relations in a multilevel governance context, where power is redistributed from the state upwards but also downwards (Tortola, 2017).

While Nordic countries and the Nordic Council display a high self-awareness on the virtues of domestic climate action, there are circumstances weighing against their climate leadership aspirations. Referring to the high per capita ecological footprint, some analysts are suspicious about the Nordic model and maintain that 'Nordic countries have some of the highest levels of resource use and CO_2 emissions in the world, in consumption-based terms, drastically overshooting safe planetary boundaries' (Hickel, 2019). The cold climate and ample natural resources available do not suffice to explain the large footprint, and critics strike a chord when contrasting rhetoric with reality.

While Nordic countries are seeking to speak with one voice, there are, as revealed in this chapter, some discrepancies in their actual strategies, fuelling doubts about their climate action. The approaches of Norway and Iceland, with an optimistic emphasis on using flexibility mechanisms including CCS, ETS, REDD+ and LULUCF differ markedly from their neighbours, and some observers regard them as followers rather than leaders on climate action (Hermansen *et al.*, 2019). Sweden, on the other hand, has pursued for many years measures with a stronger focus on domestic emissions reductions, a strategy which, after some hesitations, Finland and Denmark now seem to be inclined to. Still, Sweden, Finland and Norway have joined forces on a promotion of biofuels, which for many observers looks dubious in consideration of the short- and medium-term GHGE involved. To become successful as exemplary climate leaders, Nordic countries will have to address these doubts convincingly and they must be able to demonstrate how to reach low per capita emissions without questionable accounting and bookkeeping tricks.

Exemplary leadership has a direct as well as an indirect role; the former mainly during the agenda setting phase and the latter by underpinning credibility in negotiations where other types of leadership are exercised. Notwithstanding the pledge for exemplary leadership, when it comes to the most decisive type, structural leadership (based on power and resources), it plays an evident role in high-level international climate negotiations (Andresen and Agrarwal, 2002). The architecture of the Paris Agreement offers, despite the emphasis on nationally determined commitments, a range of opportunities to exercise influence in negotiations over the detailed mechanisms associated. The Nordic countries have some possibilities to act as a 'middle power' in these negotiations, if they understand to connect and coordinate well their positions in the various regional and European organisations, while accelerating deep emissions reductions in the real world. In other words, if they make successful use of the multilevel governance structures this may help them to increase their structural leadership capacities. Whether in the end they will succeed in having a significant or systemic impact would seem to depend furthermore on a good appreciation of the differences in governance structures and financial mechanisms at home and abroad.

Acknowledgements

Financial support (grant #82841) from Nordforsk, Nordic Energy and Nordic Innovation to the project New Nordic Ways to Green Growth (NOWAGG) (https://projects.au.dk/nowagg/).

Notes

1 UN data from the July 2018 World Development Indicators.
2 The self-ruling regions of Greenland, the Faroe Islands and the Åland Islands are associate members of the Nordic Council. The Sami Parliamentary Council has observer status. The Nordic Council of Ministers has offices in Estonia, Latvia, Lithuania and Schleswig-Holstein (Germany).
3 https://www.bloomberg.com/news/articles/2019-10-04/here-s-the-secret-list-of-norway-s-oil-stock-divestments [Last accessed 22 June 2020].
4 All sectors and indirect CO_2 (excluding LULUCF and memo items including international aviation).
5 Based on the Sustainable Development Index which includes CO_2 embedded in imports, Sweden maintains its lead with 9 tonnes per capita, against 13–14 tonnes for Norway, Iceland and Finland, while Denmark stands at 10 tonnes (2015 data) (Hickel, 2020). However, these figures exclude international shipping for which Denmark has 7 tonnes GHGE/capita, mainly due to the Maersk container fleet (cf. Statistics Denmark: https://www.dst.dk/Site/Dst/Udgivelser/nyt/GetPdf.aspx?cid=27511).
6 In the Netherlands, the Bank Nederlandse Gemeenten (BNG) was established in 1914 as a specialised financial institution for the public sector, owned jointly by the Dutch State, provinces and municipalities. Its market share is 60% of the Dutch municipal sector. The Netherlands' Waterschapsbank (NWB) was established in 1954 as a specialised lending institution to provide the regional water boards with funding (Andersson, 2014).
7 https://norden.diva-portal.org/smash/get/diva2:1098386/FULLTEXT01.pdf [Last accessed 22 June 2020].
8 http://www2.compareyourcountry.org/oda?cr=20001&cr1=oecd&lg=en&page=0 [Last accessed 22 June 2020].

Bibliography

Andersen, M.S. (2019) 'The politics of carbon taxation: how varieties of policy style matter', *Environmental Politics*, 28(6): 1084–1104.
Andersen, M.S. and Ekins, P. (eds) (2009) *Carbon-energy taxation: lessons from Europe*, Oxford: Oxford University Press.
Andersen, M.S. and Liefferink, D. (eds) (1997) *European environmental policy: the pioneers*, Manchester: Manchester University Press.
Andersson, L.M. (2014) *Local government finance in Europe - trends to create local government funding agencies*, conference paper, AB Mårten Andersson Productions. https://www.maproductions.se/?page_id=479 (last accessed 30 June 2020).
Andresen, S., and Agrawala, S. (2002) 'Leaders, pushers and laggards in the making of the climate regime', *Global Environmental Change*, 12(1): 41–51.
CICERO (Center for International Climate Reseach) (2019) *'Second Opinion' on MuniFin's green bond framework*, Oslo: CICERO. https://www.kuntarahoitus.fi/app/uploads/sites/2/2019/09/munifin_spo_draft_20062019_final.pdf (last accessed 2.7.2020).

Dansk Fjernvarme (2017) *Sådan fungerer finansiering med Kommunekredit*, Kolding: Handout.

Duus, F. (2017) '12 investeringsfonde er som de første svanemærket', *Finanswatch*, 11 October. https://finanswatch.dk/Finansnyt/article9942611.ece (last accessed 29.6.2020).

EC (European Commission) (2019) *Taxation trends in the European Union. Data for the EU Member States, Iceland and Norway*. Luxembourg: Publications Office of the European Union.

EEA (European Environment Agency) (2019) *Country profiles - greenhouse gases and energy 2019*. https://www.eea.europa.eu/themes/climate/trends-and-projections-in-eur ope/climate-and-energy-country-profiles/copy_of_country-profiles-greenhouse-gases -and (last accessed 1 December 2019).

ENDS Europe (2019) 'Finnish EU presidency to embrace climate action', 26 June.

Enevoldsen, M. (2005) *The theory of environmental agreements and taxes: CO_2 policy performance in comparative perspective*, Cheltenham: Edward Elgar.

Finnson, P.T. (2016) 'Quinze solutions climatiques nordiques pour réduire les émissions mondiales de 4Gt', *AfriMag Magazine*, 15 December.

Han, H. (2015) 'Korea's pursuit of low-carbon green growth: a middle-power state's dream of becoming a green pioneer', *The Pacific Review*, 28(5): 731–754.

Hermansen, E., Peters, G. and Lahn, B. (2019) *Climate neutrality the Norwegian way*, Oslo: CICERO. https://cicero.oslo.no/no/posts/nyheter/climate-neutrality-the-norwegi an-way-carbon-trading (last accessed 2 July 2020).

Hickel, J. (2019) 'The dark side of the nordic model', *AlJazeera*, 6 December. https://ww w.aljazeera.com/indepth/opinion/dark-side-nordic-model-191205102101208.html (last accessed 29 July 2020).

Hickel, J. (2020) 'The sustainable development index: measuring the ecological efficiency of human development in the anthropocene', *Ecological Economics*, 167: 106331.

IEA (International Energy Agency) (2017a) *Energy policies of IEA countries: Denmark 2017 review*, Paris: International Energy Agency.

IEA (International Energy Agency) (2017b) *Energy policies of IEA countries: Norway 2017 review*, Paris: International Energy Agency.

IEA (International Energy Agency) (2018) *Energy policies of IEA countries: Finland 2018 review*, Paris: International Energy Agency.

IEA (International Energy Agency) (2019) *Energy policies of IEA countries: Sweden 2019 review*, Paris: International Energy Agency.

Jänicke, M. (2017) 'Germany: innovation and climate leadership'. In: R. Wurzel, J. Connelly and D. Liefferink (eds) *Still taking a lead? The European Union in international climate change politics*, London: Routledge, 114–130.

KBN (Kommunalbanken Norge) (2019) *Investor presentation*. https://www.kbn.com/ globalassets/dokumenter/funding/investor-presentation/kbn-investor-presentation- 2019.pdf (last accessed 2 July 2020).

Keohane, R.O. (1969) 'Lilliputian's dilemmas: small states in international politics', *International Organization*, 23(2): 291–310.

Laine, A., Halonen, M., Mikkola, J., Lütkehermöller, K., Höhne, N., and de Villafranca Casas, M.J. (2019) 'Nordic opportunities to provide leadership in the global climate action agenda - international cooperative climate initiatives with major upscaling and impact potentials', *Nord* 22, Copenhagen: Nordic Council of Ministers.

Lunde, L.O. (2017) 'New nordic ecolabel for investment funds almost ready for launch', *AMWatch*, February 20th. https://amwatch.dk/AMNews/Ethics/article9375331.ece (last accessed 2 July 2020).

Mathiesen, B.V., Bertelsen, N., Schneider, N., García, L.S., Paardekooper, S., Thellufsen, J.Z., and Djørup, S.R. (2019) *Towards a decarbonised heating and cooling sector in Europe unlocking the potential of energy efficiency and district energy*, Aalborg: Aalborg University, Department of Planning.

MENR (Ministry of Environment and Natural Resources) (2018) *Iceland's climate action plan for 2018–2030*, Reykjavik: Ministry of Environment and Natural Resources.

Merrill, L., Bridle, R., Klimscheffskij, M., Tommila, P., Lontoh, L., Sharma, S., Touchette, Y., Gass, P., Gagnon-Lebrun, F., Sanchez, L. and Gerasimchuk, I. (2017) 'Making the switch - from fossil fuel subsidies to sustainable energy', *TemaNord 537*, Copenhagen: Nordic Council of Ministers.

Nissen, P.-O. (ed.) (2009) *From Poul la Cour to modern wind turbines*, Askov: The Poul la Cour Foundation.

Nordic Cooperation (2019) *Joint statement of the Nordic Prime Ministers and the Nordic CEOs for a sustainable future*, Reykjavik. https://www.norden.org/en/declaration/joint-statement-nordic-prime-ministers-and-nordic-ceos-sustainable-future

OECD (2014) *Environmental performance reviews: Iceland*, Paris: Organization for Economic Co-operation and Development.

OECD (2017) *Multi-level governance studies: making decentralisation work in Chile: towards stronger municipalities*, Paris: Organization for Economic Co-operation and Development.

Ollila, J. (2017) *Nordic energy co-operation: strong today – stronger tomorrow*, Copenhagen: Nordic Council of Ministers.

Oxfam (2018) *Climate finance shadow report 2018 – assessing progress towards the $100 billion commitment*, Oxford: Oxfam International. https://www.oxfam.org/en/research/climate-finance-shadow-report-2018

Parker, C.F., Karlsson, C. and Hjerpe, M. (2015) 'Climate change leaders and followers: leadership recognition and selection in the UNFCCC negotiations', *International Relations*, 29(4): 434–454.

Partanen, A. and Corson, T. (2019) 'Finland is a capitalist paradise', *New York Times*, 7 December.

Richardson, B.J. (2009) 'Climate finance and its governance: moving to a low carbon economy through socially responsible investing', *International and Comparative Law Quarterly*, 58(3): 597–626.

Rothstein, R.L. (1968) *Alliances and small powers*, New York: Columbia University Press.

Standard and Poors (2011) *New Zealand local government funding agency*, Global Credit Portal, https://www.standardandpoors.com/ratingsdirect (last accessed 5 December 2019).

Timperley, J. (2019) *Denmark adopts climate law to cut emissions 70% by 2030*. London: Climatechangenews.com.

Tortola, P.D. (2017) 'Clarifying multilevel governance', *European Journal of Political Research*, 56(2): 234–250.

WEF (2009) *Scaling-up in a downturn? Ideas for building the low carbon economy*, Geneva: World Economic Forum.

Wetterberg, S. (2010) 'The United Nordic Federation', *TemaNord 583*, Copenhagen: Nordic Council of Ministers.

12 Ireland's Citizens' Assembly on climate change

Institutional pioneership by a climate laggard?[1]

Diarmuid Torney, Laura Devaney and Pat Brereton

Introduction

In January 2018, then Irish Taoiseach (prime minister) Leo Varadkar made headlines by acknowledging Ireland's reputation as a 'climate laggard' during an appearance at the European Parliament (Sargent, 2018). This was a label that had long been used by civil society groups. Indeed Ireland[2] was ranked as the worst-performing European Union (EU) member state in the annual Climate Change Performance Index ranking in 2017 and 2018 (Burck *et al.*, 2017; Burck *et al.*, 2018). However, it was the first time an Irish prime minister had used the label and admitted the lack of progress being made in an Irish climate action context.

Ireland's laggard status makes Ireland's decision to include the topic of climate change on the agenda of a Citizens' Assembly surprising. An exceptional experiment in democratic governance, the Citizens' Assembly comprised 99 citizens drawn from all walks of life and afforded them the time, space and structure to consider complex questions of public policy in a deliberative way. Climate change was one of five topics considered by the Assembly. The 13 recommendations they agreed on the climate change topic were significantly more radical than many observers expected (Citizens' Assembly, 2018). Following on from the conclusion of the Assembly's deliberations on climate change, an all-party parliamentary committee was set up to consider the Assembly's proposals. It issued its own set of recommendations in 2019. These were similarly far-reaching and endorsed many of the citizens' proposals (Houses of the Oireachtas, 2019). In turn, the parliamentary committee report played a significant role in shaping an all-of-government *Climate Action Plan to Tackle Climate Breakdown*, published by the Irish government in June 2019 (Government of Ireland, 2019). A new programme for government, agreed in June 2020 by the centre-right Fianna Fáil and Fine Gael parties along with the Green Party, in turn went beyond the Climate Action Plan, promising stronger climate action.

A relatively long tradition of scholarship has advocated the use of deliberative forms of democracy (Fishkin, 1991; Dryzek, 1990). Such forums, it is argued, should be open, inclusive, public and include affected citizens on equal terms. Moreover, collective decisions should be based not just on simple aggregation of atomised preferences, but should be arrived at through reasoned debate.

Deliberative democracy has been criticised, however, because of the burden of time and effort placed upon those participating. Deliberative forums also do not replace the problems of voting and the potential tyranny of the majority. In more recent times, scholars have specifically considered the promise of deliberative democracy to address the challenges posed by environmental degradation, including climate change (Niemeyer, 2013; Stevenson and Dryzek, 2014). Nonetheless, the Irish experience was pioneering as a forum composed of a representative sample of the population, established by and reporting to Parliament. In 2019, both the UK and France announced the establishment of similar citizens' assemblies, focused exclusively on climate change (see Chapter 10 in this volume). Moreover, its potential to inform ongoing and future attempts to engage diverse publics and contribute to decision-making on the challenge of climate change remains largely untapped. It also demonstrates the prospect for seeming climate laggards to exhibit some degree of climate pioneership (or possibly even climate leadership), even if only for particular issues or certain periods of time.

What drove an apparent climate laggard to engage in such institutional innovation? This chapter analyses the establishment, activities and outcomes of Ireland's Citizens' Assembly with the aim of answering this question. We examine these phenomena through the analytical lens of leadership, pioneership and followership. In line with Liefferink and Wurzel (2017), we distinguish analytically between leaders and pioneers. Leaders are conceived of in the context of climate politics as actors who adopt ambitious policies, approaches or other initiatives with the intention of proactively seeking to attract followers. Pioneers, by contrast, adopt ambitious policies or approaches for primarily domestic reasons and are unconcerned with whether they attract followers (see also Wurzel *et al.*, 2017; Wurzel *et al.*, 2019a; Chapter 1 in this volume). While leaders and pioneers differ with respect to their intentionality in attracting followers, pioneers may in fact attract followers (although only unintentionally) while leaders may fail to attract followers despite the fact that they are trying to do so (Torney, 2015; 2019).

Patterns of climate leadership, pioneership and followership can play out in multilevel and polycentric governance contexts (Jordan *et al.*, 2015; Jordan *et al.*, 2018). Indeed, such contexts arguably provide fertile ground for these relationships to develop. By bringing governance actors into repeated interactions across jurisdictions and governance levels, multilevel governance systems allow for policy-makers to learn from each other, and for potential leaders and pioneers to engage with potential followers. Taken to its logical extreme, polycentric governance theory may limit the potential for leader-follower dynamics because of its emphasis on the independence of governance actors. However, most accounts of polycentric governance in fact emphasise the importance of experimentation and learning among formally autonomous but nonetheless interdependent governance actors (Ostrom, 2010, 2012; Dorsch and Flachsland, 2017). In climate change terms specifically, the 2015 Paris Agreement has heralded a new era that combines elements of top–down and bottom–up climate governance (see also Chapter 1 in this volume). This, as well as Ireland's and the EU's evolving responses to Paris (see also Chapter 8 in this volume), provides the broader context within which the story portrayed below has unfolded. Depending on their degree of institution-alisation in national policy-making processes, deliberative forums such as citizens' assemblies could be conceptualised as sites within a polycentric governance framework. However, in the Irish case, the Citizens' Assembly was established by, and reported to, the national parliament for a set time period and purpose. As such, it would be a stretch to classify it as contributing to the development of polycentric governance more broadly.

In this chapter we set out the institutional innovations encapsulated in Ireland's Citizens' Assembly and follow-on parliamentary committee. We find very limited and sporadic evidence that the process was aimed explicitly at providing a model of climate governance for others to follow. Evidence of such leadership was not found in the framing of the Citizens' Assembly model set-up or execution, nor did the Irish government explicitly seek to coach or inform other nations about the process, for instance, as evidenced in the framing of EU climate policy since the 1990s (Torney, 2015) or the promotion by the UK government of its Climate Change Act internationally (see, for example, Torney, 2017). However, the Citizens' Assembly chairperson, the Honourable Mary Laffoy, stated at the conclusion of the process, 'It is hoped that they will be of benefit not only to the political system, but to others involved in exercises such as those in other jurisdictions' (McGreevy, 2018). Since its conclusion, this Irish experiment in deliberative democracy has gained attention in other jurisdictions and across other policy areas that are considering emulating the Citizens' Assembly model. For these reasons, we suggest that the use of a Citizens' Assembly to deliberate on climate change policy in Ireland is best conceptualised as pioneership rather than leadership. As we argue below, what makes the Irish case particularly interesting is the fact that an acknowledged climate laggard succeeded in engaging in climate pioneership in a particular domain. This echoes the findings of Wurzel *et al.* (2019b) who argue, using the cases of Bremerhaven and Hull, that structurally disadvantaged cities can nonetheless exhibit climate pioneership.

The next section sets the context by outlining the recent history of Ireland's policy response to climate change and places this in comparative perspective. The following section discusses the work of the Irish Citizens' Assembly, paying particular attention to the origins and evolution of this institutional innovation. The penultimate section takes the story one step further by discussing the work and findings of the all-party committee on climate action that was established in summer 2018 to consider the recommendations of the Assembly. The final section concludes by reflecting on the empirical narrative from the analytical perspective of leadership, pioneership and followership.

Ireland's climate performance in comparative perspective

Ireland's first national climate strategy was published in 2000 (Department of Environment and Local Government, 2000). It set out a policy pathway towards achieving Ireland's target under the EU Burden-Sharing Agreement[3] of limiting the growth in greenhouse gas emissions (GHGE) to 13% above 1990 levels by

2008–2012. However, rapid economic growth combined with a failure to implement much of the contents of the 2000 national climate strategy, among other factors, meant that Ireland's emissions grew rapidly over the course of the 2000s. A second national climate change strategy was published in mid-2007, just prior to a general election (Department of Environment Heritage and Local Government, 2007). As with the 2000 strategy, this was aimed at providing a roadmap for achieving Ireland's emissions targets under the EU Burden-Sharing Agreement.

Overall, Ireland complied with its Burden-Sharing Agreement targets though not primarily due to policy decisions. Rather, Ireland's compliance was brought about by severe economic recession. GHGE declined by 15% from 2008–2011 with the onset of a severe economic crash. Subsequently, emissions rose again in line with economic recovery between 2011 and 2016 (EPA, 2018). According to projections by the Environmental Protection Agency (EPA), Ireland will miss its non-emissions trading GHGE target by a wide margin (EPA, 2019). According to these projections, in the best-case scenario – that is, 'with additional measures' – Ireland's GHGE in the non-emissions trading sector will decline by just 0.4% (EPA, 2019). This is relative to a 20% decarbonisation target, as set out under the EU Effort-Sharing Decision[4] (EPA, 2018).

In parallel and separate to its EU commitments, the Irish government committed in 2014 to decarbonisation of 80% relative to 1990 levels by 2050 across electricity generation, the built environment and transport, and an approach to carbon neutrality in the agriculture and land use sector (Government of Ireland, 2014). Moreover, a government white paper on energy, published in December 2015, committed to transforming Ireland's energy sector into a clean, low-carbon system by 2050, stating that 'eventually, we will have to generate 100% of all our energy needs – not just electricity – from clean sources' (DCENR, 2015: 4).

In December 2015, Ireland's climate law – the Climate Action and Low Carbon Development Act – entered into force. Eight years in the making, the final climate law enacted was significantly weaker than earlier versions that had been proposed (Torney, 2017). The climate law provided a climate policy planning framework for mitigation and adaptation that requires government to produce a National Mitigation Plan and a National Adaptation Framework every five years. Relevant ministers were required to report annually to both houses of parliament on progress in combating climate change. The law also established an independent Climate Change Advisory Council tasked with reporting annually on progress towards national, EU and international goals and objectives. However, arguably the most striking feature of Ireland's climate law was that it contained no quantified targets for emissions reduction for the medium term or even for 2050. The 80% decarbonisation target mentioned above was agreed by government but not enshrined in the climate law. This marked Ireland's climate law out as somewhat unusual by international standards and can be characterised as climate laggardship (Duwe *et al.*, 2017).

Ireland has a distinctive GHGE profile by international standards. Agriculture constitutes an unusually high share of national emissions due to the lack of a heavy industry profile, making up one-third of the total in 2017 (EPA, 2018). Among industrialised countries, New Zealand is the only other country with a similar emissions profile (see Chapter 5 in this volume), though it is much more common among developing countries with low emissions from industrial sectors (see Part 1 in this volume). Although emissions from agriculture in Ireland in 2017 were marginally below the 1990 level, emissions from the sector increased by 9.5% between 2009 and 2017, and by 2.9% in 2017 alone (EPA, 2018). A significant driver for this increase over the past decade is a substantial expansion of the beef and dairy sectors. This has been driven to a large extent by national policy signals, including the Food Harvest 2020 and Food Wise 2025 government strategies. These prioritised the expansion of beef and milk exports in particular. Furthermore, the abolition of EU milk quotas in 2015 was another significant contributory factor.

Transport is another problematic sector from the perspective of GHGE. Ireland's transport emissions remain stubbornly wedded to economic growth, registering significant decline during the financial crisis followed by strong rebound as economic growth returned. Between 1990 and 2017, GHGE from transport in Ireland increased by 133.2% – more than any other sector in the national inventory – with emissions from road transport increasing by 140% (EPA, 2018). Moreover, Irish transport governance is characterised by significant complexity and fragmentation, impeding decarbonisation progress (Devaney and Torney, 2019). This includes internal tensions between public and private actors, rural and urban divides, and the competing role of special interests as well as complex external interactions with broader policy systems, including planning, health and education. Moreover, contestation between institutional priorities has shaped the development of a carbon-intensive transport system, and low carbon transition has yet to be embedded in governing regimes.

By contrast, there has been moderate to good progress in renewable electricity deployment. Although Ireland is likely to miss its target for renewable energy development, performance in this area has been significantly better than reducing GHGE in the non-emissions trading sector. It is somewhat difficult to compare performance across member states, however, as member states have different renewable energy targets for 2020 and also come from different starting points. Gross final energy consumption from renewable sources in Ireland more than tripled during the period 2005–2016. Ireland registered the fifth highest growth in percentage terms of any member states (EEA, 2018). In 2017, renewable energy made up 10.6% of gross final consumption relative to a 2020 target of 16%. While significant progress has been made in scaling up renewable electricity, progress in the transport and heat sectors has been much more limited (SEAI, 2018).

At 13.2 tonnes of carbon dioxide (CO_2) equivalent per capita in 2018, Ireland's per capita emissions are the fourth highest in the EU and significantly higher than the EU-27 average of 8.7 tonnes (Eurostat, 2020). At the end of 2018, the Germanwatch/Climate Action Network Climate Change Performance Index gave Ireland a very low/very poor rating on climate policy for the fifth year in a row, ranking it 48th out of 60 countries, the lowest ranking of any EU state (Burck *et al.*, 2018).

Progress towards a decarbonised economy and society has thus been slow in Ireland, hampered by often haphazard, conflicting, unambitious and poorly implemented climate policies to date. Overall, it is fair to characterise Ireland as a climate laggard. Ireland's profile as a laggard developed in a European multi-level governance context that set legally binding targets for GHGE and renewable energy at member state level. Outside of renewable electricity, the limited impact of this multilevel governance framework on Ireland's domestic policy development is striking (Torney and O'Gorman, 2019). Against this backdrop, the Irish Citizens' Assembly emerged in 2016, holding the potential to address democratic failures and deliberate on some of the most important issues facing Irish society at the time, including climate change. Its evolution, structure and progress is explored next.

The Citizens' Assembly

The first materialisation of a deliberative democracy approach in Ireland came with the experimental 'We the Citizens' project in 2011. This participatory project, led by Irish academics and funded by The Atlantic Philanthropies, aimed to test the potential for giving citizens a greater role in democracy between elections. It hosted a pilot Citizens' Assembly in June 2011 where participants had access to objective information, had the chance to deliberate and arrive at informed decisions. The research found that after taking part in a pilot Citizens' Assembly, participants expressed more willingness to discuss and become involved in politics. They experienced opinion shift and felt more positive about the ability of ordinary people to influence politics (We the Citizens, 2015).

Building on the 'We the Citizens' model, the Irish government established the 'Convention on the Constitution' in 2012. The Constitutional Convention, as it became more commonly known, brought together 66 citizens with 33 legislators from both Norther Ireland and the Republic of Ireland to deliberate on proposed amendments to the Constitution of Ireland. Citizens were randomly chosen to reflect the age, regional and gender balance of the electorate. A chairperson was appointed, Tom Arnold, then CEO of Concern Worldwide, Ireland's largest humanitarian aid agency. Topics for deliberation in the Constitutional Convention included mandating same-sex marriage, reducing the age of eligibility for presidency, increasing female participation in politics and removing the offence of blasphemy from the Constitution (Constitutional Convention, n.d.).

While the Irish government was not obliged to proceed with any proposed amendment from the Constitutional Convention, it committed to formally responding to each recommendation and debating it in parliament. The most significant impact of the deliberations included feeding into the referendum on same-sex marriage in Ireland, held in May 2015. With almost 2 million votes cast, the referendum was passed by a significant majority (62% Yes to 38% No), building on recommendations of the Constitutional Convention and signalling a new era of societal and democratic reform in the Irish context. This includes the unique foundations of this referendum in a deliberative democracy experiment, its 'unusually vigorous and active' campaign and voting pattern results that 'point to a significant value shift along the deep seated liberal conservative political cleavage of Irish politics' (Elkink *et al.*, 2017: 361).

Some academic research has been carried out looking at the predecessors to the Citizens' Assembly. Positioning 'deliberation as a sign of democracy in transformation', Farrell *et al.* (2013: 100), for example, outline the process, structure and impact of the 'We the Citizens' assembly, including the impact on those taking part and potential for the model to be replicated by government. This included enhanced ability on behalf of the participants to account for hard trade-offs in policy-making (in this instance related to balancing tax increases with spending cuts), as well as the potential of involving citizens in critical public policy issues.

Suiter *et al.* (2016) similarly revealed the power of deliberative processes in encouraging opinion change and the importance of heterogeneous groups for deliberation to be successful (exposing people to views that are different to their own). More widely, Farrell *et al.* (2019: 113) praise Ireland as 'something of a trail-blazer in the use of deliberative methods in the process of constitutional review', representing one of the first countries internationally to not only host two constitutional mini-publics in quick succession, but also link deliberative (mini-publics) and direct democracy (referendums) (see also Chapter 13 in this volume). In this sense they suggest an increasing 'systemisation of deliberation' in the Irish policy process as a result of these organised 'mini-publics', defined by Goodin and Dryzek (2006: 220) as democratic innovations involving ordinary citizens in 'groups small enough to be genuinely deliberative, and representative enough to be genuinely democratic'.

Following on from these early initiatives, the centre-right Fine Gael party, which had been the senior partner in the 2011–2016 coalition government, included in its manifesto for the February 2016 general election a commitment to establishing, within six months, a Citizens' Assembly. They endeavoured for this to be composed solely of members of the public and tasked with considering reform of the senate, climate change and possible repeal of Ireland's constitutional ban on abortion (Fine Gael, 2016: 99). Following lengthy negotiations with the centre-right Fianna Fáil party after the general election, agreement was reached on a 'confidence and supply' arrangement, under which Fianna Fáil agreed to support a minority Fine Gael government from the opposition benches.

The agreement was codified in a May 2016 Programme for Partnership Government, which committed to 'the establishment of a Citizens' Assembly, within six months and without participation by politicians, with a mandate to look at a limited number of key issues over an extended time period' (Government of Ireland, 2016: 84). The Assembly was established by resolution of both houses of parliament in July 2016, and was tasked with deliberating on a number of topics of public policy (see Table 12.1).

Despite the commitment in the Fine Gael manifesto to include climate change as a topic for the Assembly, the draft resolution introduced in July 2016 contained no reference to climate change. However, as a result of a Green Party amendment, one of the topics included was 'How the State can make Ireland a leader in tackling climate change' (Citizens' Assemby, 2018). There was no requirement under the EU framework to establish a Citizens' Assembly to engage and enable the public to deliberate on the topic of climate change. Its emergence in a climate laggard state like Ireland was thus groundbreaking and novel for its time, and can be characterised as an instance of climate pioneership.

Unlike the Constitutional Convention, no politicians were involved in the Citizens' Assembly, largely attributed to political desires to distance themselves from the controversial abortion issue (Suiter, 2018). Using stratified random sampling, 99 citizens were recruited to take part in the deliberations. A polling company, Red C, was selected by public tender to recruit participants. Citizens did not receive remuneration for their participation but their travel and accommodation expenses were covered, along with any additional expenses incurred as a result of participation (e.g. contribution towards childcare). A chairperson was also appointed by government, retired Supreme Court Judge the Honourable Mary Laffoy. The chair was supported by a secretariat and by an expert advisory group for each of the topics on which it deliberated.

The process of deliberation in the Citizens' Assembly included 12 weekend-long meetings in a hotel in Dublin between October 2016 and April 2018. Different topics received different time allocations as detailed in Table 12.1. Deliberations concerning the possible removal of Ireland's constitutional ban on abortion, for example, received the most time, given mounting pressure from both top–down international human rights movements and bottom–up local campaigns to create change in this health arena.[6] The topic of climate change originally was designated one weekend of deliberations but, given the amount of material to be covered, was later granted an extension to proceed into a second weekend of deliberations.

For every topic under consideration, citizens received background material in advance of each meeting before being exposed to a series of expert and advocate presentations and facilitated roundtable discussions. Participants were able to question speakers on the content of their papers. Ballot paper voting took place on the final day of deliberations for each topic which reflected the preceding debates and culminated in a number of recommendations for each topic. Similar to the Constitutional Convention, the government was not obliged to pursue each recommendation but committed to respond to each one, particularly in cases when the recommendation was not taken forward. As with the Constitutional Conventions, however, some topics received more traction than others to date, as described in Table 12.1.

Deliberations on 'How the State can make Ireland a leader in tackling climate change' was originally scheduled to be considered last by the Citizens' Assembly. However, in January 2017, the members decided that this topic should be considered third. Deliberations took place over two weekends: 30 September – 1 October 2017 and 4–5 November 2017. The Assembly also invited submissions from members of the public (including individual citizens and representative groups). A total of 1,185 submissions were received on the topic of climate change. This was the second highest number of submissions received across the

Table 12.1 The Irish Citizens' Assembly: Topics, Meetings, Submissions, Output and Impact.

Topic	Number of weekends dedicated	Number of public submissions[5]	Output	Government response and impact
The Eighth Amendment of the Constitution (related to abortion)	5	12,200	1 key recommendation (with various parts) plus 5 ancillary ones	Considered by a special parliamentary committee. Government accepted proposal for a referendum. Referendum passed in May 2018.
How we best respond to the challenges and opportunities of an ageing population	2	122	15 recommendations plus 6 ancillary ones	No reaction to date.
How the State can make Ireland a leader in tackling climate change	2	1,185	13 recommendations plus 4 ancillary ones	Considered by a special parliamentary committee and their report published in March 2019.
The manner in which referenda are held	1	206	10 recommendations	No reaction to date.
Fixed-term parliaments	1	8	7 recommendations	No reaction to date.

five topics, though with large variance as evidenced in Table 12.1. The secretariat of the Assembly produced a summary 'signpost document' to provide an overview of the submissions received on climate change (Citizens' Assembly, 2017). Citizens could access this document online in advance of the deliberations in their own time. It was not presented formally as part of the expert or advocate presentations.

Twenty-one speakers in total presented on the topic of climate change. This included 15 experts (e.g. climate scientists, academics, government agency employees and international experts) and six advocates (e.g. farmers, a firefighter who had reduced carbon emissions from his fire station and a start-up entrepreneur engaged in sustainability-orientated businesses). Encompassing 26 hours of listening, discussion and deliberation, topics included an overview of climate science; an overview of the national policy framework; international perspectives on climate leadership; and sectoral snapshots of energy, transport and agriculture impacts and potential solutions (some of the top greenhouse gas emitting sectors in Ireland) (EPA, 2018). Speakers were chosen by the chair of the Assembly, with input from an expert advisory group of five academics with expertise across a variety of climate change disciplines including climate, political and social sciences.

On the final Sunday of climate change deliberations, 13 questions were put to the citizens for ballot paper voting on how Ireland's response to climate change could be strengthened. The 13 recommendations they agreed upon were significantly more radical than expected for a seeming climate laggard nation. A high degree of consensus was also obvious amongst the 99 citizens, with 80% or more in favour of all proposals. These recommendations are detailed in Table 12.2.

Recommendation 2, that the state should take a leadership role in addressing climate change through mitigation measures, was supported by 100% of members. This is particularly interesting in the context of this volume's focus on climate leadership, though arguably it came about – and received unanimous support – because of the strong view among members that the state was a climate laggard rather than a leader. From a polycentric governance perspective, the demand that the *state* should take a leadership role points towards centralised rather than polycentric governance. However, the unanimous support for Recommendation 6, that the state should act to ensure the greatest possible levels of community ownership of renewable energy, perhaps points towards the desire for more polycentric approaches to climate governance.

The Citizens' Assembly process highlights the potential and power of mobilising ordinary citizens in the climate crisis. When engaged and informed in this way, it appeared that citizens were able to overcome any information deficits and move beyond their own self-interests to reach collective decisions for the greater public good (criticisms often levied at deliberative democracy processes and climate engagement fields) (Fung, 2003; Smith, 2009; Suldovsky, 2017).

A Final Report and Recommendations from the Citizens' Assembly on climate change was laid before parliament on 18 April 2018. An all-party parliamentary committee was established in July 2018, charged with reviewing and responding

Table 12.2 Recommendations of the Citizens' Assembly on Climate Change.

	Topic area	Recommendations by members of the Citizens' Assembly
1	Climate governance	97% recommended that to ensure climate change is at the centre of policy-making in Ireland, as a matter of urgency a new or existing independent body should be resourced appropriately, operate in an open and transparent manner and be given a broad range of new functions and powers in legislation to urgently address climate change.
2	Leading by example	100% recommended that the State should take a leadership role in addressing climate change through mitigation measures, including, for example, retrofitting public buildings, having low carbon public vehicles, renewable generation on public buildings and through adaptation measures including, for example, increasing the resilience of public land and infrastructure.
3	Carbon tax	80% said they would be willing to pay higher taxes on carbon-intensive activities.
4	Adaptation	96% recommended that the State should undertake a comprehensive assessment of the vulnerability of all critical infrastructure (including energy, transport, built environment, water and communications) with a view to building resilience to ongoing climate change and extreme weather events. The outcome of this assessment should be implemented. Recognising the significant costs that the State would bear in the event of failure of critical infrastructure, spending on infrastructure should be prioritised to take account of this.
5	Energy: micro-generation	99% recommended that the State should enable, through legislation, the selling back into the grid of electricity from micro-generation by private citizens (for example energy from solar panels or wind turbines on people's homes or land) at a price which is at least equivalent to the wholesale price.
6	Energy: community ownership	100% recommended that the State should act to ensure the greatest possible levels of community ownership in all future renewable energy projects by encouraging communities to develop their own projects and by requiring that developer-led projects make share offers to communities to encourage greater local involvement and ownership.
7	Energy: peat and just transition	97% recommended that the State should end all subsidies for peat extraction and instead spend that money on peat bog restoration and making proper provision for the protection of the rights of the workers impacted with the majority 61% recommending that the State should end all subsidies on a phased basis over five years.
8	Transport: shared and active mobility	93% recommended that the number of bus lanes, cycling lanes and park-and-ride facilities should be greatly increased in the next five years, and much greater priority should be given to these modes over private car use.

(Continued)

Table 12.2 (Continued)

	Topic area	Recommendations by members of the Citizens' Assembly
9	Transport: electric vehicles	96% recommended that the State should immediately take many steps to support the transition to electric vehicles.
10	Transport: public transport	92% recommended that the State should prioritise the expansion of public transport spending over new road infrastructure spending at a ratio of no less than 2:1 to facilitate the broader availability and uptake of public transport options with attention to rural areas.
11	Agriculture	89% recommended that there should be a tax on GHGE from agriculture. There should be rewards for the farmer for land management that sequesters carbon. Any resulting revenue should be reinvested to support climate friendly agricultural practices.
12	Food waste	93% recommended the State should introduce a standard form of mandatory measurement and reporting of food waste at every level of the food distribution and supply chain, with the objective of reducing food waste in the future.
13	Land use diversification	99% recommended that the State should review and revise supports for land use diversification with attention to supports for planting forests and encouraging organic farming.

to these recommendations of the Assembly. Its formation, process and outcomes are discussed next.

Impact of the Citizens' Assembly on climate policy

An all-party parliamentary committee – the Joint Oireachtas Committee on Climate Action (JOCCA) – was established in mid-2018 to respond to the Citizens' Assembly recommendations on climate change. Its establishment gave momentum to those recommendations, which otherwise risked falling into obscurity once the Assembly concluded its work. The JOCCA followed broadly a template laid down by an all-party parliamentary committee established in 2017 to consider the Assembly's recommendations on Ireland's restrictive abortion regime. A report by that committee was seen as essential in building political consensus to remove the constitutional prohibition on abortion which was successfully passed at a referendum in 2018.

The JOCCA report, published in March 2019, provided roadmaps for policy development in the key sectors of agriculture, energy, transport and buildings (Houses of the Oireachtas, 2019). Building on Recommendation 7 of the Citizens' Assembly, it set out important recommendations for a just transition towards decarbonisation.

One of the most important elements of the report is its recommendations for a new framework for governing the overall response to climate change. This built on Recommendation 1 of the Citizens' Assembly that 'climate change is at the centre of policy-making in Ireland'. It proposed a fundamental revision of

Ireland's 2015 climate law that would bring it into line with the UK's Climate Change Act (see also Chapter 10 in this volume). This is particularly interesting because, during the framing of Ireland's existing climate change law, the UK model was considered but rejected because of objections by political parties and interest groups (Torney, 2017). The JOCCA's proposed revisions to the climate law would set a target of net zero GHGE by 2050, an interim 2030 target, and a 70% renewable electricity target by 2030. The proposed law would also provide for successive five-yearly 'carbon budgets'. These would place an overall cap on emissions and would be set up to a decade in advance.

While there was cross-party consensus on most elements of the report, members of the JOCCA could not find agreement on the issue of raising the carbon tax, which was originally introduced in December 2009 and aims to disincentivise high-emitting activities through a price mechanism. The point of contention at the JOCCA was whether to recommend increasing the rate of the carbon tax above the existing level of €20/tonne CO_2. Fine Gael, Fianna Fáil, the Green Party and Labour Party members of the Committee agreed to recommend increasing the carbon tax to €80/tonne by 2030 (in line with the repeated recommendation of the Irish Climate Change Advisory Council). Sinn Féin and Solidarity–People Before Profit members of the Committee, however, voted against the report as a whole, principally because of their objection to the carbon tax recommendations.

Nonetheless, the JOCCA succeeded in giving momentum to the majority of the Citizens' Assembly recommendations on climate change. Each chapter of the JOCCA report explicitly responded to the Assembly's recommendations. In most cases, the JOCCA's recommendations developed and expanded upon the Assembly's work. However, not all of the Assembly's recommendations were endorsed by the JOCCA. Notably, the JOCCA's response to the Assembly's recommendation for a tax on GHGE from agriculture was to indicate that further consideration should be given to the subject by the JOCCA in the future (Houses of the Oireachtas, 2019: 122). This highlights the continued complexity and difficulty for politicians to develop robust, just and meaningful climate action solutions, particularly in politically sensitive arenas such as – in the Irish context – agriculture.

In turn, the JOCCA report to a significant extent shaped the development of an all-of-government *Climate Action Plan to Tackle Climate Breakdown*, published in June 2019 (Government of Ireland, 2019). Among the most noteworthy elements of this plan were its proposals to reform Ireland's climate governance institutions. These proposals mirrored closely the recommendations of the JOCCA report, including amendment of the 2015 climate law that would mirror significantly the UK Climate Change Act and its system of carbon budgets (see also Chapter 10 this volume). Other institutional changes included the creation of a Climate Action Delivery Board within the Department of the Taoiseach and the establishment of a permanent parliamentary climate action committee. These reforms promised to significantly enhance Ireland's climate governance landscape. From the perspective of multilevel and polycentric governance, these institutional changes suggest a centralisation of policy-making rather than a move towards greater polycentricity.

In June 2020, a new programme for government was agreed between the centre-right Fianna Fáil and Fine Gael parties along with the Green Party, with significant climate action commitments throughout the programme (Fianna Fáil, Fine Gael and Green Party, 2020). These included many of the commitments set out in the 2019 Climate Action Plan but with significant new actions to be undertaken, including a requirement that 'every minister will make climate action a core pillar of their new departmental strategies'. This, in combination with the governance arrangements introduced by the Climate Action Plan that were to be completed by the incoming government, was in line with Recommendation 1 of the Citizens' Assembly to place climate change at the centre of policy-making. Significantly, the plan committed to a two-to-one split new transport infrastructure spending in favour of public transport, in line with Recommendation 10 of the Citizens' Assembly. The 2020 programme for government could be read as the culmination of the Citizens' Assembly's recommendations, but in the end, the Assembly's impact will be determined by the extent to which they shape policy outcomes, including, ultimately, GHGE reduction. Whether Ireland's climate pioneership is transactional or transformational will, in a similar way, be determined by the medium-term trajectory of climate action in Ireland. At this point in time, all we can say is that the Citizens' Assembly has led to governance reforms that have the potential to be transformational.

Conclusion

Ireland's Citizens Assembly represents an innovative and novel approach to governing climate change. What drove an apparent climate laggard to engage in such institutional pioneership? The ground was laid by two previous experiments in deliberative democracy, *We the Citizens* and the *Constitutional Convention*. These experiments built the case for using deliberative forums for enhancing trust in public institutions and modern representative democracies. The Citizens' Assembly was arguably seen by politicians as a way to gain support for potentially controversial or politically sensitive policies and constitutional changes such as the removal of the constitutional ban on abortion. Nonetheless, its recommendations on the topic have now been brought further into national climate policy. This arguably has changed Ireland's previous complexion as a climate laggard, making it a pioneer with respect to the use of deliberative democracy for responding to climate change.

As we have shown, the outcomes of the Citizens' Assembly deliberations were significant and far reaching. Our analysis has shown the importance of follow-up and impact to such a deliberative process. The Citizens' Assembly illustrates this clearly, with recommendations on two topics (the constitutional ban on abortion and climate change) being taken forward by parliamentary committees, while recommendations on its three other topics have been left to gather dust. Given the commitment of members of the public to participating in such processes (without remuneration, in the case of the Citizens' Assembly), it seems important that the outcomes of such a process would at minimum be given serious consideration by government. This is important also from the perspective of democratising decision-making and truly including the 'bottom–up' perspective, as advocated for in the Paris Agreement. Although under certain conditions such deliberative democratic innovations contribute to building polycentric governance, in the Irish case, the Citizens' Assembly was established by, and reported its recommendations to, parliament. Further, the overarching nature of its conclusions and impact has inspired an increased centralisation of climate policy making in Ireland. As such, it would be wrong to consider this instance of deliberative democracy as heralding a move towards polycentric governance.

In the period since it concluded its deliberations on climate change, the Citizens' Assembly has been lauded internationally, including by the 'Extinction Rebellion' movement. This campaign has among its demands that 'government must create and be led by the decisions of a Citizens' Assembly on climate and ecological justice', pointing to the example of Ireland's Citizens' Assembly (Extinction Rebellion, 2020). However, we do not find clear evidence that the process was aimed explicitly at providing an example or model of climate governance for others to follow. As such, we suggest that Ireland's innovative approach to climate governance embodied in the Citizens' Assembly is perhaps best characterised as pioneership rather than leadership. Nonetheless, the Irish case is a particularly interesting case of climate pioneership because of Ireland's acknowledged status as a climate laggard, in which we have seen the emergence of a surprising pioneer in the use of deliberative democracy in climate governance.

The Irish experience arguably provides a template for combining representative and deliberative democracy (Farrell *et al.*, 2019). These forms of democracy should not be thought of as rivals. Rather, deliberative institutions such as citizens' assemblies can be successfully embedded within structures of representative democracy. Indeed, they have further been endorsed for their ability to respond to the seeming 'crisis of democracy' perpetuating worldwide to allow for politicians to account for (and not be overwhelmed by) citizens' voices in a way that avoids polarisation and enhances decision-making (Dryzek *et al.*, 2019).

Deliberative democracy is particularly important in the case of climate action policy that requires buy-in and public support for action. Nonetheless, deliberation should not be considered a panacea for climate change policy-making. Indeed, the jury is still out in the Irish case. The test of whether Ireland's climate pioneership through the Citizens' Assembly and subsequent policy processes is truly transformational or merely transactional will be seen in the uptake and implementation – or otherwise – of the 13 recommendations in policy, and ultimately in Ireland's GHGE trajectory in the years to come.

Notes

1 The chapter draws on the Climate Citizens' research project funded by the Irish Environmental Protection Agency (research grant 2018-CCRP-DS.18). Diarmuid

Torney participated in the Citizens' Assembly process as a member of the Expert Advisory Group for the climate change topic.
2 We use the term 'Ireland' to refer to the Republic of Ireland throughout this chapter. Where clarity requires, we differentiate between the Republic of Ireland and Northern Ireland.
3 The EU Burden-Sharing Agreement is an agreement between EU member states on how to allocate the 'burden' of achieving the EU's collective GHGE reduction target for the period 2008–2012 under the Kyoto Protocol of 8% reduction relative to 1990 levels. Under this agreement, Ireland's allocated target was to limit the increase in emissions to no more than 13% above 1990 levels.
4 The EU Effort-Sharing Decision, which succeeded the EU Burden-Sharing Agreement, is an agreement between member states on how to allocate the 'effort' of achieving the EU's collective GHGE reduction target of 10% in sectors not covered by the EU ETS for the period 2013–2020. Under this agreement, Ireland was allocated a target of 20%.
5 Total number of submissions declared valid and published on the Citizens' Assembly website: https://www.citizensassembly.ie/en/.
6 Shortly after the Assembly concluded its deliberations, a referendum was held and passed by a significant majority (with 66.4% in favour and over 2 million votes cast) to repeal the constitutional ban on abortion and pave the way for legalisation of abortion in Ireland.

Bibliography

Burck, J., Hagen, U., Marten, F., Höhne, N. and Bals, C. (2018) *The Climate Change Performance Index: Results 2019*, Berlin: Germanwatch.

Burck, J., Marten, F., Bals, C. and Höhne, N. (2017) *The Climate Change Performance Index: Results 2018*, Berlin: Germanwatch.

Citizens' Assembly (2017) Submissions to the Citizens' Assembly on the third topic for consideration 'How the state can make Ireland a leader in tackling climate change', Signpost document for Assembly Members key issues raised and themes covered. Available at: https://www.citizensassembly.ie/en/Meetings/Signpost-Document.pdf (Accessed: 30 April 2019).

Citizens' Assembly (2018) *Third Report and Recommendations of the Citizens' Assembly: How the State Can Make Ireland a Leader in Tackling Climate Change*, Dublin: Citizens' Assembly.

Constitutional Convention (n.d.) 'The convention on the constitution', Archive Website. Available at: http://www.constitutionalconvention.ie/ (Accessed: 30 April 2019).

DCENR (2015) *Ireland's Transition to a Low Carbon Energy Future, 2015–2030*, Dublin: Department of Communications, Energy & Natural Resources.

Department of Environment and Local Government (2000) *National Climate Change Strategy Ireland*, Dublin: Department of Environment and Local Government.

Department of Environment Heritage and Local Government (2007) *National Climate Change Strategy 2007–2012*, Dublin: Department of Environment Heritage and Local Government.

Devaney, L. and Torney, D. (2019) *Advancing the Low Carbon Transition in Irish Transport*, Dublin: National Economic and Social Council.

Dorsch, M.J. and Flachsland, C. (2017) 'A polycentric approach to global climate governance', *Global Environmental Politics*, 17(2): 45–64.

Dryzek, J.S. (1990) *Discursive Democracy: Politics, Policy and Science*, Cambridge: Cambridge University Press.

Dryzek, J.S., Bächtiger, A., Chambers, S., Cohen, J., Druckman, J.N., Felicetti, A., Fishkin, J.S., Farrell, D.M., Fung, A., Gutmann, A. and Landemore, H. (2019) 'The crisis of democracy and the science of deliberation', *Science*, 363(6432): 144–1146.

Duwe, M., Freundt, M., Iwaszuk, E., Knoblauch, D., Maxter, M., Mederake, L., Ostwald, R., Riedel, A., Umpfenbach, K. and Zelljadt, E. (2017) *"Paris Compatible" Governance: Long-term Policy Frameworks to Drive Transformational Change*, Berlin: Ecologic Institute.

EEA (2018) *Trends and Projections in Europe 2018 Tracking Progress towards Europe's Climate and Energy Targets*, Copenhagen: European Environment Agency.

Elkink, J.A., Farrell, D.M., Reidy, T. and Suiter, J. (2017) 'Understanding the 2015 marriage referendum in Ireland: context, campaign, and conservative Ireland', *Irish Political Studies*, 32(3), 361–381.

EPA (2018) *Ireland's Provisional Greenhouse Gas Emissions, 1990–2017*, Wexford: Environmental Protection Agency.

EPA (2019) *Ireland's Greenhouse Gas Emissions Projections, 2018–2040*, Wexford: Environmental Protection Agency.

Eurostat (2020) *Greenhouse gas emissions per capita*, https://ec.europa.eu/eurostat/datab rowser/view/t2020_rd300/default/table?lang=en (accessed 18 September 2020).

Extinction Rebellion (2020) *Citizens' Assembly*, https://extinctionrebellion.uk/go-beyond -politics/citizens-assembly/ (accessed 18 September 2020).

Farrell, D.M., O'Malley, E. and Suiter, J. (2013) 'Deliberative democracy in action Irish-style: the 2011 we the citizens pilot citizens' assembly', *Irish Political Studies*, 28(1): 99–113.

Farrell, D.M., Suiter, J. and Harris, C. (2019) '"Systematizing" constitutional deliberation: the 2016–18 citizens' assembly in Ireland', *Irish Political Studies*, 34(1): 113–123.

Fianna Fáil, Fine Gael and Green Party (2020) *Programme for Government: Our Shared Future*, https://www.greenparty.ie/wp-content/uploads/2020/06/2020-06-15-Progra mmeforGovernment_Corrected-Final-Version.pdf (Accessed: 29 June 2020).

Fine Gael (2016) *Fine Gael General Election Manifesto 2016: Let's Keep the Recovery Going*, Dublin: Fine Gael.

Fishkin, J. (1991) *Democracy and Deliberation: New Directions for Democratic Reform*, New Haven: Yale University Press.

Fung, A. (2003) 'Survey article: recipes for public spheres: eight institutional design choices and their consequences', *Journal of Political Philosophy*, 11(3), 338–367.

Goodin, R.E. and Dryzek, J.S. (2006) 'Deliberative impacts: the macro-political uptake of mini-publics', *Politics & Society*, 34(2): 219–244.

Government of Ireland (2014) *Climate Action and Low-Carbon Development: National Policy Position Ireland*, Dublin: Department of Communications, Climate Action & Environment.

Government of Ireland (2016) *Programme for a Partnership Government*, Dublin: Government of Ireland.

Government of Ireland (2019) *Climate Action Plan to Tackle Climate Breakdown*, Dublin: Government of Ireland.

Houses of the Oireachtas (2019) *Report of the Joint Committee on Climate Action. Climate Change: A Cross-Party Consensus for Action*, Dublin: Houses of the Oireachtas.

Jordan, A., Huitema, D., Hildén, M., Van Asselt, H., Rayner, T., Schoenefeld, J.J., Tosun, J., Forster, J. and Boasson, E.L. (2015) 'Emergence of polycentric climate governance and its future prospects', *Nature Climate Change*, 5: 977–982.

Jordan, A., Huitema, D., van Asselt, H. and Forster, J. (eds) (2018) *Governing Climate Change: Polycentricity in Action?* Cambridge: Cambridge University Press.

Liefferink, D. and Wurzel, R.K.W. (2017) 'Environmental leaders and pioneers: agents of change?', *Journal of European Public Policy*, 24(7): 651–668.

McGreevy, R. (2018) 'Citizens' Assembly is an example to the world, says chairwoman', *Irish Times*, 21 June, https://www.irishtimes.com/news/ireland/irish-news/citizens-assem bly-is-an-example-to-the-world-says-chairwoman-1.3539370 (Accessed: 3 May 2019).

Niemeyer, S. (2013) 'Democracy and climate change: what can deliberative democracy contribute?', *Australian Journal of Politics and History*, 59(3): 429–448.

Ostrom, E. (2010) 'Polycentric systems for coping with collective action and global environmental change', *Global Environmental Change*, 20(4): 550–557.

Ostrom, E. (2012) 'Nested externalities and polycentric institutions: must we wait for global solutions to climate change before taking actions at other scales?' *Economic Theory*, 49(2): 353–369.

Sargent, N. (2018) *Taoiseach tells EU he is not proud of Ireland's role as Europe's climate 'laggard'* [online], 18 January 2018. Available from: https://greennews.ie/taoiseach -tells-eu-not-proud-ireland-climate-laggard-role/ [Accessed: 2 April 2019].

SEAI (2018) *Energy in Ireland: 2018 Report*, Dublin: Sustainable Energy Authority of Ireland.

Smith, G. (2009) *Democratic Innovations: Designing Institutions for Citizen Participation*, Cambridge: Cambridge University Press.

Stevenson, H. and Dryzek, J.S. (2014) *Democratizing global climate governance*, Cambridge: Cambridge University Press.

Suiter, J. (2018) 'Deliberation in action – Ireland's abortion referendum', *Political Insight*, 9(3): 30–32.

Suiter, J., Farrell, D.M. and O'Malley, E. (2016) 'When do deliberative citizens change their opinions? Evidence from the Irish Citizens' Assembly', *International Political Science Review*, 37(2): 198–212.

Suldovsky, B. (2017) 'The information deficit model and climate change communication. Climate science, *Oxford Research Encyclopaedias*, September 2017. doi: 10.1093/ acrefore/9780190228620.013.301

Torney, D. (2015) *European Climate Leadership in Question: Policies toward China and India*, Cambridge, MA: MIT Press.

Torney, D. (2017) 'If at first you don't succeed: The development of climate change legislation in Ireland', *Irish Political Studies*, 32(2): 247–267.

Torney, D. (2019) 'Follow the leader? Conceptualising the relationship between leaders and followers in polycentric climate governance', *Environmental Politics*, 28(1), 167–186.

Torney, D. and O'Gorman, R. (2019) 'A laggard in good times and bad? The limited impact of EU membership on Ireland's climate change and environmental policy', *Irish Political Studies*. 34(4), 575–594.

We the Citizens (2015) 'We the citizens: speak up for Ireland'. Available at: http://www .wethecitizens.ie/ (Accessed: 30 April 2019).

Wurzel, R.K.W., Connelly, J. and Liefferink, D. (eds) (2017) *The European Union in International Climate Change Politics: Still Taking a Lead?* London: Routledge.

Wurzel, R.K.W., Liefferink, D. and Torney, D. (2019a) 'Pioneers, leaders and followers in multilevel and polycentric climate governance', *Environmental Politics*, 28(1): 1–21.

Wurzel, R.K.W., Moulton, J.F.G., Osthorst, W., Mederake, L., Deutz, P. and Jonas, A.E.G. (2019b) 'Climate pioneership and leadership in structurally disadvantaged maritime port cities', *Environmental Politics*, 28(1): 146–166.

13 Switzerland

International commitments and domestic drawbacks

*Marlene Kammerer, Karin Ingold and
Johann Dupuis*

Introduction

Switzerland is a 'small state' when it comes to climate change, contributing less than 0.11% to global greenhouse gas emissions (GHGE) (Schenkel, 2000; Carter *et al.*, 2019; Crippa *et al.*, 2019). Nevertheless, Switzerland is often perceived as a leader on climate change policies. Reputed for its clean environment and ambitious policies, Switzerland was indeed one of the first countries, together with Norway (see also Chapter 11 in this volume), to have introduced a domestic carbon dioxide (CO_2) reduction target at the very beginning of the international climate negotiations in 1990. It advocated a global tax on CO_2 and insisted that industrialised countries should support and finance climate adaptation in developing countries. In order to increase its political influence, Switzerland also initiated the 'environmental integrity group' in 2000, which includes non-Annex-1 countries to the Kyoto Protocol such as Mexico and South Korea (Ingold and Pflieger, 2016; Lehmann and Rieder, 2002). The behavioural pattern that Switzerland exhibited during the early 1990s, approximates to what Liefferink and Wurzel (2017) have described as a 'climate pusher', which is a state that innovates with regard to its domestic climate policy while lobbying others to follow its lead (see also Chapter 1 in this volume).

Since the 2000s, it has become less clear whether Switzerland still acts as pusher on global warming issues. The Climate Performance Index 2016 issued by Germanwatch ranked Switzerland at the 14th position, behind European countries like France, Sweden, Belgium, Denmark and Portugal (Burck *et al.*, 2015). This somewhat disappointing ranking position is expressed through Switzerland's unambitious renewable energy policy, its tendency to buy CO_2 certificates instead of focusing on domestic reduction measures and its failure to effectively regulate emissions from the transport sector. Different studies have concluded that the modest climate policy output and implementation is caused by Swiss direct democracy: strong and resourceful target groups – mainly the transport sector – constantly threatened to block the process via referendum (Lehmann and Rieder, 2002; Steffen and Vatter, 2002; Ingold, 2011).

One consequence is that even though Switzerland still is a country that genuinely pushes for stronger climate policies in international negotiations, it tends to wait for and align itself to the EU positions rather than taking the lead by proposing more ambitious and innovative policy goals or instruments (see also Chapter 8 in this volume). Based on these more recent developments, Switzerland does not seem as a climate 'pusher' anymore. It would be more accurate to describe Switzerland as a 'conditional leader' aligning its climate policies to those of the EU or even as a 'symbolic leader', i.e. a country with high external, but low internal ambitions (Wurzel, Liefferink and Torney, 2019).

Switzerland is a small, federalist and consensus-democratic country (Lijphart, 2012). But in contrast to this ideal-type, the elaboration of Swiss climate change mitigation policy has mostly happened at the national level, in a quite centralised fashion. Moreover, the decision-making style has so far been fairly conflictual rather than consensual. The importance of peak economic associations in this process conforms, however, to the conventional picture of the Swiss pattern of democracy drawn up by Lijphart (Ingold, 2011; Sciarini, Fischer and Traber, 2015). In this context, we analyse the so-called political elite, namely all those collective public and private actors involved in the making of the climate change mitigation policy.

In this chapter, we provide an explanation why Switzerland seems to be less at the forefront of climate protection than in the past. We argue that the Swiss case offers an adequate setting to identify the key factors which could explain why climate leaders can deviate from ambitious mitigation pathways. To do so, we focus on the introduction of a CO_2 tax, which Switzerland pushed at the international level since the 1990s. Interestingly, however, Switzerland was able to implement it domestically only in 2008 and merely in partial fashion, namely in the form of a tax on fossil combustibles which excluded motor fuels.

International commitments and domestic drawbacks

Political debates around the need for a CO_2 tax have a long history in Switzerland. The topic goes back to the 1980s, when it was first mentioned in relation to clean air and energy policies (Lehmann and Rieder, 2002). The oil crises of the 1970s had the effect of raising awareness that Switzerland needed to reduce its dependency on fossil energy imports. In the aftermath of the 1986 Chernobyl catastrophe, a national programme on energy efficiency was introduced, which set the goal of stabilising fossil fuel consumption and hence CO_2 emissions by 2000. The idea of reducing CO_2 with the introduction of a tax remained on the political agenda. A turning point in the history of the tax was arguably when the Minister for Home Affairs (who was responsible for environmental issues at the time) Flavio Cotti returned from the 1992 United Nations Rio conference. He was convinced that Switzerland, which had largely invested in hydropower from the 1950s to the 1970s, could act as a pioneer in the fight against climate change. Cotti was willing to introduce one of the first domestic taxes on CO_2 worldwide in order for Switzerland to exert a form of exemplary leadership in the international climate debate (for carbon taxes in Nordic countries, see Chapter 11 in this volume). However, he failed to convince policy-makers and stakeholders back home. Although some policy actors already sensed that the CO_2 tax could be an efficient and effective instrument, economic actors harshly criticised and opposed that idea.

In response to this internal opposition, the Federal Council (FC) had to switch to an 'entrepreneurial leadership' strategy domestically. Instead of introducing a CO_2 tax, the FC developed a general programme to reduce GHGE, which targeted a wider range of sectors (such as traffic, buildings and industry) and relied on a greater number of policy instruments (Lehmann and Rieder, 2002). This new strategy resulted in the CO_2-Act that entered into force in 2000. The main objective of the new act was to implement the Kyoto Protocol requirements. Switzerland agreed to reduce its total GHGE by 8% (compared to 1990) and its CO_2 emissions from combustibles and motor fuels by 10% (FOEN, 2010). The act comprised a policy instrument mix made up of different voluntary instruments and a subsidiary carbon tax on both motor fuels to regulate traffic and combustibles to regulate industry and the building sector.

Already in 2002, CO_2 inventories pointed at Switzerland not reaching the emission targets based on voluntary instruments alone. Hence, the FC decided to initiate a partial revision of the CO_2-Act to be able to introduce a carbon tax on both motor fuels and combustibles. However, skilful lobbying by the energy and transport sector under the auspices of the oil association (*Erdöl-Vereinigung*) stopped the introduction of a carbon tax on motor fuels (Niederberger, 2005). Instead, the revised 2007 CO_2-Act only introduced a carbon tax on combustibles. To regulate the transport sector, the oil association, while deploying an ingenious mix of both structural and cognitive leadership, negotiated a compromise solution which brought about the introduction of a 'climate cent' instead of a carbon tax on motor fuels. The climate cent was widely supported by the Swiss business community and set-up as a private and voluntary tax of 1.5 Swiss cent per litre of petrol and diesel levied by oil importers. Revenues from this charge were mostly allocated to acquire the certified emission reduction credits introduced by the Kyoto Protocol and subsidiary into emission reduction projects in Switzerland, such as a scheme to subsidise energy-saving renovations in the building sector (Stiftung Klimarappen, 2013).

Only a year passed before a second revision of the CO_2-Act had to be initiated in 2008, almost simultaneously with the start of the first commitment period of the Kyoto Protocol. This major revision was triggered by a popular initiative: 'For a healthy climate'. It was launched by the Climate Alliance of NGOs together with left and green parties. The initiative demanded the adoption of a much stricter emission reduction target of 30% by 2020 (compared to 1990). Interestingly, the discourse and strategy of this Climate Alliance solely used the repertoire of cognitive leadership, notably because their capacity in terms of structural leadership was lower than that of business and some centre and right-wing parties (see Ingold and Fischer, 2014).

This led the government to react with a counter proposal, namely a novel revision of the CO_2-Act. The new version of the act also created the legal framework for the second commitment period (2013–2020) of the Kyoto Protocol. Again, the government proposed to introduce a carbon tax on motor fuels which, however, was rejected by parliament. Specifically, economiesuisse, which is the peak business association, retrenched to a structural leadership strategy by threatening to initiate a referendum, which would have endangered the whole revision process. Therefore, the new CO_2-Act, which entered into force in 2013, excluded again the tax on motor fuels. The new reduction target was set at 20% by 2020 (compared to 1990) and relied on a quite complex mix of old policy instruments (such as the continuation of the carbon tax on combustibles and voluntary agreement) and novel instruments (e.g. energetic subsidies in the building sector and a technology fund). The transport sector was to be regulated through CO_2 emissions standards for newly registered vehicles, which Switzerland borrowed from the EU, and the obligation for oil importers to partially compensate CO_2 emissions from fuels through domestic instead of international reduction projects (FOEN, 2014). While the new CO_2-Act has proven to efficiently regulate CO_2 emissions in the building and industry sector for which emissions continue to decrease, the rather lenient regulation of the transportation sector and the withdrawal of a tax on motor fuels was not able to provide comparable results. This is illustrated in Figure 13.1, which clearly shows that in contrast to the emissions from the combustibles, emission from motor fuels did not decrease.

On the contrary, CO_2 emissions of the transport sector have continued to rise since 1990, with Swiss privately used cars emitting on average 25% more CO_2 than their European counterparts (FOEN, 2019). The main reasons for this phenomenon are the increase in driven kilometres per person by more than 30% and the increased demand of big and heavy cars (i.e. sport utility vehicles – SUVs) (FOEN, 2018). As the traffic sector in Switzerland is also responsible for the largest share of GHGE today (see Figure 13.2), this can be considered as one of the main weaknesses of the Swiss climate policy.

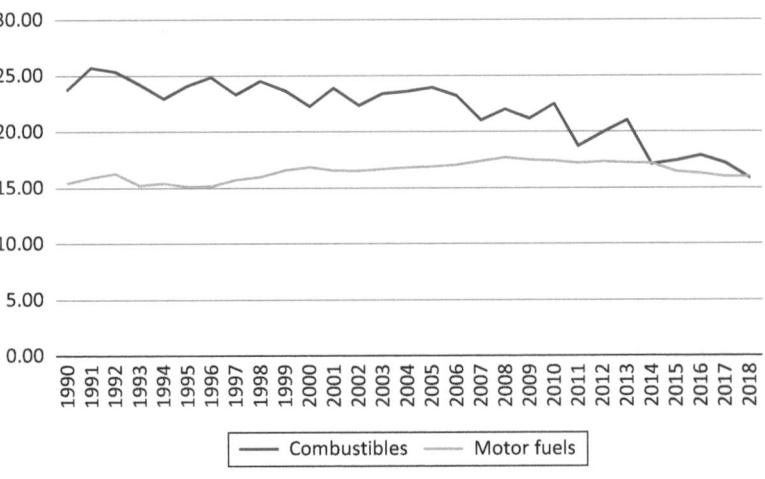

Figure 13.1 Development of GHGE from combustibles and motor fuels in Switzerland from 1990–2018. *Source:* FOEN (2019).

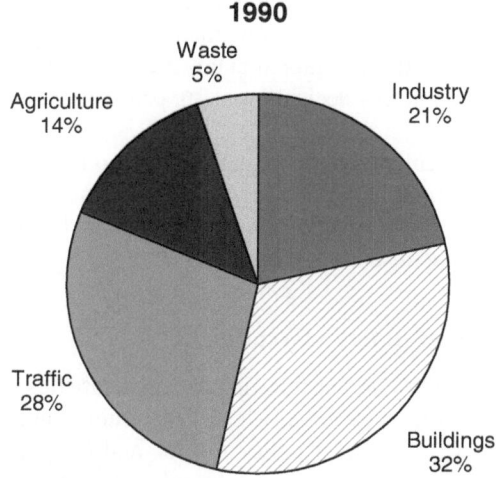

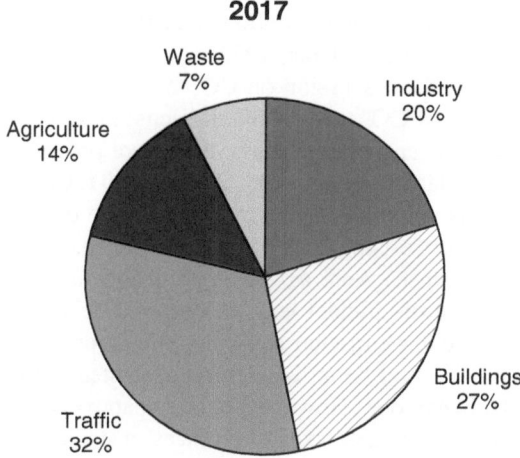

Figure 13.2 GHGE by sector in percentages in 1990 and 2017.

Hence, one of the main questions to be answered in this chapter is: despite its intention to act as a climate leader on international arenas, why has Switzerland not been able to implement stricter policies – such as a CO_2 tax – in the transport sector? To answer this question, we undertake a policy network analysis of the political elite, thus of all those public and private actors involved in national decision-making about policy instruments to mitigate climate change.

Different from many other policy fields in Switzerland, climate policy is not within the competency of the subnational units (i.e. cantons). Instead, policy instruments are mainly elaborated at the national level with only a few energy-related policies (mainly building codes) having been delegated to the cantons. Existing studies have also shown that in contrast to other national subsystems like the energy sector (see Kriesi and Jegen, 2001), cantonal authorities and regional stakeholders were not strongly involved in this decision-making process (Ingold, 2008).

After the adoption of the Paris Agreement, another round of revisions implemented Switzerland´s international commitments. The new act comprises innovations like a climate fund and a levy on flight tickets. Since the impact of the new policy is uncertain, these new developments are not part of this analysis.

Factors limiting the introduction of policy instruments

According to Majone (1975: 261), a policy is feasible 'insofar as it satisfies all the constraints of the problem which it tries to solve, where "constraint" means any feature of the environment that (a) can affect policy results, and (b) is not under the control of the policy maker'. Often, these constraints arise from an unequal distribution of costs and benefits among the affected targets groups and contrasting preferences on how a policy problem should be solved (Stavins, 1997; Carter, 2007; Bresser and O'Toole, 1998). Hence, in many cases we observe a gap between the theoretically desirable, effective policy options and the politically feasible policy options (Skodvin *et al.*, 2010; Meltsner, 1972).

This gap is, in particular, visible in the discussion on CO_2 taxes to mitigate climate change. While the introduction of CO_2 taxes is often discussed by high-level officials, environmental economists and other experts, they have proven to be unpopular and difficult to introduce (Jänicke *et al.*, 2003; Stavins, 1997; Carter, 2007; see, however, Chapter 11 in this volume). Such a tax puts a price on CO_2 emissions to internalise the external costs arising from emissions. Hence, reducing emissions is rewarded with lower costs. And, as the target groups are free to reduce their emissions in the most cost-efficient way, the desired behaviour is encouraged without formulating explicit emission reduction standards. From a societal perspective, emission taxes would be an efficient way to steer target groups towards a climate-friendly behaviour. However, from what we know about political practice, the introduction of such taxes often faces obstacles and difficulties. For example, concerns about risings costs and reduced international competitiveness create opposition by consumers and producers alike.

The political feasibility of policy instruments in general and emission taxes in particular strongly depends on the predominant constellation and policy preferences of important actors in a policy domain (Sabatier and Jenkins-Smith, 1993; Bresser and O'Toole, 2005; Majone, 1975). But the balance of power in a policy domain is usually not static and by 'the tactical manipulation of power differences in different policy processes or at different times … a shift in the balance may be obtained' (Bresser and O'Toole, 2005: 133). Altering actor constellations that shift the balance of power might open or close a 'window of opportunity' (Kingdon, 1995) for introducing an emission tax. However, the political feasibility of emission taxes is also path-dependent and is considerably dependent on the predominant political culture (i.e. the understanding of how a society should be organised) and the predominant tradition in a policy subsystem (i.e. existing policy beliefs on how a specific policy domain should be regulated) (Sabatier and Jenkins-Smith, 1993).

In consequence, the feasibility of emission taxes is connected to the general acceptance of different approaches to regulate collective action problems (Thalmann, 2004; Wüstenhagen *et al.*, 2007). For example, in a subsystem where the redistribution of wealth via taxes is a generally accepted norm, emission taxes are very likely perceived in a much more positive light than in subsystems that favour a lower level of state intervention. Likewise, in a policy subsystem in which eco-taxes have a long history the introduction of a tax usually raises far less opposition than in a policy domain with a longstanding tradition of regulations or subsidies (see also Chapter 11 in this volume). In short, the political feasibility of emission taxes depends on both the predominant actor constellation and the political context of a policy subsystem. On a more general level, Bresser and O'Toole (2005) explain this phenomenon with what they call 'contextual-interaction' theory. The political context is usually rather stable and conditions the portfolio of policy instruments to be discussed in the first place (Metz, 2017). The policy instrument or mix of instruments with the highest acceptance is then a function of the dominant actor and preference constellation. Hence, to understand why individual policy instruments (or mixes) are introduced or not it is important to study (1) the role of policy actors in the decision-making process and (2) their policy preferences (Howlett, 2014; Ingold *et al.*, 2018). This line of argument corresponds to earlier studies on the determinants of policy output and change (Fischer, 2014; Adam and Kriesi, 2007; Howlett, 2002).

According to this literature, two dimensions precondition the likelihood of the adoption of a specific policy instrument within a political subsystem: the level of actor involvement and the level of belief conflict as outlined below. In order to understand the logic of these two dimensions, we have to explain briefly what is meant by actor relations as the unit of analysis (see Dermont *et al.*, 2017). In line with the literature on policy studies or policy networks, the introduction of policy instruments is not a decision by single actors, but the result of interactions among different public and private actors (Kenis and Schneider, 1991; Knill and Tosun, 2012; Kenis and Schneider, 1991). Actors are mostly organisations and thus collective entities, such as political parties, peak and non-governmental associations, trade unions, scientific groups, civil society, industry and the public administration. These actors come together in a policy subsystem as soon as they are interested in or concerned by the production of a public policy to solve a problem (here climate change). They form a network of interactions and venues of participation (Knoke *et al.*, 1996).

The *first dimension* (level of actor involvement) can best be understood by thinking about such a policy network and the actor constellation in a given policy process, i.e. when and how often as well as what actors do participate in what types of important venues of the decision-making process? If a policy network is dominated by a small number of key actors, policy formulation and implementation is concentrated in the hands of a few, and thus, happens frequently in a top-down fashion. While this might simplify the adoption of policies in particular – when key actors share policy preferences – it also tends to lower acceptance.

In contrast, when many actors or groups of actors compete for influence, decision-making processes are (due to higher numbers of involved stakeholders) more complex, but offer opportunities for alternative policy options to reach the agenda, thus potentially facilitating compromise solutions, in particular, in settings with high disagreement between key actors. According to the literature on polycentric (climate) governance, the acceptance of specific policy instruments increases if they emerge in a bottom-up fashion or, if this is not the case, then a wide variety of concerned actors (rather than merely the 'iron triangle') will need to support them (Ostrom, 2010; Jordan *et al.*, 2015). Moreover, 'such multi-level, multi-actor systems offer important benefits in terms of fostering innovation and learning' (Jänicke and Quitzow, 2017: 123). Note that we investigate in this chapter a national subsystem, which is why we restrict the multilevel or polycentrism characteristics of the policy network assessed to the horizontal dimension, e.g. the involvement of public and private actors. In a subsystem characterised by the interplay of different levels, this first dimension could definitely also include the vertical aspect where different decisional levels or jurisdictions would be involved too.

The *second dimension* reflects the level of belief conflict in the policy network. A high level of conflict exists when most actors, and especially the key actors, hold different views on policy instruments. In this case, the policy network shows a clear-cut cleavage with respect to the preferred policy option. Conversely, a low level of conflict prevails when actors predominantly share preferences on the discussed policy options.

Table 13.1 presents a typology that combines these two dimensions: they precondition the likelihood of policy instruments (or mixes) to be introduced in four ways. First, in a policy network in which only a few key actors are actively involved in the policy process (low level of actor involvement) and agree on the policy option (low level of belief conflict), policy adoption is very likely (upper left cell in Table 13.1). Second, in a policy network that involves a variety of public and private actors in which key actors agree on the policy option (lower left cell in Table 13.1), we expect an increased likelihood for innovative policy options to reach the political agenda, because such a setting enables entrepreneurial or cognitive leadership. Indeed, the literature points out that the interactions of multiple actors in non-conflictive contexts favour the emergence of novel and alternative solutions. Third, the likelihood of policy adoption decreases in a policy network dominated by a small number of key actors who largely disagree (upper right cell in Table 13.1). Due to hardened positions, the respective political process is in danger of ending in deadlock or with the lowest common denominator compromise solutions (Nohrstedt and Weible, 2010).

Finally, in a policy network, where many policy actors are involved, showing a high level of belief conflict (lower right cell in Table 13.1), we expect a high

Table 13.1 Typology of influence of actor constellation on policy output.

| | | Level of belief conflict | |
		Low [Key actors mostly agree on policy issues]	High [Key actors do not agree on policy issues]
Level of actor involvement	Low [Clear domination of the policy network by one or more actors]	Introduction of policy or policy mix is very likely	Danger of deadlock
	High [No clear domination of the policy network by one or more actors]	Potential for innovations	Potential for compromises

Source: *Adapted from Kammerer (2018: 133)*

potential for compromise solutions: key actors, so-called policy brokers, might be able to connect different actors with different preferences and compromise solutions might be produced (see Angst *et al.*, 2018; Ingold and Varone, 2012). This is because conflictual policy context leads actors to bargain for alternative solutions that are acceptable to the opposite side. The presence of multiple actors favours the emergence of compromise solutions, compared with deadlock situations that are often characterised by lower actor involvement (Ingold, 2011).

From earlier research we know that the Swiss climate policy subsystem is conflictual, i.e. the level of belief conflict among the involved actors is rather high in comparison to other policy areas, like energy or water (Ingold *et al.*, 2020). Hence, we expect for a higher likelihood for policy deadlock or compromise solutions to emerge in Switzerland's national climate policy.

Data and methods

We focus on two subsequent revision processes explaining the drawbacks in adopting a CO_2 tax on motor fuels in Switzerland, i.e. the partial revision (first revision) of the CO_2-Act between 2005 and 2007 and the total revision (second revision) of the CO_2-Act from 2008 to 2012.

Network data collection

Our analysis is based on data gathered on both revision processes, which can be divided into a sequence of linked political events, such as committee sessions, political statements, consultations and hearings, parliamentary debates, decisions, etc. The collection of information on the sequence of these linked events and core actors allows for a comparative analysis of the two policy processes.

Based on the detailed narration of these processes, we systematised the information on event participation (Widmer *et al.*, 2008). The result produced two two-mode matrices showing actors and political events, which can be represented as affiliation networks (Borgatti *et al.*, 2013). The cells in these two-mode matrices contain a value of "1" if an actor (i.e. governmental agencies, legislative entities, political parties, business groups or civil society actor) has participated in an event and "0" if not. The network relations reflect the participation of national political actors in an event. To reconstruct the two national climate policy processes, we analysed official policy documents issued by the Swiss parliament (Curia Vista) to identify our actors and events. For revision 1 (2005–2007), we identified a total of 35 actors involved in 48 events. For the revision 2 (2008–2012), we identified a total of 39 actors involved in 41 events.[1]

That is, as a data base we have two different affiliation networks, each representing the policy-making process through the respective period. For this analysis, we used standard transformation routines, as implemented in UCINET (Borgatti *et al.*, 2002) to convert the actor-event matrix into a one-mode actor–actor matrix, which links actors when they have participated in the same event, i.e. the network relation is joint event participation. Hence, the cells in the one-mode matrix contain a value as high as the number of jointly participated events. The resulting matrix serves as valuable approximation of the network of political interactions that have taken place throughout the two-revision processes (Widmer *et al.*, 2008). It is possible to illustrate graphically constellations (compare Figures 13.3 and 13.4) and to analyse the influence of different actors with the help of network indicators, which will help us to explain the policy output in the Swiss context over time.

Operationalising of typology

To operationalise the two dimensions of the typology presented in Table 13.1, we rely on overall network centralisation measures for dimension 1 (level of actor involvement) and a combination of actor centrality and policy preferences for dimension 2 (level of agreement).

Dimension 1: level of actor involvement

To assess actor involvement as the first dimension of the typology presented in Table 13.1, we rely on network centralisation measures on the network level (Borgatti *et al.*, 2013; Wasserman and Faust, 1994). In general, it can be stated that a network is more inclusive and thus less dominated by few actors if many political actors participate to a similar degree. Conversely, network centralisation (or monocentric as an extreme) is indicated by a core-periphery structure, when a large number of actors is less connected and a smaller number of actors sits on

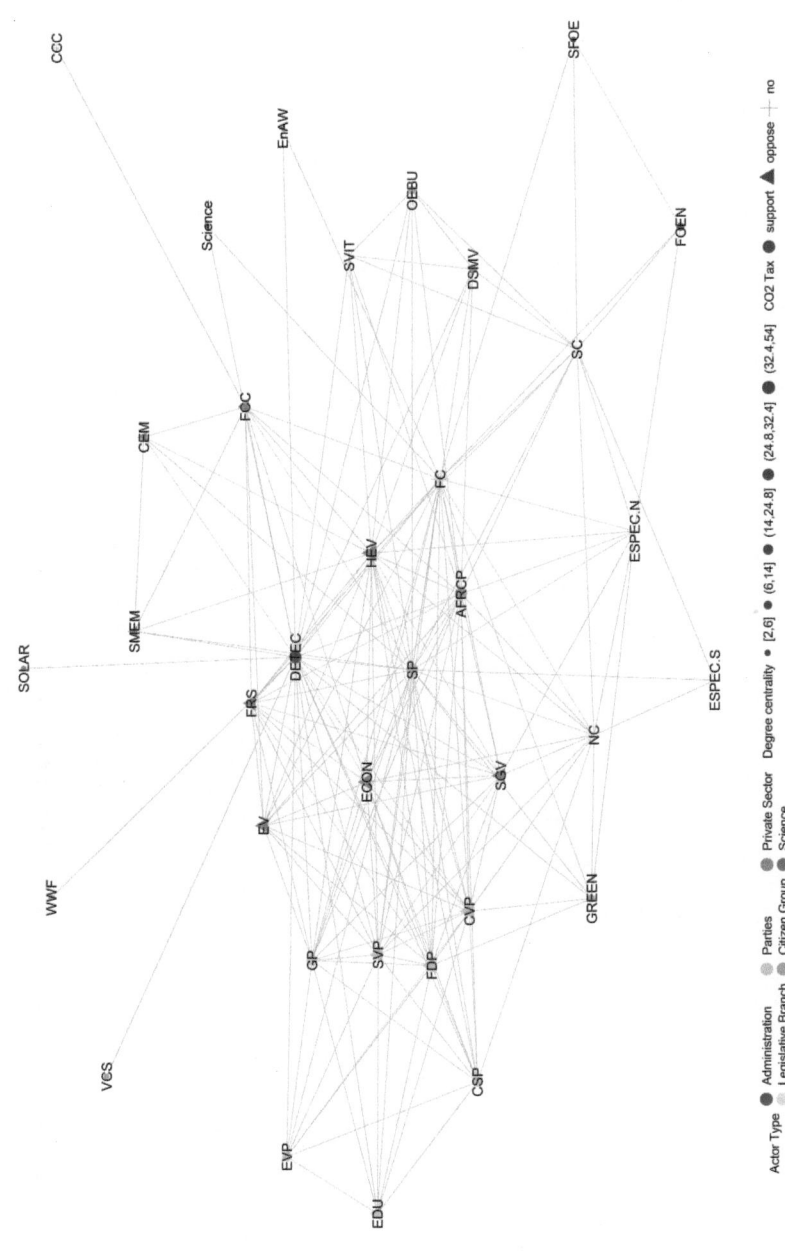

Figure 13.3 Policy network, first revision, 2005–2007, CO_2 tax.

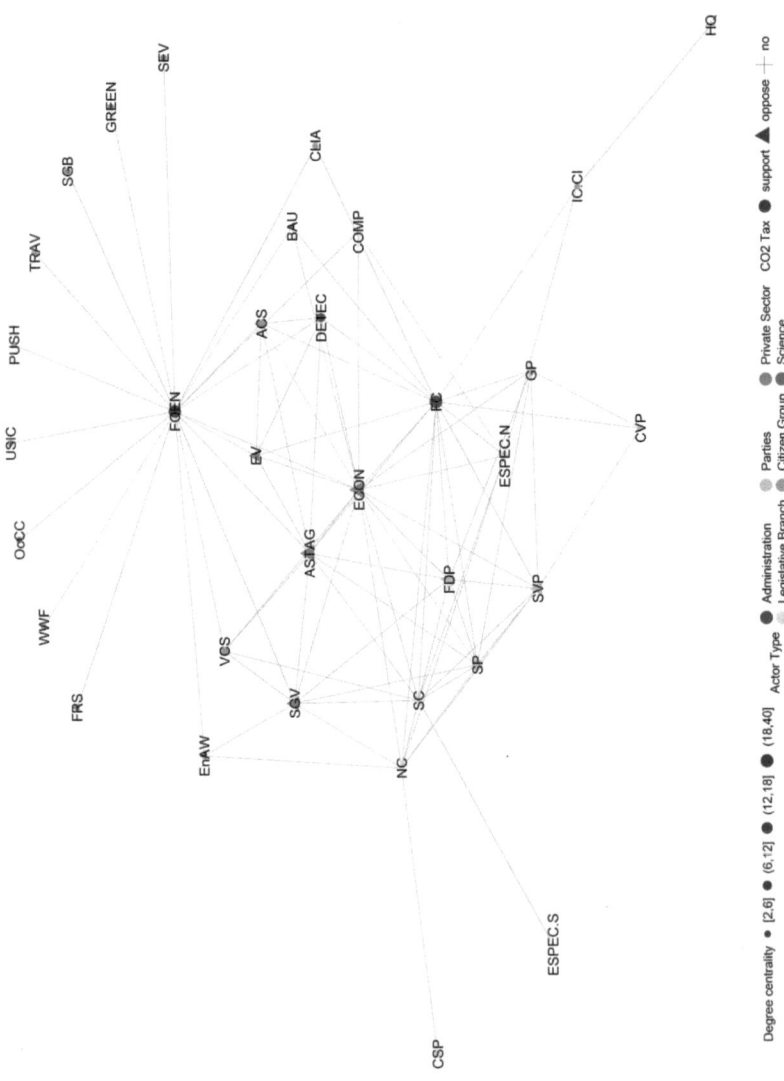

Figure 13.4 Policy network, second revision, 2009–2013, CO_2 tax.

central positions. To assess this, we use two statistics that measure actor involvement on the macro-level.

First, network cohesion measures the degree to which the network is knitted. We used density and the average degree of the network to assess this. Density measures the number of observable ties in a network relative to all possible ties. We used a normalised version of density. Moreover, we also used the average degree, as this is more reliable to compare networks of different sizes. A higher density or average degree indicates stronger network cohesion as more actors are tied to each other. A low density or average degree reflects weaker network cohesion, because fewer actors are tied to each other. This points to a more centralised actor structure, as only a small number of actors share numerous ties and dominate the network. The contrary would show evidence for high involvement of many actors.

Secondly, network centralisation 'refers to the extent a network is dominated by a single node' (Borgatti *et al.*, 2013: 159). Complete centralisation implies that one actor is connected to all other actors, there are no connections otherwise, and hence the network would resemble a star. The contrast is a network in which all nodes are connected to each other. The link to the policy network is rather straightforward: a highly centralised network is dominated by a small number of dominant actors, whereas in a network of high actor involvement, centralities among actors are more equally distributed among all those who are tied to each other.

Dimension 2: level of belief conflict

In a combined assessment of the centrality of individual policy actors and the preference structure of the actors (micro level), we are able to evaluate the level of belief conflict between key actors. The data on policy preferences were coded for both policy networks using the statements made by policy-makers during consultations related to the revision processes in 2004 and 2009, respectively, as well as policy documents for governmental actors. For the first revision (2005–2008), we coded positions on four contested policy instruments, namely the overall CO_2 tax (combustibles and motor fuels), the climate cent, tradeable permits and voluntary instruments. For the second revision, we coded the position towards the carbon tax on motor fuels (support, neutral, oppose) and two further critical aspects, namely the strictness of emission reduction targets (higher than proposed by the federal government, i.e. >20%), medium (governmental target, i.e. 20%) and low (lower than the governmental target, i.e. <20%) and the degree to which these reduction should be reached through compensation measures abroad (support, neutral, oppose). We selected these preferences, as they were the most contested in the respective decision-making processes (Ingold, 2011; Kammerer, 2018). For all four instruments, we followed the following coding scheme of support, neutral, oppose. Furthermore, we also included the actor type as a control variable.[2]

To determine the level of belief conflict, we ran ordinary least squares (OLS) regression models for both phases as an auxiliary analysis. For this purpose, we used degree centralisation as independent variable and the preference on the CO_2 tax as dependent variable. *Degree centrality* shows which actors hold the most central positions. It is calculated by counting the number of ties an actor has in a network. For this analysis, we use the statistic as a proxy for the importance of an actor in a network. In this context, a high degree of centrality might indicate an influential position in the policy network, which gives the actor power over the decision-making process. Significantly, positive or negative parameter estimates related to specific policy preferences indicate that influential actors tend to jointly agree (positive) or disagree (negative) on the climate tax. Conversely, non-significant parameter estimates indicate that central actors do not agree over the CO_2 tax.

Actor involvement and belief conflict

To assess the level of feasibility of the CO_2 tax in the Swiss climate policy network throughout both revision processes, we look at both dimensions of the typology.

Level of actor involvement

Table 13.2 gives a first impression on the findings related to the level of actor involvement (dimension 1). The different actor types have similar levels for all centrality statistics, except science. Thus, many actors from different types were active and influential during the first revision process. For revision 2, we see a clear divide of the level of actor involvement. On the one hand, there is a group of most active and influential actors that are from the administration, followed by the legislation and the private sector. On the other hand, the rest of the actors

Table 13.2 Centrality scores for first revision by actor group.

Actor type	Centrality scores		
	Degree	Betweenness	Closeness
Revision 1			
Administration	0.33	0.08	0.57
Citizen group	0.21	0.01	0.54
Legislative branch	0.28	0.01	0.54
Parties	0.44	0.02	0.63
Private sector	0.29	0.01	0.58
Science	0.10	0.00	0.50
	Degree	Betweenness	Closeness
Revision 2			
Administration	0.28	0.14	0.14
Citizen group	0.07	0.01	0.14
International	0.01	0.00	0.08
Legislative branch	0.18	0.03	0.14
Parties	0.09	0.00	0.09
Private sector	0.12	0.02	0.14
Science	0.03	0.00	0.14

played a subordinated role, both in terms of their presence in the process (degree centrality),being in powerful positions (betweenness centrality), or their low connection to important actors (closeness centrality). Overall, there is a general tendency of actors to be at the periphery of the network. In sum, the resultspoint to a policy network with high degree of actor involvement in revision 1 and a more centralised policy network in revision 2.

The macro-level statistics point at a similar direction. During the first revision, the policy network shows an average degree of 24.6 and a density of 0.73. The values indicate that actors tend to be tied to about two-thirds of other actors in the policy network. This is a rather high value, as usually social networks show values around 0.30 (Wasserman and Faust, 1994). Concerning actor constellations, this implies that many actors are tied to each other. Therefore, the policy network is inclusive with a high involvement of many actors. Accordingly, the degree centralisation of 0.23 is relatively low. For our analysis it shows that at least during the first revision, the policy network shows a rather decentralised subsystem structure with many actors highly involved in the decision-making process. In contrast, during the second revision of the act, the policy network shows an average degree of 15.85 and a density of 0.40. Hence, these values are much lower as compared to the first phase. The policy network has thus a core-periphery structure with only few actors being very central and dominating the process. This is also highlighted by the lower centralisation score of 0.41.

In sum, this implies these findings point to a higher level of involvement in revision 1 than in revision 2.

Level of belief conflict

With respect to the level of belief conflict (dimension 2), our analysis shows that during both the first and second revision of the CO_2-Act the policy network was rather conflictive with respect to the CO_2 tax. As illustrated in Figures 13.1 and 13.2, actors that were most central in the network had contrasting views on the CO_2 tax. Both one-mode network graphs represent the actor constellation during the two revisions. The nodes show the involved policy actors per actor type. In the first revision, key actors were the FC, the environmental department (DETEC), the Climate Alliance (AFRCP, Alliance for a Responsible Climate Policy) and several business and traffic organisations, such as the economiesuisse, the House Owner's Association and the oil association (Lehmann and Rieder, 2002). Whereas the FC, the DETEC and the Climate Alliance supported the tax, the other key actors held opposite positions. In the second revision, these actors were the FC, the Federal Office for the Environment (FOEN) and the economiesusisse.

This finding is also supported by the auxiliary (OLS) regression, which we used to assess the level of belief conflict during the two revisions. The regression models indicate whether key actors tend to have the same preferences with respect to the tax. If they support the same preferences, we infer that the level of belief conflict over the tax is low and the feasibility of the instrument is higher. Conversely, if key actors tend to disagree over the tax, the level of belief conflict is high and the feasibility of the instrument is lower. We used the degree centrality score as independent variables. The related parameter estimates should not be significant to indicate a high level of belief conflict. Conversely, if the related parameters were significant, more central or influential actors tend to commonly support (positive) or commonly oppose (negative) the CO_2 tax. Moreover, we controlled for the position on other policy preferences (revision 1: climate cent, tradeable permits and voluntary measures; revision 2: emission targets and compensation abroad) to see if the support for specific instruments goes together with the support or opposition of the tax. Also, we tested if specific actor types agree on the CO_2 tax.

In Table 13.3, model 1A and 1B show the results for revision 1 and models 2A and 2B show the results for revision 2. The results across all models display no significant parameter estimates related to the degree centrality variable. This implies that neither during revision 1 or 2 did key actors agree on the CO_2 tax. We may thus infer that the level of belief conflict with respect to the tax was high during both revision processes. This suggests a low level of feasibility with respect to the CO_2 tax.

We also controlled for actor type and preferences towards other policy preferences. For the first revision, it shows that citizens groups are more likely to be in favour of the CO_2 tax, so are actors that also support tradable permits. In the second revision, no systematic effect related to a specific actor type can be shown. But, the results indicate that actors, who prefer buying CO_2 credits abroad, tend to oppose a domestic CO_2 tax.

Bringing both dimensions together

The above analysis has shown that during both revisions the level of belief conflict (dimension 2) with respect to the CO_2 tax was high. The two revision processes differ with respect to the network level of actor involvement (dimension 1) that changed over time. During the first revision more actors from different types and levels were active, i.e. the policy network showed a higher level of actor involvement. So, the involvement of different actors from different sectors produced a more inclusive policy environment where different opinions and preferences, also priorities and interests, come together. In fact, skilful lobbying by the oil association prevented the CO_2 tax on fuels by replacing it with the climate cent, but finally the FC introduced a combination between the tax and the climate cent. This situation confirms what we expected in our typology and through the bottom right cell in Table 13.1: a large actors' involvement associated with belief conflict tends to enable policy compromise.

Conversely, in the second revision process fewer private actors from both the business sector and citizen organisations played a central role and the policy network was dominated by a few actors. Indeed, during the second revision, only two actors from the private sector can be highlighted: the Climate Alliance (AFRCP)

Table 13.3 Regression results for first (2005–2007) and second revision (2008–2012).

	Model 1A	Model 1B	Model 2A	Model 2B
Intercept	2.23*** (-0.30)	1.05* (0.44)	2.52*** (0.42)	0.32 * (0.14)
Degree centrality	-0.08 (0.49)	0.19 (0.46)	-0.08 (-0.89)	-0.26 (0.49)
Actor Types (baseline group: administrative agencies)				
Citizen group	0.79* (0.34)	0.94** (0.32)	0.34 (0.46)	-0.03 (0.24)
Legislative branch	-1.20**(0.38)	-0.59 (0.37)	-1.01* (0.48)	-0.21 (0.27)
Parties	0.31 (0.32)	-0.14 (0.32)	0.08 (0.43)	0.04 (0.22)
Private sector	-0.20 (0.30)	-0.31 (0.33)	-0.26 (0.41)	-0.17 (0.21)
Science	0.78 (0.63)	0.70 (0.60)	0.48 (0.77)	-0.20 (0.40)
Instruments				
Permits		0.64* (0.23)	-0.03 (-0.07)	
Climate cent		-0.42 (0.28)	-0.08 (-0.12)	
VA		0.27 (0.19)		
Targets				-0.80***(0.12)
Flexibility				-0.04 (0.08)
R^2	0.56	0.71	0.30	0.82
Adj. R^2	0.47	0.60	0.16	0.78
Num. obs.	35	35	38	38
RMSE	0.56	0.49	0.66	0.34

Notes $p < 0.001$ ***, $p < 0.01$ **, $p < 0.05$ *; standard error in parentheses

and economiesuisse This indicates that lobbying was less intensive in the revision period 2009–2012, a finding that is also stressed by a business representative during the interviews who explained that the revision process was not taken too seriously, as it was not directly connected to any international developments. As a result, lobbying organisations opposing the introduction of the CO_2 tax sleep-walked through this decision-making process until they finally realised the need to intervene at a stage when it was too late. So again, this is in line with what we expected through our typology (see upper-right cell of Table 13.1): the strong opposition between key actors and the low level of actor involvement during the decision-making process led to a deadlock.

Conclusion

During the first revision between 2005 and 2008, a political environment that involved many actors with divergent policy preferences enabled the adoption of the climate cent on motor fuels, instead of a CO_2 tax on combustibles. From a national standpoint, this is an innovative solution: some key interest groups managed to suggest a new voluntary policy instrument that has so far not been foreseen in the standard portfolio of policy instruments by the government and its administration. However, from an international perspective that includes the Kyoto commitment and the post-Kyoto targets, the Swiss portfolio of instruments was not sufficient to reach its promised emission reduction target. This situation did not change during the second revision, when the tax on motor fuels was not introduced either. But in contrast to the first revision, the failure to introduce the tax on fuels can now be explained with policy conflict between a small number of key policy actors dominating the network.

Switzerland is generally perceived as a climate pusher internationally including the UNFCCC Conferences of the Parties (COPs). In international fora, Switzerland tries to sell innovative ideas to fight climate change while offering entrepreneurial and especially 'cognitive leadership'. The speeches and proposals by Swiss environmental ministers in the international climate negotiations clearly demonstrate this point. But when considering the domestic policy output, and especially the policy instruments introduced to fight climate change, one has to conclude that Switzerland is only a 'conditional' or 'symbolic leader' because national policy-making is unable to follow-up on the promises and proposals made on the international level and is clearly orientated to the level of ambition in EU climate policy. In this chapter, we were able to show that this is mainly due to the fact that some effective climate policy instruments, such as CO_2 taxes, are difficult to introduce because of belief conflict between actors and the need to find compromises between the multiple players involved in the domestic climate politics. Conflict leads the Swiss government to introduce compromises, which most often do not consist of the best but only second-best solutions in respect of target-effectiveness and innovative commitments.

Acknowledgements

We acknowledge financial support from the Oeschger Centre for Climate Change Research and Swiss National Science Foundation (Project number: 137808).

Notes

1 For data gathering, we relied on the Actor-Process-Event-Scheme (APES, Widmer *et al.*, 2008). APES is a non-technical method for systematising qualitative information as provided by texts and documents into quantitative data. The source material of this analysis is a detailed narration of the two revision processes (Kammerer, 2018: 133, Chapter 3). For the two revision processes, data was collected systematically on the pre-parliamentary phase and parliamentary-phase, e.g. public initiatives, stakeholder meetings, etc. For this purpose, the written documentation of the Curia Vista – database of parliamentary proceedings – was used. (See https://www.parlament.ch/en/ra tsbetrieb/curia-vista). It contains detailed information on parliamentary proceedings such as Federal Council dispatches, procedural requests, etc. In addition, a simple media analysis was carried out to compliment the Curia Vista data, i.e. to account for political events that have not been covered by the official reporting. For this pur-

pose, the Swiss news wire service, SDA, was searched for articles on the CO_2-Act in the respective periods. We searched for articles in the two respective revisions (i.e. 2005–2007 and 2008–2012). For revision 1, we assessed 58 SDA media press releases and 29 releases for period 2. Finally, to validate the information from the document analysis and to enrich the primary empirical evidence, ten semi-structured interviews with a business organisation, an energy association, a road traffic organisation, an industry organisation, the economics department, an environmental organisation, the homeowners' organisation and the climate cent foundation. All interviews were conducted in January 2017.

2 Compare Kammerer (2018: 133, Chapter 3) for a detailed description of the coding and variables.

Bibliography

Adam, S. and Kriesi, H. (2007) 'The network approach'. In: P. Sabatier (ed.) *Theories of the Policy Process*, Boulder: Westview Press, 129–154.

Angst, M., Widmer, A., Fischer, M. and Ingold, K. (2018) 'Connectors and coordinators in natural resource governance: insights from Swiss water supply', *Ecology and Society*, 23(2):1. doi: 10.5751/ES-11087-240327

Borgatti, S.P., Everett, M. and Freeman, L.C. (2002) *UCINET for Windows: Software for Social Network Analysis*, Harvard, MA: Analytic Technologies.

Borgatti, S.P., Everett, M. and Johnson, J. (2013) *Analyzing Social Networks*, London: SAGE.

Bresser, H. and O'Toole, L. (1998) 'The selection of policy instruments. A network-based perspective', *Journal of Public Policy*, 18(3): 213–239.

Bresser, H. and O'Toole, L. (2005) 'Instrument selection and implementation in a networked context'. In: P. Eliadis, M. Hill and M. Howlett (eds) *Designing Government: From Instruments to Governance*, Québec: McGill-Queen's University Press, 133–153.

Burck, J., Marten, F. and Bals, C. (2015) *The Climate Change Performance Index: Results 2016*, Bonn: Germanwatch Nord-Süd Initiative e.V.

Carter, N. (2007) *The Politics of the Environment: Ideas, Activism, Policy*, 2nd edition, Cambridge: University Press.

Carter, N., Little, C. and Torney, D. (2019) 'Climate politics in small European states', *Environmental Politics*, 28(6): 981–996.

Crippa, M., Oreggioni, G., Guizzardi, D., Muntean, M., Schaaf, E., Lo Vullo, E., Solazzo, E., Monforti-Ferrario, F., Olivier, J.G.J. and Vignati, E. (2019) *Fossil CO2 and GHG Emissions of All World Countries: 2019 Report, EUR*, Vol. 29849, Luxembourg: Publications Office of the European Union.

Dermont, C., Ingold, K., Kammermann, L., Stadelmann-Steffen, I. (2017) 'Bringing the policy making perspective in: a political science approach to social acceptance', *Energy Policy*, 108: 359–368.

Fischer, M. (2014) 'Coalition structures and policy change in a consensus democracy', *Policy Studies Journal*, 42(3): 344–366.

FOEN (2010) *Schweizer Klimapolitik auf einen Blick: Kurzfassung des klimapolitischen Berichts 2009 der Schweiz und das UNO-Klimasekretariat*, Bern: Federal Office for the Environment.

FOEN (2014) *Schweizer Klimapolitik auf einen Blick: Stand und Perspektiven auf Grundlage des Berichts 2014 der Schweiz an das UNO-Klimasekretariat*, Bern: Federal Office for the Environment.

FOEN (2018) *Klimapolitik in der Schweiz: Umsetzung des Übereinkommens von Paris, Umwelt-Info Nr.1803*, Bern: Federal Office for the Environment.

FOEN (2019) *Emissionen von Treibhausgasen nach revidiertem CO2-Gesetz und Kyoto Protokoll*, 2. Verpflichtungsperiode (2013–2020), Bern: Federal Office for the Environment.

Howlett, M. (2002) 'Do networks matter? Lining policy network structure to policy outcomes: evidence from four Canadian policy sectors 1990–2000', *Canadian Journal of Political Science. Revue canadienne de science politique*, 35(2): 235–267.

Howlett, M. (2014) 'From the "old" to the "new" policy design. Design thinking beyond markets and collaborative governance', *Policy Sciences*, 47(3): 187–207.

Ingold, K. (2008) *Analyse des mécanismes de décision: Le cas de la politique climatique suisse. Vol. 8 of Analyse des politiques publiques*, Zürich: Rüegger.

Ingold, K. (2011) 'Network structures within policy processes. Coalitions, power, and brokerage in Swiss climate policy', *Policy Studies Journal*, 39(3): 435–459.

Ingold, K. and Fischer, M. (2014) 'Drivers of collaboration to mitigate climate change: an illustration of Swiss climate policy over 15 years', *Global Environmental Change*, 24: 88–98.

Ingold, K., Moser, A., Metz, F., Herzog, L., Bader, H.-P., Scheidegger, R. and Stamm, C. (2018) 'Misfit between physical affectedness and regulatory embeddedness: the case of drinking water supply along the Rhine River', *Global Environmental Change*, 48: 136–150.

Ingold, K. and Pflieger, G. (2016) 'Two levels, two strategies: explaining the gap between Swiss national and international responses toward climate change', *European Policy Analysis*, 2(1): 20–38.

Ingold, K. and Varone, F. (2012) 'Treating policy brokers seriously: evidence from the climate policy', *Journal of Public Administration Research and Theory*, 22(2): 319–346.

Ingold, K., Varone, F., Kammerer, M., Metz, F., Kammermann, L. and Strotz, C. (2020) 'Are responses to official consultations and stakeholder surveys reliable guides to policy actors' positions?', *Policy & Politics*, 48(2): 193–222. doi: 10.1332/030557319 X15613699478503

Jänicke, M., Kunig, P. and Stitzel, M. (2003) *Umweltpolitik*, Bonn: Dietz.

Jänicke, M. and Quitzow, R. (2017) 'Multi-level reinforcement in European climate and energy governance: mobilizing economic interests at the sub-national levels', *Environmental Policy and Governance*, 27(2): 122–136.

Jordan, A., Huitema, D., van Asselt, H. and Forster, J. (eds) (2018) *Governing Climate Change: Polycentricity in Action?* Cambridge: Cambridge University Press.

Jordan, A., Huitema, D., Hildén, M., van Asselt, H., Rayner, T., Schoenefeld, J., Tosun, J., Forster, J. and Boasson, E. (2015) 'Emergence of polycentric climate governance and its future prospects', *Nature Climate Change*, 5(11): 977–982.

Kammerer, M. (2018) *Climate Politics at the Intersection between International Dynamics and National Decision-Making. A Policy Network Approach*, Dissertation, Universität Zürich, Zürich.

Kenis, P. and Schneider, V. (1991) 'Policy networks and policy analysis: scrutinizing a new analytical toolbox'. In: B. Marin and R. Mayntz (eds) *Policy Networks: Empirical Evidence and Theoretical Considerations*, Frankfurt am Main: Campus-Verlag, 26–59.

Kingdon, J.W. (1995) *Agendas, Alternatives and Public Policies*, New York: Addison-Wesly Educational Publishers.

Knill, Christoph, and Tosun, Jale (2012) *Public Policy: A New Introduction*, Basingstoke: Palgrave Macmillan.

Knoke, D., Pappi, F., Broadbent, J. and Tsujinkaka, Y. (1996) *Comparing Policy Networks: Labor Politics in the U.S., Germany, and Japan. Cambridge Studies in Comparative Politics*, Cambridge: Cambridge University Press.

Kriesi, H. and Jegen, M. (2001) 'The Swiss energy policy elite: the actor constellation of a policy domain in transition.' *European Journal of Political Research*, 39(2): 251–287.

Lehmann, L. and Rieder, S. (2002) *Wissenschaftliches Wissen in der politischen Auseinandersetzung: Fallstudie zur Genese des CO2-Gesetzes im Auftrag der Arbeitsgruppe Transdisziplinarität der Energiekommission der Schweizerischen Akademie der Technischen Wissenschaften (SATW)*, Luzern: Interface. https://www.int erface-pol.ch/app/uploads/2018/09/Be_CO2-Studie.pdf

Liefferink, D. and Wurzel, R. (2017) 'Environmental leadersand pioneers: agents of change?', *Journal of European Public Policy*, 24(7): 951-968, DOI:10.1080/135017 63.2016.1161657

Lijphart, A. (2012) *Patterns of Democracy: Government Forms and Performance in Thirty-Six Countries* (2nd edition), New Haven: Yale University Press.

Majone, G. (1975) 'On the notion of feasibility', *European Journal of Political Research*, 3(3): 259–274.

Meltsner, A.J. (1972) 'Political feasibility and policy analysis', *Public Administration Review*, 32(6): 859–867.

Metz, F. (2017) *From Network Structure to Policy Design in Water Protection: A Comparative Perspective on Micropollutants in the Rhine River Riparian Countries, Springer Water*, Cham: Springer International Publishing.

Niederberger, A.A. (2005) 'The Swiss climate penny: an innovative approach to transport sector emissions', *Transport Policy*, 12(4): 303–313.

Nohrstedt, D. and Weible, C.M. (2010) 'The logic of policy change after crisis: proximity and subsystem interaction', *Risk, Hazards & Crisis in Public Policy*, 1(2): 1–32.

Ostrom, E. (2010) 'Polycentric systems for coping with collective action and global environmental change', *Global Environmental Change*, 20(4): 550–557.

Ostrom, E. (2012) 'Nested externalities and polycentric institutions: must we wait for global solutions to climate change before taking actions at other scales?', *Economic Theory*, 49(2): 353–369.

Sabatier, P. and Jenkins-Smith, H. (1993) *Policy Change and Learning: An Advocacy Coalition Approach*, Boulder: Westview Press.

Schenkel, W. (2000) 'From clean air to climate policy in the Netherlands and Switzerland: how two small states deal with a global problem', *Swiss Political Science Review*, 6(1): 159–184.

Schreurs, M. and Tiberghien, Y. (2007) 'Multi-level reinforcement. Explaining European leadership in climate change mitigation', *Global Environmental Politics*, 7(4): 19–46.

Sciarini, P., Fischer, M. and D. Traber (2015) *Political Decision-Making in Switzerland: The Consensus Model under Pressure. Challenges to Democracy in the 21st Century*, London: Palgrave Macmillan UK.

Skodvin, T., Gullberg, A.T. and Aakre, S. (2010) 'Target-group influence and political feasibility. The case of climate policy design in Europe', *Journal of European Public Policy*, 17(6): 854–873.

Stavins, R.N. (1997) 'Policy instruments for climate change. How can national governments address a global problem', *University of Chicago Legal Forum*, January 1997: 293–329.

https://www.belfercenter.org/sites/default/files/files/publication/Policy%20Instrumen ts%20for%20Climate%20Change%20-%20E-96-03.pdf

Steffen, I. and Vatter, A. (2002) *Erfolgsfaktoren Klimapolitischer Abstimmungsvorlagen*, Bern: Büro Vatter.

Stiftung Klimarappen (2013) 'Abschlussbericht 2005–2013'. http://www.klimarappen.ch/ en/.4.html (accessed 3 May 2016).

Thalmann, P. (2004) 'The public acceptance of green taxes: 2 million voters express their opinion', *Public Choice*, 119(1): 179–217.

Wasserman, S. and Faust, K. (1994) *Social Network Analysis: Methods and Applications*, Cambridge: Cambridge University Press.

Widmer, T., Hirschi, C., Serdült, U. and Vögeli, C. (2008) 'Analysis with APES, the actor process event scheme'. In: M.M Bergmann (ed.) *Advances in Mixed Methods Research*, Los Angeles: SAGE, 150–171.

Wurzel, R., Liefferink, D. and Torney, D. (2019) 'Pioneers, leaders and followers in multilevel and polycentric climate governance', *Environmental Politics*, 28(1): 1–21.

Wüstenhagen, R., Wolsink, M. and Bürer, M. (2007) 'Social acceptance of renewable energy innovation: an introduction to the concept', *Energy Policy*, 35(5): 2683–269.

Part 4

Conclusion

14 Conclusion

Pioneers, leaders and followers in multilevel and polycentric climate governance reassessed

*Paul Tobin, Rüdiger K.W. Wurzel and
Mikael Skou Andersen*

Introduction

As the world's gaze has turned to focus increasingly upon the climate change crisis, there has been a growing clamour – both within academia and beyond – for leaders to marshal resources and guide us towards an effective response to this complex global challenge. In acknowledging not only the need for financial and technological support to developing countries, but also how 'developed country Parties shall continue taking the lead' (article 4.4), the architecture of the Paris Agreement is explicit about the need for leadership. Indeed, the commitment under the Paris Agreement to make 'rapid reductions' in greenhouse gas emissions (GHGE) (article 4.1) is complemented by a mechanism to facilitate 'the exchange of information, experiences and best practices'. This mechanism thus creates an institutionalised channel for leading states to influence other countries' strategies. Indeed, all states share a commitment to the 'highest possible ambition, reflecting [their] common but differentiated responsibilities and respective capabilities, in the light of different circumstances' (article 4.3) when preparing their Nationally Determined Contributions (NDCs).

Previous work has theorised what we mean by an environmental 'leader' or 'pioneer' (e.g. Young, 1991; Underdal, 1994; 1998; Andersen and Liefferink, 1997; Andresen and Agrawala, 2002; Jänicke, 2006; Liefferink and Wurzel, 2017; see also Chapter 1 in this volume) while an increasing number of studies has used such concepts to assess climate governance (e.g. Oberthür and Roche Kelly, 2008; Wurzel, Connelly and Liefferink, 2017; Wurzel, Liefferink and Torney, 2020). The existing literature on environmental leaders and pioneers has almost exclusively focused on economically highly developed countries (i.e. the Global North) while largely neglecting the emerging economies in the Global South as well as the role of followers across the globe. This edited volume deliberately explores both the global South and North, and traverses pioneers, leaders and followers alike. The outcome, we hope, is a collection of chapters that provides a more comprehensive exploration of the actions of the key players in global climate governance, especially those pushing for and demonstrating greater ambition, whilst also enabling greater analytical clarity through a guiding theoretical framework that is applied throughout the book.

As outlined in Chapter 1, the core analytical themes of this book are the conceptualisation of pioneers, leaders and followers within multilevel governance (MLG) and polycentric (climate) governance structures. These conceptual framings are overlapping and mutually supportive in the quest for greater analytical purchase. Specifically, as most cases exhibit different forms of leadership and pioneership – and even, perhaps simultaneously, followership and possibly also laggardness – MLG and polycentricity permit such complex identities to be located and examined in detail, by enabling the multifaceted 21st-century state to be examined from multiple angles. The theoretical insights and empirical findings obtained across this book suggest that while pioneership and leadership may be more commonly associated with the Global North – especially following the explicit allocation of primary responsibility for climate action to developed 'Annex-I' states via the 1997 Kyoto Protocol – they may be increasingly found across the globe. Indeed, as the chapters in this volume show, there are instances of climate leadership and pioneership within the Global South and followership within the Global North, as well as the other way round. Although the 2015 Paris Agreement emphasises again the principle of Common But Differentiated Responsibilities (CBDR), it requires all parties to put forward voluntary pledges in the form of NDCs. Climate leadership and pioneership from countries in both the Global North and South will therefore be important for achieving the Paris Agreement's goal of keeping global temperatures to well below 2°C and to pursue efforts to limit it to 1.5°C. In order to find instances of ambition, the book's use of MLG and polycentricity as guiding themes enables contributing authors to find climate leadership and pioneership beyond the 'usual suspects', and to acknowledge both the guidance of the state and the importance of non-state actors.

In this concluding chapter, we summarise and build upon the preceding chapters as follows. We begin by exploring the examples of pioneership and leadership identified by the chapter authors, focusing explicitly on the different leadership types explained in Chapter 1 (structural, entrepreneurial, cognitive and exemplary). From here, we turn to the followers, exploring the factors that led to such stances and their implications for global climate action. Second, we analyse the role of MLG and polycentricity in enabling new actors to shape policy-making, as well as their capacity to interrogate the actions of those that have previously evaded the analytical spotlight. Third, we draw together the innovations developed within this book including the 'emotional leadership' sub-type of cognitive leadership (see below) introduced by Hall in Chapter 5, and the theorisation by Lederer *et al.* in Chapter 6 on the application of leadership types to pioneership, and the significance of the vertical dimension within their usage. Our penultimate section compares Global North and South actors, before highlighting those actors and processes that merit further exploration. Finally, we conclude the book. Looking to the pursuit of the 2030 and other targets, we call for further research on the important role of leaders, pioneers and followers during this most pivotal of decades – which has started so tragically with the COVID-19 pandemic – in the struggle to mitigate and adapt to climate change effectively.

Pioneers, leaders and followers

In this volume we have followed the distinction provided by Liefferink and Wurzel (2017: 952–953), whereby pioneers are 'ahead of the troops', while leaders explicitly seek to lead or obtain followers. Furthermore, the chapters differentiated between the following four types of leadership (Liefferink and Wurzel, 2017; Wurzel, Connelly and Liefferink, 2017). *Structural leadership* draws from an actor's economic and/or military power, the latter of which is usually of little relevance for environmental governance. *Entrepreneurial leadership* reflects the use of diplomatic or negotiation skills to broker new agreements. *Cognitive leadership* encompasses the promulgation of new ideas or concepts that alter understandings or approaches in response to challenges. Finally, *exemplary leadership* occurs when an actor provides an example that others may emulate. Wurzel, Liefferink and Torney (2019: 11) note that leaders can combine combinations of these four manifestations of leadership, as, indeed, we have found in this book. Indeed, while leaders have often been identified in the literature as affluent states, this leadership status has been hindered somewhat during the challenging global context following the 2008 Global Financial Crisis (Burns, Tobin and Sewerin, 2019; Burns, Eckersley and Tobin, 2020), while Global South countries have exhibited numerous instances of leadership in this volume. Yet, the structural challenges these states face remain real. Indeed, as Underdal (1998: 107) claimed, 'All being equal, therefore, the smaller and poorer the country, the more rarely can it (afford to) mobilise the amount of expertise and diplomatic activity needed to play a leading role' even in purely cognitive environmental leadership terms. However, Underdal uses the term instrumental leadership to capture analytically what we have divided conceptually into cognitive and entrepreneurial leadership types (see also Liefferink and Wurzel, 2017; Wurzel, Liefferink and Torney, 2019).

The COVID-19 crisis is likely to make it even more challenging for countries in the Global South – and, quite possibly, the Global North as well – to offer cognitive climate leadership/pioneership, which is often resource-intensive and usually takes time to generate (e.g. on the basis of scientific findings). Below, we examine the primary instances of pioneership and each form of leadership in turn, noting that new locations for ambition are arising across the world, but also the difficulty of becoming a leader in an arena that comprises every state, business and individual. We then reflect upon the role of followers.

Pioneers

The chapters in this book identify numerous examples of pioneering climate action across the globe. Pioneers take actions that endeavour to address collective action problems that are hindering a wider community from reaping potential joint benefits (cf. Young, 1991). Pioneers differ from leaders in that only the latter explicitly try to attract followers although the former may nevertheless be emulated by others (Liefferink and Wurzel, 2017; Wurzel, Connelly and Liefferink, 2017). Development and dissemination of solutions and strategies at the national level is often a precondition for successful transfer to the international level and/or diffusion to other countries (Jänicke, 1995; Andersen and Liefferink, 1997). Without detailed research, it is often challenging to establish the motivations behind the actions of leaders and pioneers. It may even be the case in some situations that what appeared to be 'pioneership' would have been 'leadership' had the actor had greater resources to encourage other actors to follow, particularly in the case of those based in the Global South. Here, the chapter by Urban *et al.* (Chapter 4) is illustrative, as they find both Costa Rica and Vietnam to be pioneers, despite their being less economically developed than those cases that had previously been seen as leaders (e.g. Liefferink *et al.*, 2009). Relatedly, the parallel conceptual focus upon MLG enables us to make further distinctions between the two states; while Vietnam's (authoritarian) top–down approach is almost entirely resultant from the actions of government and party officials, Costa Rica's approach is more bottom–up, reflecting a more polycentric approach that involves civil society actors. Hall also finds civil society actors to have been pioneers in New Zealand, with Māori tribal organisations and activist groups hindering the development of fossil fuel extraction, via the cultural concept of *kaitiakitanga*.

In addition, our contributing authors frequently identified cities and municipalities as providing sites of pioneership within states, especially those that have otherwise not been so ambitious on the global stage. Li (Chapter 2) highlights cities in the southeast coastal areas, especially Shanghai, as playing a pivotal role in China's emissions trajectory. Similarly, Jörgensen (Chapter 3) posits that Gujarat in India was regarded as a pioneer having created a Department of Climate Change. Finally, Lederer *et al.* (Chapter 6) show how cities in Brazil assumed pioneering roles in the C40 Cities Climate Leadership Group, up until the election of President Bolsonaro who has shown total disregard for climate change.

The election of Brazil's populist right-wing leader is mirrored in the Global North case of the USA. There, Selin and VanDeveer (Chapter 7) explain that polycentric activity need not be one of collaboration and cooperation, but can in fact manifest as contestation and conflict between local-level actors, such as the State of California, and the national government. In contrast to car emission regulations, for which California has consistently been able to set the pace for other USA states, creating the so-called California effect (Vogel, 1997), its influence on climate governance seems much weaker. However, Selin and VanDeveer show that the states in the USA have been able to offer cognitive and entrepreneurial leadership, as well as some structural leadership. This leadership has been especially pronounced when states have teamed up, as has been the case, for example, with the Regional Greenhouse Gas Initiative (RGGI), which links up regional emissions trading schemes (ETS).

In other Global North states, climate action has been less actively contested but ambition nonetheless has plateaued – in part due to the impacts of the Global

Financial Crisis (Burns *et al.*, 2019; Burns *et al.*, 2020). It is too early to say what impact the COVID-19 crisis will have on efforts to mitigate and/or adapt to climate change throughout the world and at different levels of governance. However, it is likely to hit poorer countries harder than more affluent ones. In the meantime, the negative (differentiated) impact of the 2008/2009 financial crisis is becoming clearer. For instance, although Ireland was hit hard, Torney *et al.* (Chapter 12) find that the recent introduction of a Citizens' Assembly on climate governance is an example of pioneering behaviour. The Nordic states and Germany were less heavily afflicted by the financial crisis, and have continued to develop pioneering activities throughout the 2010s. Municipalities in the Nordic states benefit from availability of long-term and affordable credits for green investments from local government financing agencies (Chapter 11), while Germany's state-owned development bank's targeted support for energy-efficiency in the building sector (Chapter 9). Thus, from a multilevel perspective, the capacity of the EU at the global scale to influence negotiations has been galvanised by such actions by its member states. Tobin and Schmidt (Chapter 8) argue that around the time of the 2015 Paris Climate Change Conference (COP25), despite reductions in influence, the EU was still closer to being a leader than a pioneer, for example as a result of its active shepherding of states via the High-Ambition Coalition.

Leaders

While some countries act as pioneers others have positioned themselves very firmly as climate leaders at least in terms of their ambitions. For example, as Moulton (Chapter 10) explains, the UK 'wants to be a leader much more than it wishes to be a pioneer'. Moulton also points out that especially post-Brexit the UK or, to be more precise, pro-Brexit UK governments have been keen to 'go it alone' on climate action and in other aspects of international collaboration. Thus, leadership may be pursued due to a commitment to see a certain outcome realised, and also because a state wishes to be seen as a leader and have followers as part of its perceived status in the world.

Structural leadership

The geopolitical landscape has transformed since the 1992 United Nations Framework Convention on Climate Change (UNFCCC). Brazil, India and particularly China have since experienced the rapid growth of their economies, GHG emissions and structural power. However, so far, they have rarely used their power to offer structural leadership in international climate governance. As Lederer *et al.* (Chapter 6) argue damningly regarding Brazil's recent facilitation of deforestation despite possible carbon market opportunities: 'The central government [of Brazil] thus provided structural leadership, but of the wrong kind.' Thus, this manifestation contravenes the understanding assumed in this book and elsewhere, following Underdal (1998: 101), that leadership should be 'positive' to be considered thus. Jörgensen (Chapter 3) highlights India's capacity for structural leadership, but in contrast to the climate-damaging actions of Brazil, focuses upon the former state's increasing structural leadership in the field of solar power. Finally, while China is regularly identified as a key player at UNFCCC negotiations, Li (Chapter 2) analysed the oft-neglected internal leadership of the central government, noting that 'preferential policies and resources allocation … [ensure that] pilot cities or provinces have been allocated structural leadership to implement innovative low-carbon practices'. Thus, this book's usage of structural leadership as a guiding concept has enabled the authors who focused on these increasingly influential states to analyse them with greater nuance as to the exact manifestation of their power. Hall (Chapter 5) notes, as we might expect, that New Zealand lacks structural leadership, except when dealing with Pacific Islands. In so doing, he reminds us that structural leadership need not be global, but can be applicable to actors within a more local context. As such, New Zealand's actions on climate change, although small from a global perspective, can influence surrounding actors, reflecting leadership. China, in contrast, holds the power resources to underpin its ambitions for structural leadership (see Dong, 2017).

The expectation that power may be a necessary but not sufficient condition for structural leadership (Burns, 1978; Young, 1991; Liefferink and Wurzel, 2017) has been confirmed in the chapters on China and the USA (Chapters 2 and 7), which are both very powerful countries and also major GHG emitters. As China is now the largest GHG emitter, it has become an actor of systemic relevance for global climate governance. In contrast to the growing structural leadership potential of Brazil, India and China, and the low potential of New Zealand, the chapters on the Global North countries highlight how structural leadership is diminishing for traditionally influential actors including certain larger European countries. Tobin and Schmidt (Chapter 8) highlight the paradox that if the European Union (EU) succeeds in reducing significantly its GHGE, its structural leadership capacity in international climate governance will simultaneously be reduced. The authors use MLG and polycentric theory to focus upon the key actors within the EU – the European Parliament, Germany and Sweden – that strengthened its capacity to exert structural leadership nonetheless. Germany and Sweden are explored in further detail by Steuwer and Hertin (Chapter 9) and Andersen (Chapter 11), respectively. Yet, as Selin and VanDeveer (Chapter 7) posit, a state with the potential to exert structural leadership will not necessarily do so, or at least, not in a consistent manner at the national governance level, depending on the individuals shaping central policy decisions. Most notably, the decision of President Donald Trump to withdraw from the Paris Agreement undermined the USA's capacity to demonstrate structural climate leadership, despite its enormous latent power to do so.

Entrepreneurial leadership

Liefferink and Wurzel (2017) suggest that entrepreneurial leadership, which involves the use of negotiating and/or diplomatic skills and resources, usually occurs in conjunction with other leadership types. For example, New Zealand has combined entrepreneurial leadership with its usage of 'soft power' in its foreign policy, partly in order to compensate for its lack of structural leadership capacity. Another widely-recognised soft power, the EU, demonstrated entrepreneurial leadership in the run-up to the 2015 Paris Climate Change Conference (COP21), having submitted its voluntary pledge (Intended NDCs) to reduce GHGE second only to Switzerland (see Chapter 13), followed four months later by New Zealand, which was sooner than the majority of states. Tobin and Schmidt (Chapter 8) build on this point by identifying the EU's Climate Commissioner Miguel Arias Cañete as being 'the figurehead of the EU's entrepreneurial leadership in Paris', due to his work in liaising with other states, particularly via the creation of the High-Ambition Coalition. Moreover, the EU's capacity to exert entrepreneurial leadership was strengthened by both its status as a *de facto* host, and through the large number of highly-connected Member States, such as France, Germany and Sweden, that could simultaneously push the EU's narrative. Indeed, Steuwer and Hertin (Chapter 9) highlight Germany's entrepreneurial proficiency, such as its many international energy dialogues and partnerships.

Increasingly, Global South countries seem to offer entrepreneurial climate leadership, especially at the subnational governance level. In China (Chapter 2), cities have been important drivers of entrepreneurial climate leadership by, for example, joining international city networks such as the C40 and Local Governments for Sustainability (ICLEI) networks. Jörgensen (Chapter 3) points out that India's vibrant NGO sector and think tanks have been able to offer some entrepreneurial climate leadership. Finally, Urban *et al.* (Chapter 4) isolate the development of renewables within Vietnam and Costa Rica as being instances of such leadership in the two states. Thus, entrepreneurial activities in these instances need not elicit a large number of followers, but can make important contributions nonetheless.

Cognitive leadership

While the theorisation of cognitive leadership is relatively straightforward, the identification of cognitive leadership within our cases is a more nebulous challenge because it is hard to identify empirically those states that have expressed cognitive leadership, which manifests itself often only over longer time periods. In contrast, structural leadership, for example, can be engaged more or less instantly, at least by powerful states. It can often take years or longer for ideas to alter behaviour, meaning that any study on cognitive leadership will struggle to identify with certainty which ideas merit the label. Young (1991: 298) argued 'that new ideas generally have to triumph over the entrenched mindsets or worldviews held by policymakers', which usually takes time. Similarly, Liefferink and Wurzel (2017: 595) postulated that 'scientific expertise and experiential knowledge is usually generated on the domestic level only over a longer time period'.

Moreover, with a policy challenge as complex as climate change, multiple actors within a state produce numerous policy ideas at once, meaning that it is especially challenging to demonstrate where an idea came from, and thus where the agency behind the activity was located. As such, future research on cognitive leadership may benefit from using a framework such as Schmidt's (2008)

Discursive Institutionalism, which demarcates ideas across three levels in order to describe their status, including an upper 'paradigm' level that corresponds to the kind of cognitive shift implied as being connected the moniker of 'leader'. From this perspective, it is unsurprising that our chapter authors ascribed cognitive leadership to several cases in general terms, but the exact machinations of such behaviour were complicated to track.

Underlining the importance of taking a long-term perspective when examining this leadership type, Andersen (Chapter 11) highlights Finland's introduction of a carbon tax in 1990 as being pivotal in the state's subsequent emissions reductions, not least as the state was followed by Sweden one year later. Tobin and Schmidt (Chapter 8) observe a more recent example of cognitive leadership, applying such status to the EU's championing of a 1.5°C maximum temperature increase at the Paris COP. This idea was simultaneously advocated by a large number of other states within the High-Ambition Coalition and also environmental NGOs, again underlining the importance of viewing such activities from a polycentric viewpoint. Lederer *et al.* (Chapter 6) observed numerous examples of cognitive leadership in Brazil and Indonesia as a result of their sub-state focus, as well as the international linkages sometimes underpinning these leaps; the German development agency was found to play an influential role in capacity building within Indonesia via cognitive leadership, as was the Norwegian government within Brazil.

However, due to the lack of resource opportunities, it was harder to find instances of cognitive leadership that gained influential status within the Global South. Hall (Chapter 5) suggests that 'the Ardern Government has mostly been a taker of ideas, adopting existing frameworks rather than devising its own', be it the legacy of previous New Zealand governments, or the ideas pushed by more influential global actors. While cognitive leadership was identified as 'emerging' in the chapter on China (Chapter 2), a decade or so from the time of writing we may consider that the state has demonstrated even more cognitive leadership than we realise, requiring the benefit of hindsight to be seen. Similarly, we may then be able to discern cognitive leadership from other states that are not yet identified as cognitive leaders. Jörgensen (Chapter 3) points out that 'India exhibited cognitive leadership by introducing the equity principle to the international climate negotiations, which was met with strong approval by fellow industrialising countries'. Global South countries have championed the CBDR principle and other internationally accepted principles that emphasise the importance of justice and equity issues (see Chapters 2 and 3).

Exemplary leadership

In Chapter 11, Andersen posits that exemplary leadership is especially important during international climate negotiations, as such behaviour signals to other actors that a state is committed to acting on climate change. Exemplary leadership is similar to the directional leadership formulated by Grubb and Gupta (2000), except that it may be either intentional or not, and will commonly be combined with entrepreneurial leadership (Liefferink and Wurzel, 2017). For instance, the decision of the EU to submit its Paris Intended NDC early is an indication of entrepreneurial leadership, while its formatting of the target in the exact format preferred by the UNFCCC is identified by Tobin and Schmidt (Chapter 8) as intentional exemplary leadership. Within the EU, Steuwer and Hertin (Chapter 9) highlight Germany as providing examples to other states through its successful *Energiewende* (energy transition) and also, from a multilevel perspective, via its enthusiastic (but non-binding) uptake of the EU's Energy Performance of Buildings Directive. These examples show the importance of considering MLG within conceptualisations of leadership, particularly within the EU. Relatedly, and as is discussed in more detail in the section on 'theoretical innovations' below, Chapter 4 by Urban *et al.* emphasises vertical exemplary leadership, finding many such examples in Costa Rica and Vietnam as a result of the authors' explicit MLG perspective. Indeed, the local level is repeatedly found to be a source for exemplary leadership, as Li shows regarding the Low-Carbon Pilot Cities (Chapter 2) and Lederer and colleagues (Chapter 6) likewise find via individual city initiatives, such as São Paulo's 2009 climate policy and East Kalimantan's forest governance reforms.

Combining different leadership types

Importantly, it is rare for countries to offer only one type of leadership over time. Instead, different leadership types are usually combined (Young, 1991; Underdal, 1998; Parker and Karlsson, 2014: 586; Liefferink and Wurzel, 2017). The specific mix of different types of leadership employed by a particular actor, as well as the different ways in which they may interact, varies across issues and may evolve over time (Wurzel, Liefferink and Torney, 2019) or be contradictory when examining instances across multiple levels.

We may assume that large powerful jurisdictions – such as China and the USA, as well as to some degree the EU – are at least theoretically more easily able to offer structural climate leadership compared to small countries, such as Costa Rica, Ireland, New Zealand and Switzerland or the Nordic countries. This hypothesis derives from much of the existing literature (e.g. Young, 1991; Underdal, 1998; Parker and Karlsson, 2014: 586; Liefferink and Wurzel, 2017; Wurzel, Connelly and Liefferink, 2017; Wurzel, Liefferink and Torney, 2019) and appears to be supported by several empirical findings put forward in the chapters of this volume. Wurzel, Liefferik and Torney (2019: 15–16) have argued that 'some actors which have relatively little structural power may nevertheless become relatively influential climate governance actors capable of showing leadership or pioneership'. The main reason for this is that actors such as small states may be able to compensate at least partly for their lack of structural leadership capacity by creating considerable entrepreneurial, exemplary and/or cognitive leadership capacities, although this may take a considerable amount of time. Here, we may assume that states follow a degree of path dependence; those areas in which a state is already favourably disposed may become the areas they choose to prioritise regarding their leadership efforts. For example, Sweden is a small, wealthy, export-orientated state that was already highly defossilised in its electricity prior to the ascent of climate change as a global challenge. This status lends itself to the country making exemplary climate leadership as a dominant feature of its foreign policy identity, which it then builds through further instances of cognitive (e.g. polycentric governance methods) leadership.

Followers

While the bulk of this volume is structured around the actions of leaders and pioneers, several chapters provide valuable insights also for the behavioural patterns and motives of followers. Torney (2019) provides an important conceptualisation for the otherwise nebulous idea of 'the follower', particularly regarding climate governance. There, he defines climate followership as:

> The adoption of a policy, idea, institution, approach, or technique for responding to climate change by one actor by subsequent reference to its previous adoption by another actor. Note that there must be intentionality on the part of the follower but not the leader/pioneer.
>
> (Torney, 2019: 169)

The challenge is to identify intentionality on the part of the follower, with specific reference to the actions of a preceding pioneer/leader. Yet, there is a political as well as empirical challenge in identifying such behaviour. As Urban *et al.* argue (Chapter 4), '[p]olitically, Costa Rica is rather isolated in Central America … and no other country in the region ever officially labelled Costa Rica as an example that it wants to follow'. Thus, although Costa Rica has demonstrated greater ambition than its neighbours, this activity has not produced followers. Here, we must note an important dimension in researching climate leadership and followership: just because an actor has developed an innovative policy tool that could be replicated elsewhere does not necessarily mean that others will openly acknowledge that they have followed their lead. This difficulty is particularly pronounced when researching cases at the global level, rather than focusing on relatively homogenous states that are more willing to highlight collaboration and coordination, say within the EU. As a result, Urban *et al.* once again draw from the sibling conceptual framework within this volume by emphasising the importance of MLG, as they find no clear-cut leader-follower relationship. In addition to the political challenge of states being willing to reveal that they have followed others' lead, we must also note in a volume focused on leaders and pioneers that the cases selected to be included in this volume are more likely to be ambitious and/or influential states, making the identification of followers less likely. However, Andersen's chapter on the Nordic states (Chapter 11) notes that Norway is considered a follower rather than a leader, due to its emphasis on flexible mechanisms instituted by EU rather than domestic action. Moreover, different countries may arrive independently from each other at similar policy solutions in simultaneous or sequential fashion (Wurzel, Liefferink and Torney, 2019). Establishing empirically climate followership is therefore a challenging task.

However, our authors found examples of followership in both the Global North and South. The leadership shown in the run-up to the 2015 Paris COP, and the leadership demonstrated in creating the High-Ambition Coalition, resulted in several instances of followership according to Tobin and Schmidt (Chapter 8). Although non-EU Member States, Iceland and Norway committed to fulfilling their Paris climate pledges via collective delivery with the EU. As Kammerer *et al.* (Chapter 13) echo, Switzerland 'tends to wait for and align itself to the EU positions rather than taking the lead'. Indeed, within the Intended NDCs, The Gambia (2015: 1, 5, 19) noted its gratitude to Germany in particular for its support in the development of their pledge. At this point, we may wish to reflect on the implications of states in the Global South following those in the Global North, and the attendant power differentials that exist within such relationships. At what point does the pursuit of followership become neo-colonial *realpolitik* through other means? Due to a dearth of comparative Global South–Global North studies, we also know little about whether climate leader (and pioneer) countries in the Global South are able to attract followers primarily from other Global South countries or whether they can also persuade Global North countries to follow their examples.

The chapter authors in this volume have found several instances of followership leading to increasing ambitions. Steuwer and Hertin (Chapter 9) note that although France and Flanders were previously followers, they have used this experience as a springboard to become pioneers. Likewise, Vietnam was found by Urban *et al.* (Chapter 4) to have followed the actions of South Korea and China regarding Green Growth and energy policy respectively, before becoming a pioneer in its own right. There is reason to feel cautiously optimistic that followership can lead to future climate leadership, in the right circumstances. Indeed, New Zealand has placed 'fast followership' at the heart of its climate strategy (Hall, Chapter 5), replicating vehicle emissions standards, 'feebates', and investment vehicles, amongst others. Further research is encouraged in order to trace such patterns in a comprehensive manner, particularly relating to the factors that facilitate and obstruct a follower subsequently becoming a leader.

The need for longitudinal and multi-case perspectives

Although the focus of our book is on climate leaders, pioneers and followers, several chapters have identified also empirical examples of climate laggardship. This identification is perhaps not surprising as environmental leaders and pioneers usually have some blind spots (Wurzel, 2008). Moreover, who acts as a climate leader, pioneer, follower or laggard can change over time (Liefferink and Wurzel, 2017; Wurzel, Liefferink and Torney, 2019). In our volume this reality is best illustrated by the USA, whose climate change policy has been 'erratic over time and as internally contradictory', as Selin and VanDeveer have detailed (Chapter 7). The complex nature of climate governance and the large number of states involved mean that a country's status as a leader or pioneer, particularly if understood in relative terms, may come and go over time. 'Pioneer' and 'leader' are not timeless labels, but positions that must be continuously earned over time, and identified by researchers.

Moreover, due to the limitations of a book-length project, we have not explored the majority of the 195 signatory states of the Paris Agreement. Not every state can be a leader; indeed, it may not be beneficial for every state to attempt to be so, if the outcome is a fragmented and contradictory approach to global climate governance. However, one state's laggardship may reduce ambition throughout the global community, and understanding why states drag their feet is of vital importance to the study, and policy implications, of climate governance. Moreover, these factors could be beyond the control of the states in question: for instance, the Intended NDC submitted by the state of Jordan in 2015 highlights that Syrian refugees comprise 13% of their population, creating significant pressures on the small state to meet its everyday needs, let alone transitioning to a low-carbon future. As we touch upon later, further studies are needed to explore the intricate nuances of polycentricity and ambition within the states of the world.

Multilevel governance and polycentricity

As Wurzel, Connelly and Liefferink (2017) discuss, despite the overlapping shared presuppositions of MLG and polycentric concepts – such as focusing on multiple levels of governance and sources of authority – MLG approaches usually ascribe a higher importance to government, while polycentricity focuses upon broader *governance* (see also Homsy and Warner, 2014; Jordan *et al.*, 2018). Relatedly, the national level is identified as a key locus of power with MLG theory (Marks, 1993; Hooghe, 1996), whereas, as Ostrom (2010: 552) makes clear, '[e]ach unit within a polycentric system exercises considerable independence to make norms and rules within a specific domain'. In this volume, then, scholars have sought to draw from either or both concepts, as appropriate for their cases in question.

Perhaps it is of little surprise that the chapters that have emphasised either MLG or polycentricity within this volume are federal or quasi-federal jurisdictions. In particular, the chapters on India (Chapter 3), Germany (Chapter 9), Switzerland (Chapter 13) and the EU (and Chapter 8) have each highlighted the importance of considering MLG for explaining the instances of leadership and pioneership within their borders. In addition, Selin and VanDeveer (Chapter 7) argue that a 'polycentric turn' is emerging in the USA. Yet, we can also see how more unitary governance models, such as the Nordic states explored in Andersen's chapter (Chapter 11), have pursued their own models of MLG via the creation of the Nordic Council. China's Communist government has governed via a top–down approach, whereby selected cities are encouraged to experiment with innovative climate governance approaches at the municipal or city level in a learning-by-doing fashion with the aim of finding solutions which can then be upscaled to the national level. However, the *ecological civilisation* conference in Guiyang is identified by Li (Chapter 2) as being founded in a 'bottom–up or polycentric fashion', and was subsequently given a greater status from 2013.

Polycentricity has drawn increasing attention from academic circles and policy-makers alike since the 2010s as a means of facilitating more effective climate action. However, Jordan *et al.* (2015) note that there has never been a 'monocentric' international climate regime, but rather a series of interacting regimes. As such, when making claims about the rise of polycentric governance in some jurisdictions, we are keen to emphasise that the dominant understandings of the policy-making context, against which comparisons will be made, neglects the degree of polycentricity underway. Moreover, as Rayner and Jordan (2013: 80) point out, the 'the more polycentric a governance system, the greater the likelihood that its component parts pursue different and possibly incoherent approaches'. Indeed, it may transpire that greater polycentric interaction actually enables individuals who wish to *weaken* climate policy ambition to gain a stronger foothold (Boasson, 2018: 131). As such, as we reflect upon the polycentric communities that are examined in this volume's constituent chapters, we are at pains to emphasise that we do not view polycentric governance as being a panacea. Rather, polycentricity can be a potential catalyst for facilitating the kinds of benefits – experimentation, more robust institutions, new norms, trust-building and so on – that can help to assuage cooperation difficulties (see Dorsch and Flachsland, 2017).

Theoretical innovations

The primary contribution of this volume is the creation of a body of empirical data examining the existing theorisations of leadership types and MLG/polycentricity. Yet, in the process of conducting these analyses, contributing authors have made especially the following three further theoretical innovations. First, Hall (Chapter 5) introduces the concept of 'emotional leadership' to the exploration of national climate leadership types. While 'emotional leadership' has been identified in numerous fields previously (Humphrey, 2002; Loerakker and van Winden, 2017), its introduction to climate leadership types is noteworthy as it brings back the locus of analysis onto the individual level, which has been neglected in more recent climate leadership research. As a result, the conceptualisation is especially complementary with polycentric governance, in the event that multiple 'emotional leaders' may be located within a single network. Hall identifies Jacinda Ardern as a prime example of this leadership type. We may wish to place emotional leadership 'under', or at least in association with, cognitive leadership, due to the need for emotional intelligence to achieve such leadership, which is, after all, a cognitive ability. As such, more research is encouraged in order to theorise how this conceptualisation of emotional leadership interacts with other leadership types.

Lederer *et al.* (Chapter 6) provide a second instance of theoretical innovation in the book, in their work on Brazil and Indonesia. There, they emphasise the importance of vertical interactions between different *governmental* levels within states when examining the four climate leadership types. From here, the scholars then examine the precise nature of leadership exerted in their cases. For instance, they found that Brazil and Indonesia exerted vertical cognitive leadership through their national plans to tackle deforestation, with effective results, while Indonesia also demonstrated vertical structural leadership via its REDD+ task forces to develop provincial strategies to be followed by local leaders.

Finally, Kammerer *et al.* (Chapter 13) provide a typology that combines two dimensions that precondition the likelihood of a given policy instrument's adoption. These two dimensions relate to the level of actor involvement (Dimension 1) and the level of agreement in a policy network (Dimension 2). As a result of this innovation, the authors find that the level of belief conflict with regard to the CO_2 tax was high, as shaped by a low level of political feasibility, which they hypothesise may explain why motor fuels were never included in the tax accordingly. Kammerer *et al.*'s innovation enables us to achieve a more nuanced understanding of the policy process, from which future research may in turn be able to situate the roles of pioneers, leaders and followers.

Comparing Global South and North

To date, there has been limited comparison between Global South and North countries regarding the nature of climate leadership and pioneership. In part, this lacuna has been due to the clearly demarcated role for mitigating climate change established in the 1990s and early 2000s, whereby economically developed states were allocated primary responsibility for action. The 2015 Paris COP was the first UNFCCC COP in which *all* states were expected to state their commitments towards this shared problem. Thus, until the mid-2010s, any systematic attempt to compare or contrast the leadership behaviours of all states would have been stymied by the reality that cases were operating in entirely different policy contexts. This edited volume has sought to provide one of the first attempts to track the variegated forms of pioneership and leadership in both the Global North and South. Of course, any such claims are tentative at this stage due to the small number of cases that could be analysed, but we hope that more detailed analyses will be conducted following from this early work.

Here, we note that instances of climate leadership, pioneership and followership have been found across the globe via our chapter authors. The status of a country as a climate leader need not prohibit that state from being a follower. For example, Steuwer and Hertin find that, despite Germany's apparent leadership status with regard to climate change, the state was either a laggard or at best a follower of EU regulations when it came to the building sector. Conversely, despite their relatively minor geopolitical sway in the global arena, and hindered economic development, Costa Rica and Vietnam are both found to be pioneers by Urban *et al.* (Chapter 4) due to their strong governments and effective bureaucracies. Thus, as the UNFCCC shifts to encouraging more and more polycentric climate action that includes as many actors as possible, we can expect the ascribed statuses of states to move away from being starkly divided between 'leaders' and 'laggards'. Instead, we may move towards a more nuanced research landscape in which both the instances of greater action and followership are judged simultaneously.

It is prudent to highlight cases that we could not explore within this volume due to the limitations of space, as future areas of research that merit consideration. In particular, this volume has provided analyses of the Nordic states, Germany, Ireland, Switzerland and the UK, as well as the EU as a whole. However, the complex challenge of mitigating climate change for new EU member states or those countries especially heavily affected by the Global Financial Crisis means that greater exploration of Mediterranean nations and Eastern Europe is welcome. Existing work on Eastern European states has been provided by Jankowska (2016) on Poland, for instance, and the rising significance of major continental Member States in the EU merits further exploration. Looking beyond Europe, while Brazil, India, China and Indonesia have been examined in this volume, the remaining high-profile, fast-growing state, South Africa, deserves further analysis (Fløttum and Gjerstad, 2013). Likewise, the development of green efforts across Africa, as explored by Death (2016), is increasingly overdue, not only for examination of the roles of Western states in shaping African countries' climate policies as identified in Paris in 2015 (Tobin and Schmidt, Chapter 8), but for instances of leadership that may be replicated elsewhere. We are acutely aware that African countries, many of which have supported the above-mentioned High-Ambition Coalition, are not assessed in our volume. There clearly is a need to learn more about climate pioneership and leadership in and from those countries. Finally, we suggest greater exploration of that most Janus-faced of climate actors, Canada, as a simultaneous champion of environmental action and as laggard that is increasing its emissions via tar sands exploitation while frequently stymying action internationally.

Conclusion

Elinor Ostrom (2010: 555), winner of the Nobel Prize for Economics, reminds us that '[s]elf-organised, polycentric systems are not a panacea!' Yet, this volume has sought to provide instances of polycentric governance in order to glean a more nuanced understanding of the empirics supporting this concept. Moreover, the chapters in this volume have identified and examined instances of leadership, pioneership and followership within MLG structures. Time is running out for ambitious steps on climate change that can prevent warming over 2°C, necessitating that the 2020s are a crucial decade of climate action. This volume has identified numerous instances of climate leadership and pioneership across the globe in response to this shared problem, and provides many causes for optimism. However, we also see stark reminders of how environmental concerns can be pushed down the political agenda when seemingly more urgent problems rise to the fore. The lessons of the Global Financial Crisis are that countries, cities, businesses and networks must continue to develop more and more ambitious environmental protection measures, regardless of the ongoing crises surrounding us. It is in this context that the COVID-19 pandemic that has shaken the world at the start of this new decade is even more worrying. As states rush to grow their economies following the slump that started in 2020, leaders must not forget climate change.

And so, we urge that the instances of climate leadership and pioneership identified here are emulated as widely as possible, while new innovations are pursued wherever possible. Polycentricity may not be a panacea, but this volume has shown that inspiring action can be found at all governance levels, and in any country. Mighty oaks from little acorns grow.

Acknowledgements

The support of the Economic and Social Research Council (ESRC) is gratefully acknowledged by Paul Tobin, having funded him via grant ES/S014500/1 during the writing of this chapter. Rudi Wurzel would like to thank the British Academy for grant SG 131240. Mikael Skou Andersen gratefully acknowledges financial support from Nordforsk, Nordic Energy and Nordic Innovation for grant 82841. All three authors are grateful to the Innovations in Climate Governance (INOGOV) programme of COST which funded a workshop on 'Pioneers and Leaders in Polycentric Climate Governance' in Hull, UK.

Bibliography

Andersen, M.S. and Liefferink, D. (eds) (1997) *European Environmental Policy: The Pioneers*, Manchester: Manchester University Press.

Andresen, S. and Agrawala, S. (2002) 'Leaders, pushers and laggards in the making of the climate regime', *Global Environmental Change*, 12: 41–51.

Boasson, E. (2018) 'Entrepreneurship: a key driver of polycentric governance?' In: Jordan, A., Huitema, D., van Asselt, H. and Forster, J. (eds) *Governing Climate Change: Polycentricity in Action?* Cambridge: Cambridge University Press, 117–134.

Burns, C., Eckersley, P. and Tobin, P. (2020) 'Environmental policy in times of crisis', *Journal of European Public Policy*, 27(1): 1–19.

Burns, C., Tobin, P. and Sewerin, S. (eds) (2019) *The Impact of the Economic Crisis on European Environmental Policy*, Oxford: Oxford University Press.

Burns, J. M. (1978). *Leadership*. New York: Harper and Row.

Death, C. (2016) *The Green State in Africa*, New Haven: Yale University Press.

Dong, L. (2017) 'Bound to lead? Rethinking China's role after Paris in UNFCCC negotiations', *Chinese Journal of Population Resources and Environment*, 15(1): 32–38.

Dorsch, M.J. and Flachsland, C. (2017) 'A polycentric approach to global climate governance', *Global Environmental Politics*, 17(2): 45–64.

Fløttum, K. and Gjerstad, Ø. (2013) 'Arguing for climate policy through the linguistic construction of narratives and voices: the case of the South-African green paper "National Climate Change Response"', *Climatic Change*, 118(2): 417–430.

Grubb, M. and Gupta, A. (2000) 'Climate change, leadership and the EU'. In: J. Gupta and M. Grubb (eds) *Climate Change and European Leadership. A Sustainable Role for Europe?* Dordrecht: Kluwer, 3–14.

Homsy, G.C. and Warner, M.E. (2014) 'Cities and sustainability: polycentric action and multilevel governance', *Urban Affairs Review*, 49(1), 1–28.

Hooghe, L. (ed.) (1996) *Multi-Level Governance and European Integration*, Oxford: Clarendon Press.

Humphrey, R.H. (2002) 'The many faces of emotional leadership', *Leadership Quarterly*, 13(5): 493–504.

Jankowska, K. (2016) 'Poland's clash over energy and climate policy: green economy or grey status quo?'. In: R. Wurzel, J. Connelly, and D. Liefferink (eds) *The European Union in International Climate Change Politics: Still Taking a Lead?* Abingdon: Routledge, 145–158.

Jordan (2015) 'Intended nationally determined contribution'. Available from: https://www w4.unfccc.int/sites/submissions/INDC/Published%20Documents/Jordan/1/Jordan% 20INDCs%20Final.pdf [Accessed 26 June 2020].

Jordan, A., Huitema, D., Van Asselt, H. and Forster, J. (eds) (2018) *Governing Climate Change: Polycentricity in Action?* Cambridge: Cambridge University Press.

Jordan, A.J., Huitema, D., Hildén, M., van Asselt, H., Rayner, T.J., Schoenefeld, J.J., Tosun, J., Forster, J. and Boasson, E.L. (2015) 'Emergence of polycentric climate governance and its future prospects', *Nature Climate Change*, 5: 977–982.

Jänicke, M. (1995) 'The political system's capacity for environmental policy'. In: Jänicke, M. and Weidner, H. (eds) *National Environmental Policies: A Comparative Study of Capacity Building*, New York: Springer, 1–24.

Jänicke, M. (2006) 'Trend setters in environmental policy: The character and role of pioneer countries'. In: M. Jänicke and K. Jacob (eds) *Environmental Governance in Global Perspective*, Berlin: Environmental Policy Research Centre, 51–66.

Liefferink, D., Arts, B., Kamstra, J. and Ooijevaar, J. (2009) 'Leaders and laggards in environmental policy: a quantitative analysis of domestic policy outputs', *Journal of European Public Policy*, 16(5): 677–700.

Liefferink, D. and Wurzel, R.K.W. (2017) 'Environmental leaders and pioneers: agents of change?', *Journal of European Public Policy*, 24(7): 951–968.

Liefferink, D. and Wurzel, R.K.W. (2018) 'Leadership and pioneership: exploring their role in polycentric governance'. In: Jordan, A., Huitema, D., Van Asselt, H. and Forster, J. (eds) *Governing Climate Change: Polycentricity in Action?* Cambridge: Cambridge University Press, 135–151.

Loerakker, B. and van Winden, F. (2017) 'Emotional leadership in an intergroup conflict game experiment', *Journal of Economic Psychology*, 63: 143–167.

Marks, G. (1993) 'Structural policy and multi-level governance in the EC'. In: A. Cafruny and G. Rosenthal (eds) *The State of the European Community*, Boulder: Lynne Rienner, 391–411.

Oberthür, S. and Roche Kelly, C. (2008) 'EU leadership in international climate policy: achievements and challenges', *The International Spectator*, 43(2): 35–50.

Ostrom, E. (2010) 'Polycentric systems for coping with collective action and global environmental change', *Global Environmental Change*, 20: 550–557.

Parker, C. F. and Karlsson, C. (2014). 'Leadership and international cooperation'. The Oxford Handbook of Political Leadership / [ed] Paul 't Hart and R.A.W. Rhodes, Oxford: Oxford University Press, 2014, 580–594.

Rayner, T. and Jordan, A. (2013) 'The European Union: the polycentric climate policy leader?' *WIREs Climate Change*, 4: 75–90.

Schmidt, V.A. (2008) 'Discursive institutionalism: the explanatory power of ideas and discourse', *Annual Review of Political Science*, 11: 303–326.

The Gambia (2015) 'Intended nationally determined contribution of the Gambia'. Available from: https://www4.unfccc.int/sites/submissions/INDC/Published%20D ocuments/Gambia/1/The%20INDC%20OF%20THE%20GAMBIA.pdf [Accessed 26 June 2020].

Tobin, P., Schmidt, N., Tosun, J. and Burns, C. (2018) 'Climate policy innovation: mapping states' Paris climate pledges', *Global Environmental Change*, 48: 11–21.

Torney, D. (2019) 'Follow the leader? Conceptualising the relationship between leaders and followers in polycentric climate governance', *Environmental Politics*, 28(1): 167–86.

Underdal, A. (1994) 'Leadership theory: rediscovering the arts of management'. In: Zartman, I.W. (ed.) *International Multilateral Negotiation: Approaches to the Management of Complexity*, San Francisco: Jossey-Bass, 178–97.

Underdal, A. (1998) 'Leadership in international environmental negotiations: designing feasible solutions'. In: A. Underdal (ed.) *The Politics of International Environmental Management*, Dordrecht: Kluwer, 101–127.

UNFCCC (2015) *Paris agreement*, FCCC/CP/2015/L.9. Geneva: United Nations.

Vogel, D. (1997) 'Trading up and governing across: transnational governance and environmental protection', *Journal of European Public Policy*, 4(4): 556–71.

Wurzel, R.K., Liefferink, D. and Torney, D. (2019). 'Pioneers, leaders and followers in multilevel and polycentric climate governance.' *Environmental Politics*, 28(1): 1–21.

Wurzel, R.K.W. (2008) 'Environmental policy: EU actors, leader and laggard states'. In: J. Hayward (ed.) *Leaderless Europe*, Oxford: Oxford University Press, 66–88.

Wurzel, R.K.W., Connelly, J. and Liefferink, D. (eds) (2017) *The European Union International Climate Change Politics. Still Taking a Lead?* London: Routledge.

Wurzel, R.K.W., Liefferink, D. and Torney, D. (2019) 'Pioneers, leaders and followers in multilevel and polycentric climate governance', *Environmental Politics*, 28(1): 1–21.

Young, O. (1991) 'Political leadership and regime formation: on the development of institutions in international society', *International Organization*, 45(3): 281–309.

Index

Note: Page locators in **bold** refer to tables and *italics* refer to figures.